Multifacetted Modelling and
Discrete Event Simulation

Multifacetted Modelling and Discrete Event Simulation

Bernard P. Zeigler

Department of Computer Science
Wayne State University
Detroit, USA

ACADEMIC PRESS · 1984

(Harcourt Brace Jovanovich, Publishers)

London · Orlando · San Diego · San Francisco · New York
Toronto · Montreal · São Paulo · Sydney · Tokyo

ACADEMIC PRESS INC. (LONDON) LTD.
24/28 Oval Road
London NW1

United States Edition published by
ACADEMIC PRESS INC.
(Harcourt Brace Jovanovich, Inc)
Orlando, Florida 32887

Copyright © 1984 by
ACADEMIC PRESS INC. (LONDON) LTD.

All Rights Reserved

No part of this book may be reproduced in any form by photostat, microfilm,
or any other means, without permission from the publishers

British Library Cataloguing in Publication Data

Zeigler, Bernard P.
  Multifacetted modelling and discrete event
simulation.
  1. Digital computer simulation
  I. Title
  001.4'24   QA76.9.C65

ISBN 0-12-778450-0
LCCCN 83-72316

Filmset by H Charlesworth, Huddersfield
Printed in Great Britain by St Edmundsbury Press

FOREWORD

Model-based simulation is like a gem: it is multifacetted. Some of the specialists too close to one of the facets, perceive only that single facet and the reflection of the success of their careers through it. The more they see the latter, the more, it seems, they are enamoured with that aspect of modelling and simulation instead of exploring new horizons. If it was to this attitude, nobody would have discovered the New World.

The fable of the elephant and the blindfolded men is a well-known metaphor in eastern cultures, to illustrate how easy it is to confuse the parts with the whole.

Being familiar with more than twenty definitions of the term "simulation" compiled by A. A. B. Pritsker in the August 1979 issue of *Simulation* and some others scattered in numerous publications, I often remember that fable.

Nowadays, computers are becoming more powerful and affordable; complexity of our environment and available knowledge are increasing at alarming rates; the number of simulation languages is increasing dramatically — almost proving implicitly the dissatisfaction of people with existing ones.

Zeigler, in his book "Multifacetted Modelling and Discrete Event Simulation" casts the light on the many facets of modelling and simulation and presents a systemic view of the discrete event simulation.

The methodology presented in Zeigler's book can be useful to realize advanced tools for the information society.

Zeigler's new book will, I am sure, be welcomed by many serious modellers and simulationists.

January, 1984
                Tuncer I. Ören
                University of Ottawa
                Canada

PREFACE

Modelling, in its computerized form, increasingly will take its place as the key knowledge component in all forms of decision making in modern life. Consequently, more attention should be paid to understanding the processes by which models are built and employed. More particularly, the problems facing modern society are often called "large scale" (an expression of awe at their enormous scope, perhaps) or "complex" (an expression of exasperation in dealing with them?) as are the models proposed to understand and manipulate them. Beyond these cliches of the systems approach lies a reality causing our discomfort: we shall call it "multifacetted". Indeed, we live in a world of multiplicities: our fates are thrown together with many other people, we must recognize many points of view, disciplines, and "inputs". Our systems and models, the solutions of our problems, must begin to reflect this relativity, this multiplicity of objectives and perspectives.

This book claims to be the first to lay its finger on this sense of unease, letting the multifacetted beast out of the closet, and recognizing it for the virtue that it is.

So we are concerned with multifacetted modelling. What kinds of concepts are needed to make sense of such an ambitious form of modelling? What methodologies of simulation modelling are needed to bring some conviction to its results? What kinds of computer systems are needed to make the approach feasible at all? These are the questions that I address and answer (expecting the answers to have a very short life-time!).

Multifacetted modelling methodology should, in principle, not be limited to any one form of model expression. Why then "discrete event" simulation in the title? Only because we must have some substrate to work on, and the discrete event concept is a good one to start with. Discrete event simulation is widely employed in systems engineering, operations research, and management science. It is the modelling child of this century (where differential equations are the child of the last) and its full potential has yet to be realized.

"Multifacetted Modelling and Discrete Event Simulation" therefore explores the formalism of discrete event systems with some intensity, and employs this understanding to formulate the issues and approaches of multifacetted system modelling.

January, 1984 *B. P. Zeigler*

CONTENTS

Foreword v
Preface vii
How to read and teach this book 1

PART 1 Motivation and perspective

1. MULTIFACETTED MODELLING: MOTIVATION AND PERSPECTIVE

1. Models in Management, Control and Design 3
2. Multifacetted System Modelling 8
3. Structures for Multifacetted Modelling 14
4. Summary 19
References 19

PART 2 Fundamentals and formalism

2. UNDERSTANDING FORMALISMS

1. Abstraction: Sets and Set Operations 21
2. Formalisms and Objects 22
3. Useful Abstractions 33
References 34

3. SYSTEM SPECIFICATIONS AND MORPHISMS

1. Levels of System Specification 37
2. Association Mappings for Moving Down the Hierarchy . . . 42
3. Hierarchy of System Specification Morphisms 46
4. Moving Down the Morphism Hierarchy 51
5. Climbing up the Hierarchy 53
6. Modelling Formalisms 54
7. Closure of Formalisms under Coupling 60
8. Summary 60
References 61

ix

4. FORMALISM FOR DISCRETE EVENT SYSTEMS

1. The DEVS Formalism. 62
2. The System Specified by a DEVS 63
3. Interpreting the DEVS 65
4. Example: A Generalized Queue 66
5. Special Classes of DEVS Models. 71
6. Summary 76
References 77

5. A WORKING FRAMEWORK

1. System Formulation 79
2. Knowledge of Internal Structure. 81
3. Verification 82
4. Validation 82
5. Experimental Frames 82
6. Experimental Frames and Validity 84
7. Varieties of Validity/Structural Inference 85
8. Valid Simplification 86
9. Implication for Computer Assistance to Modelling and Simulation . 87
10. Summary 89
References 89

PART 3 Multicomponent models

6. MULTICOMPONENT MODELS OF SPATIALLY DISTRIBUTED SYSTEMS

1. The Cellular Automaton Formalism 92
2. Next Event Cell Space Models 97
3. Formalism for General Discrete Event Cell Space Models . . 106
4. Formalism Adequacy: Representing Real World Relationships . 107
5. Summary 115
Problems. 115
References 116

7. DEVS MODELS AT THE STRUCTURED SYSTEM LEVEL

1. Multicomponent DEVS in Structured System Form . . . 117
2. User Oriented DEVS Language 123
3. Hierarchical Multicomponent DEVS 127
4. Summary 128
Problems. 130
References 131

8. MULTICOMPONENT DEVS: MODULARITY AND HIERARCHY

1. Modularity and Hierarchical Construction: Informal Concepts . . 132
2. Multicomponent DEVS in Modular Form 133

3. Language for Modular DEVS Specification 137
4. Translating Non-modular Formalisms in Modular Form . . . 137
5. Examples: Multicomponent Models in Modular Form . . . 141
6. Equivalence of DEVS Formalisms 145
7. Hierarchical Specification of DEVS Models 149
8. Summary: Adequacy of DEVS Formalisms. 150
Problems. 157
References 157

9. AGGREGATION AND OTHER SIMPLIFICATION PROCEDURES

1. State Space Reduction: The Congruence Relation 159
2. Aggregation: The Uniformity of Influence Principle 159
3. Random Phase-Random Space Approximation 163
4. Representing Systems in DEVS Formalism. 166
5. Parameter Correspondences 169
6. Summary 171
References 171

PART 4 Multifacetted system modelling

10. SPECIFICATION OF MODEL STATIC STRUCTURE

1. Static Structure — Introduction 174
2. Prestructures for Static Structuring 178
3. The Entity–Attribute–Set (EAS) Formalism 190
4. Summary 192
Problems. 192
References 193

11. THE SYSTEM ENTITY STRUCTURE

1. The Entity Structure Axioms 195
2. More on Specialization Hierarchies 199
3. Semantic Structure on Variables 202
4. Summary 205

12. OBJECTIVES-DRIVEN METHODOLOGY: EXPERIMENTAL FRAMES

1. Example: Experimental Frames for the University Bus System . . 206
2. Definition of Experimental Frames 210
3. Development of Experimental Frames 212
4. Input Segment Specification 216
5. Realization of Experimental Frames 224
6. Wymore's Tricotyledon Theory of System Design 226
7. Summary 232
Problems. 232
References 235

13. EXPERIMENTAL FRAMES: OPERATIONAL CONCEPTS

1. Derivability Concepts: Motivation and Approach 236
2. Experimental Frames for Isolation of Components 248
3. Formalization of Modelling Assumptions 250
4. Experimental Frames for Model Simplification 251
5. Summary 254
Problems. 256

14. MODELLING IN THE LARGE: BOUNDARY EXPANSION, INTEGRATION

1. Objectives and Expanding System Boundaries 257
2. Entity Structure as an Organizer of Models and Frames . . . 261
3. Propagation of Parameter Information 267
4. Summary 270
References 271

15. MULTIFACETTED MODEL BUILDING METHODOLOGY

1. Pruning the Entity Structure 272
2. Concepts for Model Coupling 277
3. The Composition Tree Formalism 281
4. Iterative Methodology for Model Construction 288
5. Paradigms for Frame Based Model Retrieval and Synthesis . . 290
6. Summary 294
Problems. 295
Appendix 1: Interpretations of Composition Tree Formalism. . . 295
References 298

PART 5 Multifacetted system architectures

16. MODULARITY IN MODELS AND EXPERIMENTAL FRAMES

1. Modular Realization of Model/Experimental Frame Couplings . . 301
2. State Variables and Parameters in Execution Control. . . . 309
3. Off-line Realization of Experimental Frame Application . . . 316
4. Summary 317
References 317

17. SIMULATION OF MODULAR AND HIERARCHICAL MODELS

1. Abstract Simulator of a DEVS 318
2. Simulation Languages Realization of the Modular DEVS . . . 323
3. Hierarchical DEVS Models and Simulators 326
4. Distributed, Asynchronous DEVS Simulation 327
Problems. 333
References 333

18. SYSTEM ARCHITECTURES FOR MULTIFACETTED MODELLING

   1. Architectural Framework for Support of Multifacetted
      Methodology 334
   2. Contexts For Support of Multifacetted Methodology 336
   3. Software/Hardware Steps Toward Multifacetted System
      Architecture. 341
   4. Epilogue 363
   References 364

Subject Index 366

HOW TO READ AND TEACH THIS BOOK

Three distinct kinds of material make up this presentation of multifacetted modelling concepts: motivational/conceptual, foundational, and technical. This didactic breakdown of the material is somewhat, but not fully, correlated with the logical breakdown of the subject matter which is reflected in the five parts of the book.

Motivational and conceptual material largely constitutes the first and last chapters. The reader may thus wish to read Chapter 1 and then skim Chapter 18 to obtain a sense of the directions in which the book is heading.

The bulk of the novel and technical material relating to multifacetted modelling methodology is contained in the chapters in Part 4. A reader somewhat familiar with the theory and practice of simulation modelling should be able to understand the essence of most of this development without having to read the foundational material. To help fill in gaps in background, references to this material are liberally sprinkled throughout the text.

Foundational material is largely centralized in Part 2. Except for Chapter 3, this material is of a fairly introductory nature and should be readily assimilated. Chapter 3 is highly integrative and intended to provide a rigorous basis for the technical material in the book. It can be skimmed upon first reading.

Material specialized to the discrete event formalism is concentrated in Part 3 and in Chapters 16 and 17 of Part 5. This material is important for the design of multifacetted methodology support environments in the context of discrete event simulation.

The book is suitable for intermediate and advanced courses at the graduate level. An intermediate course might be centered on the modular and hierarchical design of discrete event models and simulators. It would then be based on the following material: Chapters 2, 4 (Part 2); Chapters 6, 7, 8 (Part 3); Chapters 10, 11, 12 (Part 4); Chapters 16, 17 (Part 5). Such a course should prove intellectually challenging and rewarding for students of computer and related systems sciences. It should teach them modular and hierarchical design principles in rigorous form in an interesting and useful domain, that of discrete event simulation.

An advanced course would focus on the concepts of multifacetted methodology and its support by the computer *per se*. Such a course might center on the hierarchy of system specifications and morphisms as a means of establishing the apparatus for a theoretical approach to the subject. It might then go on to examine the issues in multifacetted modelling from this perspective (concentration on Part 4). Projects could have students going to the source literature on existing software systems and proposals, entree to which is provided in the last chapter.

In developing the concepts of multifacetted modelling, I have tried to be as concrete as necessary to bring concepts to a level where design of computer implementation is conceivable. The interested reader: theoretician, software designer, or graduate student, will find that much remains to be done to bring the multifacetted approach to fruition.

PART 1
MOTIVATION AND PERSPECTIVE

Chapter 1

MULTIFACETTED MODELLING: MOTIVATION AND PERSPECTIVE

1. Models in Management, Control and Design

The ultimate benefit of modelling is to increase decision making capability. Although this statement does not deny the role of modelling as an activity for its own sake, it does emphasize that modelling and simulation are activities to be viewed within a larger pragmatic context. We thus suppose that we are faced with a real system — an existing part of reality — into whose operation we intend to intervene so that it better suits our purposes.

As shown in Figure 1.1, we can think of this real system as being composed of *controllable* and *uncontrollable* parts. The controllable part refers to all aspects of the system which are in our power to modify, alter, transform, augment, replace, etc. The uncontrollable part refers to whatever is left — usually most of the system — namely the aspects which are not in our power to control. (In a moment it will become clear that what we take to be the real system and its controllable/uncontrollable decomposition is not fixed for all time but depends both on our perception and viewpoint as well as the objective situation.) Naturally intervention must be limited to the controllable part of the system.

We can distinguish three broad levels of intervention. In order of increasing power these are: *management*, *control* and *design*.

Management connotes a limited power to intervene. One can set goals and determine broad courses of action, but in their nature, these policies cannot be spelled out in full detail. Their execution must thus be delegated to, and interpreted at, subordinate levels; so the link between intention and implementation is ambiguous.

In the control context, however, action is deterministically related to policy. Still, constraints remain on the scope of intervention in that control action is limited to the selection of alternatives within fixed extant domains.

In contrast, design connotes greater scope for choice in that the designer

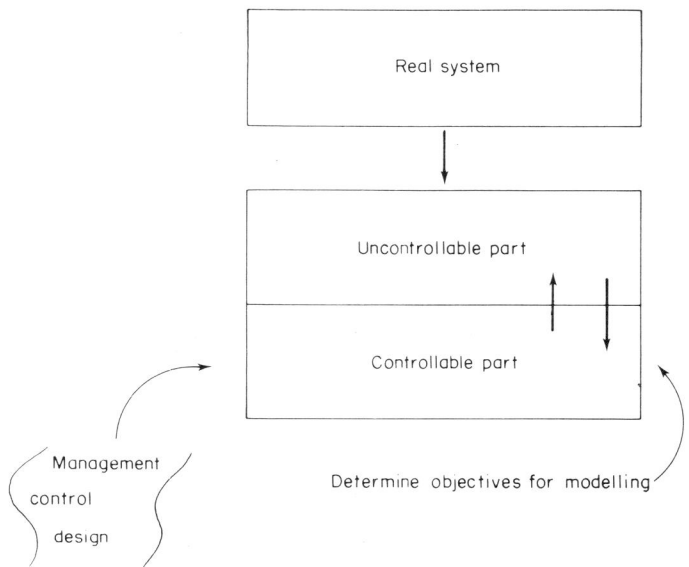

Fig. 1.1. Decomposition of a real system into controllable/uncontrollable parts.

expects to augment, or replace a part of, the existing reality. Implementation of a design is relatively expensive and infrequent while control and management are continued "on-line" activities. Thus we can think of management, control and design as lying along axes representing degree, cost and time scale of intervention on the controllable part of a real system.

The effect of such intervention, however, is uncertain due to the existence of the uncontrollable part and its interaction with the controllable part. The greater one's knowledge of the latter, the greater the certainty one can have in assessing the effect of intervention. Models encode such knowledge. Their *sine qua non* is the ability to integrate sets of disconnected relationships whose joint implications would otherwise be difficult to draw. They are especially useful to the extent that they enable extrapolation beyond the data so far acquired.

Modelling methodology may thus be viewed as a component within computer-based systems intended to aid decision making in management, control, and design. Such a view is typified in current approaches to providing more effective computerized environments for business managers coming under the rubric: decision support system [1]. A *decision support system* is an interactive computer-based system intended to help decision makers utilize data and models in the problem solving process. Sprague and

Carlson [1] view such a system as consisting of a data base, a model base, and a dialog interface with the user.

> "The data base component provides access to the raw material for decision making. It is the modelling component that gives decision makers the ability to analyze the problem fully by developing and comparing alternative solutions. *In fact, it is the integration of models into the information system that moves an MIS* [management information system] ... *into a full decision support system*" [1, page 257] (emphasis added).

While Sprague and Carlson view models as generalized tools for operating on input data to produce output helpful in decision making, we restrict the concept of *model* to that of a mathematical representation of a system intended to reproduce its responses to input stimuli (see Chapter 5). Such models serve as the basis for a host of decision support activities such as asking "what if" questions, evaluating alternative courses of action, and seeking optimum control strategies, all of which are included in Sprague and Carlson's undifferentiated conception of modelling activity. Nevertheless, the point remains that model construction is to be considered an integral component within a decision support system. Many other examples of model-based management, control, and design could have been used to make the same point. In Section 12.6 we show how the multifacetted modelling methodology to be developed here is an integral part of the formal methodology of systems design.

Thus modelling is an activity undertaken to increase decision making capability — the ability to assess the effects of interventions before they are actually carried out and to pre-select promising ones, in view of the objectives one is trying to achieve. These objectives in turn serve to orient modelling efforts as we shall see.

1.1. Multiple Objectives: Multiple Models

Let us now recognize that the same real time system may be subject to a multiplicity of management, control and design objectives. Consider for example a short list of categories relevant to the operation of a business firm as outlined in Figure 1.2. An all inclusive model would be able to provide the basis for decision making in each of these categories. However, the impediments to constructing such a comprehensive model make it a near impossibility for the forseeable future. Instead we can envisage a collection of partial models, each oriented to one or more objectives. We shall return to the systematic aspects of such a collection — indeed, this is our main concern — in a moment.

For now, let us see how objectives orient the model building process by

OBJECTIVES	NEED MODEL OF:
manpower requirements	plant operations timings, routings, etc.

location of plant	environment:
	    availability of resources,
	    energy, personnel; impact.

layout of plant	plant machines: sizes, routings, etc.
marketing	environment: consumer tastes, competition
safety assurance	material characteristics, alarms, escape routes, etc.
quality control	material characteristics, production process, etc.
customer satisfaction	customer interface: waiting, routing, etc.
inventory control	environment: customers, orders firm: purchasing, production, delivery departments
corporate planning	environment: investment capital, long term trends, etc. firm: unused capacity, growth potential
research and development	projects: interactions, likely payoffs

Fig. 1.2. The firm as a multifacetted system.

determining such factors as the *system boundaries* and the model *components* of relevance, also the *level of detail* at which the latter are represented. This is illustrated in Figure 1.2 where different choices of these elements are associated with the various objectives.

System boundaries refer to the partitioning of reality into a part to be focused on and the rest whose effect on the former is to be represented by input variables (generalized boundary conditions). Examining Figure 1.2 we see that, depending on the objective, the firm, some part of it, its environment, or the firm together with its environment may be the focus of attention. Indeed, we may as well admit that each such focus might be called a system and thus that the "real system" is a crude way of referring to all of these potential systems.

Having chosen our system boundaries, we are in a position to model the enclosed system. As is evident from Figure 1.2, the *components* we include in our model and the *abstractions* we make of them are governed by the objectives. A production process might be studied in detail for quality control or might be represented by "black box" features (such as its input resources, output products and process time characteristics) for manpower assessment. It may well not be represented at all in plant location studies.

Let us consider an example to illustrate the foregoing ideas. We shall return to this example periodically as new concepts are being introduced.

1.2. Example — A University Bus Service

A university currently provides a transportation service for its students in which a single bus shuttles between a downtown station and a university station. The area surrounding the university has been developing, however, and increasing numbers of nonstudents are also using the bus service, at least complaints to this effect have been registered by the students who sometimes cannot get on the bus and must wait for it to return from its round trip.

The university administration has set up a committee to review the situation and make recommendations. They are to assess the merits of the following options:

(1) Leave the situation as it is
(2) Permit only students on the bus
(3) Permit anyone to ride but charge nonstudents

The administration also foresees the possibility of adding a second bus on the route, but would rather not begin consideration of this until it becomes convinced of its necessity.

To help in their deliberations, the committee has asked a team of computer and management science professors to build a model on which they can test the various policies.

Let us analyze the situation in the terms we have introduced.

The *decision maker* is the university administration. Its *primary objective* is apparently to guarantee a satisfactory level of bus service for the students. However, it may also see in the bus service a potential for a much needed source of revenue. The good will afforded by providing the community with such a service would be a welcome byproduct. Thus a *second objective*, that of financial gain through expanded service, looms on the horizon.

The *system boundary* implied by these objectives is to encompass the present bus and its route. A second bus may later be added, but no further expansion is contemplated. Nor have other issues such as acceleration of local development or environmental or safety problems yet been raised which would require extending the system scope. Let the term "bus system" designate the system within the above boundaries.

The *controllable* part of the bus system implicit in the above objectives has to do firstly with the identification of potential passengers and secondly with the addition of a second bus. In the first case, possibilities for intervention are limited to providing students with identification and prohibiting entry to or charging nonstudents. In the second case, these possibilities are expanded somewhat, but most of the system such as the external demand for service and the capacity of the present bus is considered *uncontrollable*.

The kind of intervention being contemplated has management, control and design aspects. The institution of a new passenger identification policy or addition of a bus would be a *managerial level* decision — here carried out by the administration. Working out the details of the identification policy and especially of the operation of the expanded bus system would be a *design* problem in this case being handled by the adhoc review committee. Execution of the identification policy would be of a *control* nature perhaps assigned to the driver.

Although the committee has asked for a single model, it is likely that the modelling team will end up developing more than one, as we shall see.

1.3. Models for the University Bus Service Example

Within the objectives and system boundaries given them, the modelling team considers the development of three models. The first will model the bus system as it is currently and the second will model the current bus system with passenger identification possibilities. The third, to model the expanded bus system, will only be initiated under further instruction.

This development from scratch of models for specific objectives is what we shall call "modelling in the small". Even in this case, we see that good practice suggests beginning with the simplest model apparently required and adding elaboration as necessary.

Note that the second model is the one presumably that the committee has in mind to help formulate and test identification policies. But the modelling team develops the first model expecting to complete it quickly because of its relative simplicity. Moreover, it considers the knowledge that it represents, after validation, will be transferable to the more complex models required for decision making. These aspects of "modelling in the large" are already at work even in an apparently one shot modelling project.

2. Multifacetted System Modelling

The example just given to bring out multiplicities of objectives and models is not unique. Such a situation is pervasive in what are usually called large scale systems, socio-economic, ecological, urban, etc. But the large scaleness is really not the significant distinguishing sign, since if one looks, one can find the same situation pertaining to systems of medium scale (e.g., the brain) or small scale (e.g. a biological cell). The term *multifacetted* is introduced to denote an approach to modelling which recognizes the existence of multiplicities of objectives and models as a fact of life. More than this, it seeks concepts and theories to understand the potentiality of the real world to be dealt with in this way and simulation methodologies to exploit this potentiality and avoid succumbing to its pitfalls.

2.1. Multifacetted Systems as Partially Decomposable Systems

A long standing controversy in science is that between the "reductionists" and the "holists". The former claim that all phenomena can be broken down into physico-chemical elements and understood on the basis of the laws of physics. The latter deny that this is possible and assert that the most interesting systems must be understood as indivisible wholes.

Multifacetted modelling acknowledges that both could be right to some degree. It views real systems as partially decomposable and seeks to make the best of this situation.

Consider the "bowl of spaghetti" analogy. An *indecomposable* system would be one in which you could not pull on any one strand without pulling out the whole mass of spaghetti. A *fully decomposable* system, on the other hand, would allow you to extract any one strand without disturbing the rest.

Real systems seem to be somewhere between the two extremes, that is, they are *partially decomposable*. They are well cooked spaghetti bowls in which one can get some distance by tugging on a strand (how far may depend on which strand one pulls), but eventually other strands will start to interfere with further progress.

In such a bowl, the strands represent *facets* and the tangle represents the fact that ultimately "everything is linked to everything else", a commonly expressed view concerning biological and higher systems on the organized complexity scale. Tugs on the strands represent *objectives* and their ability to narrow one's interest in the system to one of its facets.

In a partially decomposable system, there is a marked utility in such a narrowing. However, only in a totally decomposable system would it be possible to forever ignore other strands. Specialized disciplines in today's sciences assume such total decomposability. Decision makers who must solve problems which cut across many disciplines are more aware of the limited ability of each speciality to contribute relevant, noncontingent information.

The multifacetted modelling approach thus seeks to capitalize on the decomposability where it exists by facilitating the multi-objective/multi-model methodology. We call such a divide-and-conquer methodology "modelling in the small". But recognizing limitations due to partial decomposibility, we also seek to facilitate the evolution of coherent models of such systems. We call such a holistic methodology "modelling in the large". In fact,

MODEL DEVELOPMENT

* in the small (model as tool)
  - top down design
  - informal $\begin{bmatrix} \text{formal} \\ \text{program} \end{bmatrix}$ specification
  - simplification/elaboration
  - validation/verification

* in the large (model as knowledge)
  - communication/documentation
  - team co-ordination
  - modularity and large model development
  - multi-model organization

Fig. 1.3. Modelling in the small and in the large.

the two methodologies are mutually supportive as we shall see.

There is an important way in which the spaghetti bowl fails as an analogy. It suggests that the strands or facets are inherent features of the real system. This may be true only in part. The situation is more like a bowl of clay in which we are free to string out our own strands subject to the constraints imposed by the clay. Each such "spaghettization" represents a decomposition of the system. What may be tangled in one decomposition may be separable in another. Thus multifacetted methodology seeks to facilitate the development of alternative and useful decompositions.

In Figure 1.3, we summarize some of the points characterizing modelling methodologies "in the small" and "in the large".

2.2. Methodology for Modelling "in the small"

In the small, we are concerned with the immediate task of constructing a model for a given decision making objective. The model is thus a tool to answer questions about the real system relevant to the objective.

Top down modelling is based on the distinction between a *model* (an abstract set of instructions for generating behavior) and its realization by a *simulation program*. As in top down program design, we encourage conceptual development of the model prior to program writing. The conceptual development starts with an *informal model description* and proceeds in stages to more *formal specifications*.

One benefit of such a procedure is that there now exists an independent model description which can be employed in both *verification* (ascertaining that the simulation program correctly implements the model) and in *validation* (ascertaining that the model faithfully represents the real system).

A second benefit is that such model descriptions are more easily manipulated, facilitating exploration of the "model space". For example, by *simplification* and *elaboration* (its inverse), the modeller can assess the sensitivity of the model behavior to his assumptions.

Because it requires abstract and formal thinking, such top down methodology is not easy. Recognizing this problem has led to the development of software tools to facilitate this methodology. We shall presently discuss both the methodology and the availability of tools.

2.3. Methodology for Modelling "in the large"

Modelling in the large concerns itself with the model as more than a tool of immediate application to be used and then thrown away. Here we conceive of the model as embodying knowledge about the real system. Thus the model should have some of the characteristics of knowledge — it should be transmittable from one person to another and open to modification, improvement and combination with other knowledge.

Methodology in this domain thus concerns good practices for *documentation* of models and their *communication* from human to human and also between human and computer. It concerns the coordination of teams of modellers, consisting, for example, of specialists from various disciplines, which must be brought to bear on policy problems in the environmental, energy or other such transdisciplinary fields.

Methodology here also concerns the development of large models whose construction may be greatly assisted by good *decomposition* and *modularity* design. A good decomposition sets up simple interfaces ("sockets") into which modules may be "plugged". This facilitates among other things "plugging in" previously developed models as components in larger models. However, to control the growth in large model complexity, simplification of component models for their new context should be considered.

Such recycling of models calls for *systematization* and *organization* of model collections. We call such an organization a *model base* and will soon stress its centrality to the modelling process.

2.4. Synergism: Modelling in the Small and in the Large

A moment's thought will suggest that good practice of modelling in the small is mutually supportive with good practice of modelling in the large. Top down design is intended to improve the development from scratch of a particular model for a particular objective. But, as a side effect, it provides the basis for good documentation and the decomposition structure required for recycling this model in the context of others. Conversely, a well organized bank of models puts them at one's disposal for use as building blocks in new model development. This obviates always having to start model building from scratch, thus reducing the tremendous expense and time this may involve. It should not be surprising then that the same software tools should be applicable both to modelling in the large and in the small. In fact, this suggests that we conceive of both methodologies as parts of the same whole. Let us, therefore, make a model (!) of the modelling and simulation process itself. Our ultimate objective: to improve the performance and management of this process and consequently to improve its products — the models.

2.5. Modelling and Simulation as a Process

Figure 1.4 depicts the activities of modelling and simulation as a process (an integrated dynamic system) in which the model base figures centrally. The process is driven by objectives being generated from outside the system boundaries over time. As new objectives come in, they initiate model synthesizing activity. The available knowledge for such activity comes from the model base and also from the data base — which stores and organizes data gathered about the real system. Model construction is followed by simulation (model experimentation) and validation phases. Validation leads to new experimentation on the real system and may require more modification or even rejection and reinitiation along different lines. In the process, new or more refined objectives may be formulated as deficiencies in the current knowledge base (model, data) become more apparent. Ultimately,

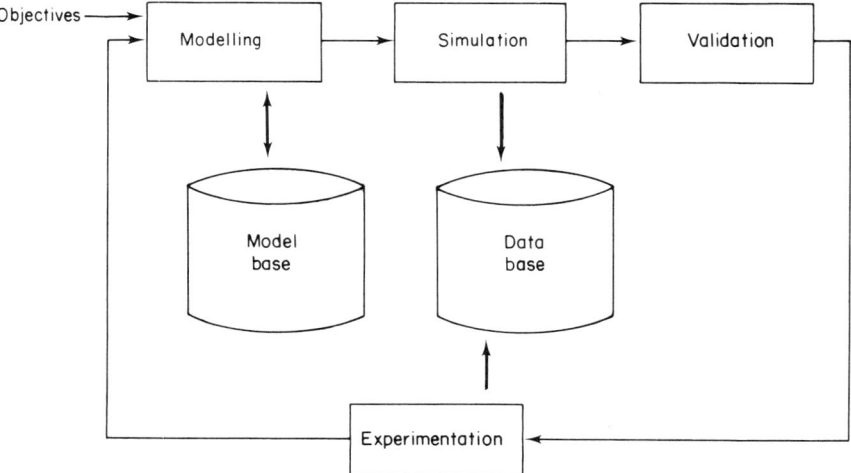

Fig. 1.4. Modelling and simulation as a process.

one or more models are produced to meet the external objectives (unless the process gets into a loop it cannot get out of). Besides being sent to the decision maker, these models are also stored in the model base and so made available for further rounds of activity stimulated by new objectives.

Let us note some implications for management of the modelling and simulation process emerging from the foregoing portrayal.

Continued and indefinite span of the process

The process is viewed as an ongoing, never ending set of activities, not as taking place only during a "modelling project". The classical view (which still governs funding allocation policies of government agencies, for example) is to initiate a modelling project with a set of objectives (more or less clearly stated), some funds and personnel and a delivery date. In this view, the process exists only between initiation and delivery dates. Better put, an *instance* of the process exists between such initiation and delivery dates, each such instance being viewed as an independent start-from-scratch, one-shot sequence of activities. With no institutionalized integration and coordination of such instances, one may expect wasteful duplication, excessive cost and shortfall of objectives, even with the best "in the small" methodology. Of course, any cumulative growth in knowledge associated with the "in the large" development would be purely accidental.

Partial data availability and partial model validity

At the initiation of modelling, there will usually be a lack of real system data for calibration and validation. Because experimentation during a project is limited, this will likely be true also at the project's termination. Thus, at the delivery date, the model will be in some state of partial validity. But if the model remains accessible in a model base, it can be tested against new relevant data whenever they are obtained, perhaps in connection with some other project. Moreover, the model is also being tested whenever it participates as a component in a composite model. Thus, while the integrated view recognizes the partialness of data and model validity, it also conceives of possibilities for efficient use of this available knowledge to sustain its continued growth.

Multiple complementary and competitive representations in the model base

Models are *competitive* when they embody different, mutually exclusive hypotheses about how the real system works. They are *complementary* when they embody the same hypotheses, but represent them in different ways. Both kinds of situations may exist in a model base. Because of the partial validity problem, we may have to maintain competitive models (those achieving a certain confidence level, say) on which to base decision making. We should be able to test the sensitivity of management, control or design decisions to the model alternatives. Complementary models, on the other hand, represent the same hypotheses about the real system, but one representation may be better than another in a particular application. For example, even though time and frequency domain representations of a linear system embody the same information, the first is better for answering steady state, the second, for transient, questions of behavior.

Models at various levels of abstraction, simplification and aggregation

We have referred previously to the role of objectives in orienting model construction and in particular the level of detail required. There is an advantage to minimizing detail. Since the coarser a model is, the easier it is to construct and the less expensive it is in computer time and space. Furthermore it is easier to calibrate since it has fewer parameters. On the other hand, a more refined model is likely to validly represent the real system. A solution to this dilemma lies in constructing families of models at different complexity levels which can be checked against each other (mutual consistency) and against the real system (validity). In fact, such a methodology lies at the heart of the systematizing of model collections mentioned earlier.

3. Structures for Multifacetted Modelling

The advantages of the multifacetted approach to modelling having been stated, it is time to start the work of developing appropriate concepts aimed at supporting the concrete practice of such methodologies. Succeeding chapters will consider these in some depth but it might help to provide a brief exposition of some of the novel concepts in illustrative form. The example we shall use is a situation where the existence of multiple objectives is blatantly obvious.

As has been indicated, multifacetted modelling arises when decision makers are faced with a multiplicity of objectives. This is even more apparent when the decision makers themselves do not share the same goals. In this case there will certainly be a multiplicity of modelling objectives which correspond to the various questions that are of interest to the different decision makers.

When do decision makers not share the same goals? Consider the annual negotiations that involve labor, management and government in an effort to reconcile the divergent proposals of the parties for the new work contracts. Each party wants to see that its own interests are furthered by any agreement and is therefore interested in answers to certain questions about it to the exclusion of others. Often negotiations snag because the bargaining representatives cannot agree not only on the desirability of the effects of a proposed policy but on just what these effects will be. Elzas [2] has proposed that the application of simulation modelling within a gaming format might do a lot to rationalize the negotiation process. Each party can use its own models to forecast the results of proposed policies and thereby estimate their value for the realization of its particular interests. Use of the models is to be structured by a pattern of negotiations between parties (horizontal games) and within parties (vertical games) in which attempts are made to bridge the gaps between policies acceptable to the parties.

What can multifacetted modelling methodology contribute to this proposed decision support system? It must be ready to provide models capable of responding to the objectives of the many participants. Such models must have a high degree of credibility, must be improveable, and must be compatible across levels of aggregation. In view of the state of our modelling capability for socio-economic systems this is no small requirement. Several economic modelling systems exist or are under development which provide tools for model construction, calibration, validation, storage and retrieval.[1] But beyond this there should be an integrated framework capable of organizing the possibilities for system boundaries, decompositions and objectives that reflect the interests of the various parties.

At this point our aim is to introduce two key concepts of such a framework: the *entity structure*, and the *experimental frame*.

The *entity structure* is based on a tree-like graph embodying the system boundaries and decompositions which have been conceived for the system. An *entity* signifies a conceptual part of the system which has been identified as a component in one or more decompositions. Figure 1.5 presents an appropriate entity structure for our work contract example. Note that the top most entity is the NATIONAL ECONOMY and thus the national border constitutes the widest system boundary of interest. A widened boundary which would allow explicit consideration of interaction with other countries would require extending the entity structure accordingly. Note that there are several ways of decomposing the NATIONAL ECONOMY — into labor, business and government components; into economic sectors, into regions such as states or provinces. Each such decomposition is called an *aspect*. Actually, both entities and aspects are better thought of as types of components and decompositions, respectively, since (subject to constraints) they may appear more than once in the structure. In fact anywhere it is placed, an entity (or aspect) carries with it the same decomposition sub-

---

[1] Examples: the IAS (Interactive System) developed by the Institute for Advanced Studies, Vienna, the SIM (System for Interactive Modelling) being developed at the Institute for System Studies, Moscow; and the MBS (Model Base System) developed at the GMD, Bonn.

16   1. *Motivation and perspective*

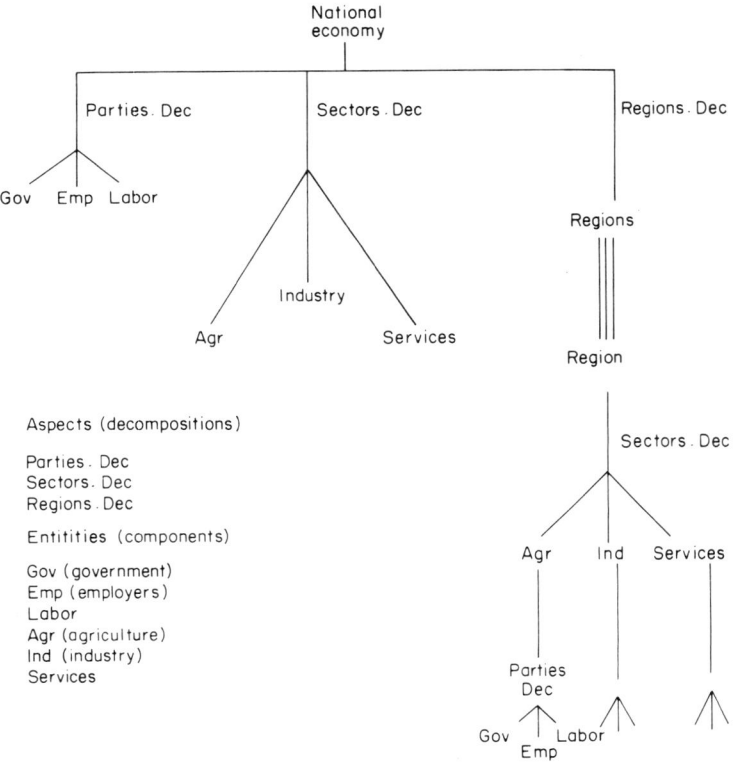

Fig. 1.5. Entity structure for Elzas negotiation methodology.

structure. For one thing this means that new refinements of decompositions may be specified by re-applying existing ones. For example in Figure 1.5, each REGION is further decomposed into economical sectors by attaching to it the same SECTORS aspect that has been attached to the NATIONAL ECONOMY. More complex decompositions may be built up by successive re-applications of the same principle. Of course a naming convention must be given to distinguish multiple occurrences of the same entity (or aspect). For example, the AGRICULTURAL sector of the NATIONAL ECONOMY is to be distinguished from the AGRICULTURAL sector of a particular REGION. But once this is done we can see that the entity structure provides a compact means of displaying the components, and generating the various decompositions, that have been formulated for a given system.

A second important concept is the *experimental frame*. This concept characterizes modelling objectives by specifying the form of the experimentation that is required in order to obtain answers to the questions of interest. At minimum an experimental frame states the *input* and *output* variables that are of interest. *Input* variables are variables that are supposed not to be under the control of the model (or counterpart real system), hence must be determined by the experimenter (or the real system environment). *Output* variables are variables existing in the model (or measureable in the real system) which are determined in response to the input behavior. A frame thus sets up a space in which data may be collected in the form of pairs of input and associated output time series. It is by examining, aggregating and otherwise processing such data that answers to the motivating questions may be obtained.

In our example, each of the parties has its characteristic experimental frames as shown in Figure 1.6. The performance indexes of particular concern to the party constitute its output variables. Thus LABOR is interested in the behavior of REAL WAGES while PROFIT is in the EMPLOYERS experimental frame. The input variables of a frame are comprised of variables controllable by the party itself, those controllable by other parties, and those determined exogenously (in the case of the national economy this means by other countries, in the case of a region, this means by other regions as well). A policy is expressed as determining the time behavior of the controllable input variables. In order to test the policy with a model, assumptions, often called a *scenario*, must be made concerning the behavior of the exogenous part of the input series. A party is particularly concerned with manipulating its own controllable variables to achieve acceptable levels of the indexes reflecting its own interests. Thus GOVERNMENT will be examining the effect of different taxation levels while EMPLOYERS study the effect of varying the distribution of its investments.

The concept of experimental frame also makes provision for *run control* variables and constraints. That is, the frame asserts that it is only interested in the model (or real system) behavior so long as the constraints specified are being satisfied. For example, such constraints may check that the assumptions of a model are being realized. Also model trajectories may be required to satisfy feasibility conditions, for example that rates of change are kept below given bounds. Finally each party may try to optimize the performance indexes of interest to it while acknowledging that minimum levels of other indexes must be attained if a policy favorable to it is to have a chance of being accepted by the other policies. In this case, the performance indices of the other parties appear as run control variables in the frame of a party.

One of the important functions of the entity structure is to organize all models and experimental frames of interest. The basic idea is to associate models and frames with the entities that they concern. Thus each entity represents a component of the real system and therefore ought to point to the experimental frames which characterize questions about it and the models

18   *1. Motivation and perspective*

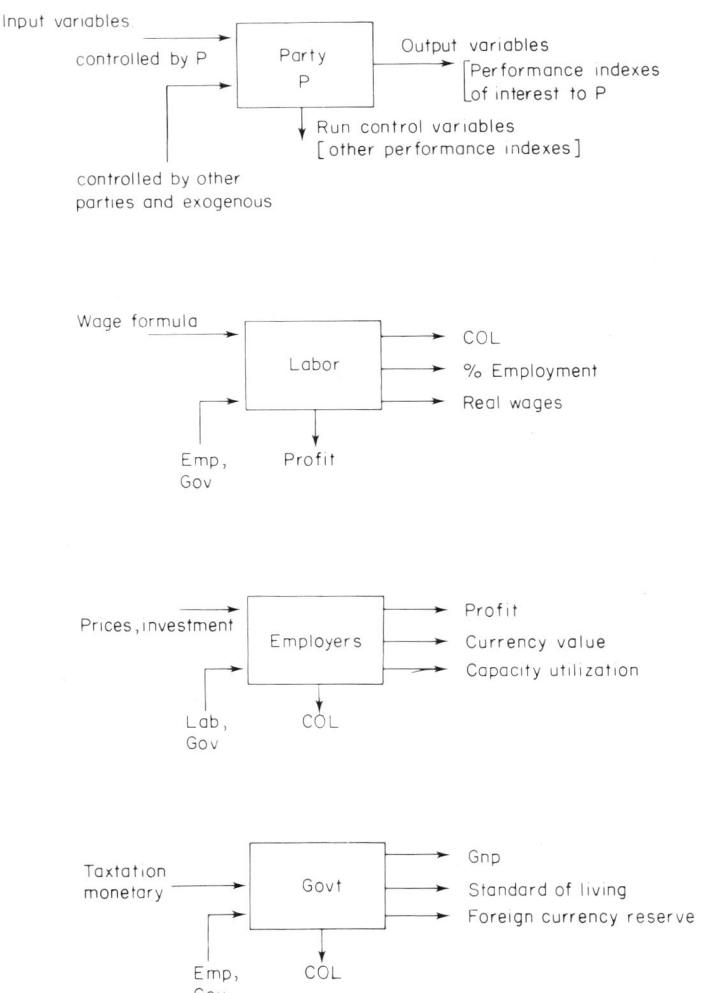

Fig. 1.6. Experimental frames for Elzas negotiation methodology.

which can be used to answer such questions. In our example, the prototypic experimental frame may be attached to many of the entities in the structure. Each of the parties at the NATIONAL ECONOMY level has such a frame, but so do each of the parties at the REGIONal levels, as well as each of the subdivisions of the parties representing the SECTORS. For example, the AGRICULTURal workers at both the NATIONal and REGIONal levels have their own experimental frames. Elzas' gaming methodology provides both for horizontal games in which NATIONal performance indexes are reconciled, as well as vertical games in which nationally proposed policies are examined for their regional implications.

4. Summary

This chapter has shown how objective oriented modelling arises naturally in decision making for management, control or design. This multiple objectives/multiple models approach is justified to the extent that real systems are partially decomposable, or more picturesquely, multifacetted. Methodologies for modelling "in the small" and "in the large" are mutually supporting and unifiable in a framework which sees modelling and simulation as an ongoing, purposeful, integrated and cumulative process. In the coming chapters, we explore this process in more depth, seeking concepts and software tools to facilitate it.

References

[1] Sprague, R. H. and E. D. Carlson (1982). *Building Effective Decision Support Systems*, Prentice Hall, N. J.
[2] Elzas, M. S. (1982). "The Use of Structured Design Methodology to Improve Realism in National Economic Planning". In *Model Adequacy* (Proceedings of the Intl. Conf. on Model Realism, Bonn) (ed, H. Wedde), Pergamon Press, London.

Other Relevant Literature
Zeigler, B. P. (1976). *Theory of Modelling and Simulation*, Wiley, NY. This is the framework in which this approach of this book is set.
Shannon, R. E. (1975). *Systems Simulation: The Art and Science.* Prentice Hall, NJ. Provides a comprehensive treatment of simulation methodology from the classical project-oriented viewpoint.
Zeigler, B. P., M. S. Elzas, G. J. Klir and T. I. Oren (1979). *Methodology in Systems Modelling and Simulation*, North Holland, Amsterdam. The first collection of articles seriously concerned with methodology of modelling and simulation, its relation to design methodology, and its computerization.
Oren, T. I., M. S. Elzas and B. P. Zeigler (1984). *Simulation and Model-Based Methodologies: An Integrative View*, Springer-Verlag, NY. Lectures presented at the NATO sponsored Advanced Study Institute, Ottawa, Canada, Summer, 1982. Successor to above.
Vansteenkist, G. C. and J. Spriet (1982). *Computer Aided Modelling and Simulation*, Academic Press, London. A preliminary version of the present work appears in Chapters 2 and 5 of this book based upon lecture notes I developed for a special Chair in simulation sponsored by IBM Belgium.
Greenberg, H. J. and J. S. Maybee (1981). *Computer-assisted Model Analysis and Model Simplification*, Academic Press, NY.

PART 2

FUNDAMENTALS AND FORMALISM

Chapter 2

UNDERSTANDING FORMALISMS

Discussion of modelling methodology is greatly facilitated by adhering to certain conventions of communication called *formalisms*. A formalism specifies a class of objects under discussion in an unambiguous and general manner. Its key advantage is that it permits one to focus on features of objects which are salient to a discussion, a process called *abstraction*. Manipulating abstractions is a potent means of formulating and solving real problems. Indeed, progress not only in mathematics, but in science and engineering, is dependent to a significant extent on finding the right abstractions for the objects under study. However, because abstraction removes one from the more familiar world of the concrete, it is also more difficult to grasp. This chapter is intended to introduce the concepts of formalism and abstraction underlying the approach of succeeding chapters. The basis of abstraction is set theory. We turn to a brief discussion of set theory and abstraction.

1. Abstraction: Sets and Set Operations

The essence of the set concept is *abstraction*. Set operations allow us to manipulate abstractions without having to specify their detailed nature. Set theory is so useful to the modelling enterprise precisely because abstraction is fundamental to this enterprise.

Some commonly employed sets are finite sets such as the digits $\{1,2,3,4,5,6,7,8,9\}$ or the alphabet $\{a,b,...,z\}$ and infinite sets such as the integers $I = \{0, \pm 1, \pm 2,...\}$ and the real numbers, R. Subsets of these sets are also important: $I^+ = \{1,2,3,...\}$ is a subset of positive integers; $I_0^+ = I^+ U \{0\} = \{0,1,2,3...\}$ is the subset of non-negative integers. Adding the infinity concept, we get: $I_\infty^+ = I^+ U \{\infty\}$, the set of non-negative integers together with the symbol ∞, R^+, R_0^+ and R_∞ denote the analogous sets in the case of the reals.

A fundamental set operation is a *crossproduct*. Let A and B be arbitrary sets. Then the crossproduct, $A \times B = \{(a,b) | a \in A, b \in B\}$, the set of all pairs of elements from A and B in that order. Note how the detailed nature of the elements in A and B need not be known in order to form pairs of these elements. This is the power to manipulate abstractions that we just referred to. Similarly, one can define triples, quadruples, ..., and in general, n-tuples. $A = \{(a_1,a_2,...,a_n)|a_i \in A, i \in I\}$ is the set of all n-tuples of elements of A. We may also identify the tuple $(a_1,a_2,...,a_n)$ with the sequence or string $a_1 a_2 ... a_n$. The set of all such sequences is denoted A^+, while $A^* = A^+ \cup \{\Lambda\}$ includes the empty sequence, Λ.

Manipulating such abstractions becomes interesting when one forms *structures* by not accepting all the combinations in a crossproduct. The simplest such structure is a *binary relation* $R \subset A \times B$. A *function* or *mapping* is a relation $f \subset A \times B$ in which there is at most one pair involving any member of A, i.e., $(a,b) \in f$ and $(a,b') \in f \rightarrow b = b'$. In this case, we also write $f:A \rightarrow B$ and $f(a)$ to mean the unique b for which $(a,b) \in f$. In a moment, we shall be forming systems and other structures on these principles.

Thus *theory construction* starts with abstraction and builds compound structures from these abstractions. But eventually to apply such structures to the real world we must connect the abstractions with real objects. Let us think of a real object as being an abstract one with much detail. Then we go in the direction of connecting abstractions to real objects by adding detail to these abstractions. We call this process *concretization* (making the abstract more concrete). Now, adding detail amounts to replacing an abstract set by a structure. So both theory construction and its concretization involves building structures — but in opposite directions.

2. Formalisms and Objects

Set theory provides the means to construct formalisms which specify objects. Let us now see how this is done. The formal objects will be set-theoretical structures such as finite state machines, directed graphs and trees. Each class of objects is represented by a formalism which prescribes its parameters and any govering constraints. To specify a particular object from the class, the "user" assigns values to the parameters which satisfy the constraints.

Thus, a *formalism* prescribes a list of *parameters* which are set-theoretic constructs, and list of *constraints*. An *object* in the class signified by the formalism is specified by an assignment of values to the parameters which satisfy the constraints. The *structure* of a formalism refers collectively to its parameters and constraints. Its elements provide a source of complexity measures as we shall see.

Consider the following example:

Finite State Machine (FSM) Formalism [1,2]

    Parameters: inputs, states, outputs,
                   transition function, output function

    Constraints: inputs, states, outputs are finite sets;
                   the transition function maps
                   states × inputs into states;
                   the output function maps
                   states into outputs (Moore version),
                   or states × inputs into outputs
                   (Mealy version).

Directed Graph (Diagraph) (DIG) Formalism (See Figure 2.1)

    Parameters: vertices, directed edges

    Constraints: vertices and directed edges are finite sets;
                   directed edges are ordered pairs of vertices.

Tree Formalism (See Figure 2.2)

    Parameters: nodes, root, successor function

    Constraints: nodes is a finite set;
                   root is a member of nodes

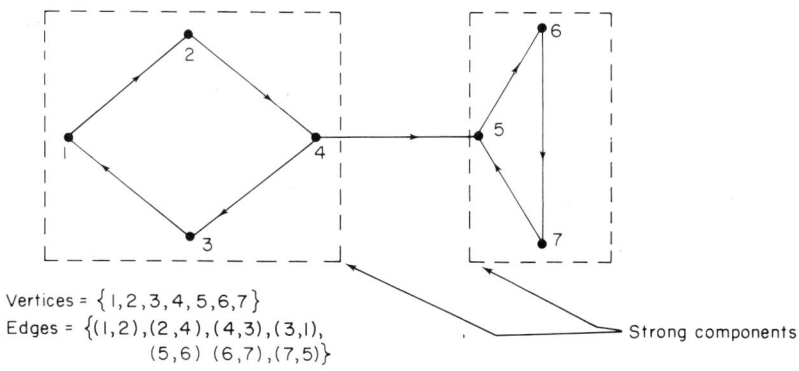

Fig. 2.1. Diagraph example showing strong components.

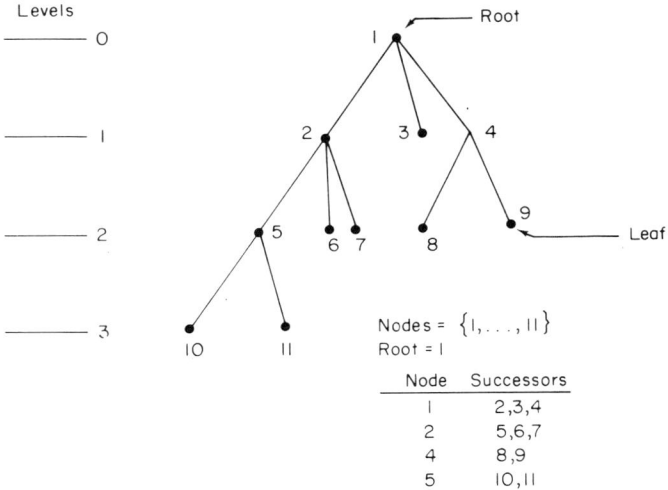

Fig. 2.2. Tree example showing root, leaf and levels.

> successor function assigns to each node a subset of nodes called its *successors*
> root is not the successor of any node, every other node is a successor of some (unique) node
>
> the successors of distinct nodes are disjoint
> a node is not in its own successor set

A *leaf* is a node with no successors, other nodes are called *interior nodes*.

The structure of a formalism provides a basis for measuring the complexity of objects. A *complexity measure* for a formalism class assigns to each object in the class a non-negative integer called its *complexity* in this measure. A complexity which can be computed directly from an object's formal structure is called a *formal complexity* (or *structural complexity*: using the term *formal* is clearer in that it leaves no doubts that the complexity is relative to some formal description).

Examples of formal complexity measures are:

Finite State Machines
   Complexity Measures: number of inputs, number of states, number of outputs, total number of inputs, states, outputs.

Directed Graphs
> Complexity Measures: number of vertices, number of directed edges, maximum fan-in, maximum fan-out, number of strong components, maximum size of strong component, maximum feedback indegree, minimum cycle break set size.

Tree
> Complexity Measures: number of nodes, maximum number of successors, number of levels.

Thus each formalism offers a variety of possibilities for measuring the complexity of its objects. Some of these measures are immediately computable (e.g. number of states), some are relatively easy to compute (e.g. number of levels) and some are computable only with great computational effort (e.g. minimum cycle break set size). Some are *local* measures, i.e. refer to properties of individual parameters of the structure (e.g. maximum number of successors); some are *global* measures referring to properties of the structure as a whole (e.g. number of levels).

2.1. Formalisms for Hierarchical Structuring

The preceding discussion of finite state machines enables us to set forth some formalisms for multicomponent models in a rather straightforward way. After the general systems formalism has been introduced we shall be in a position to discuss more useful and common means of connecting together model components. A *coupling* or *network* of finite sequential machines represents an interconnection of stand alone systems to form a composite system (in software terminology the component machines are called modules and the interconnection is called interfacing). When each component is further represented as a coupling of component machines, and this decomposition is continued recursively to some finite level, the result is a *hierarchical* specification. Consider now formalisms to represent these concepts.

FSM Network (Coupling of Finite State Machines) (See Figure 2.3)

> Parameters: index set, D
> set of FSMs indexed by D
> influencers indexed by D
> interface maps indexed by D
>
> Constraints: D is a finite set (of names for the component FSMs)

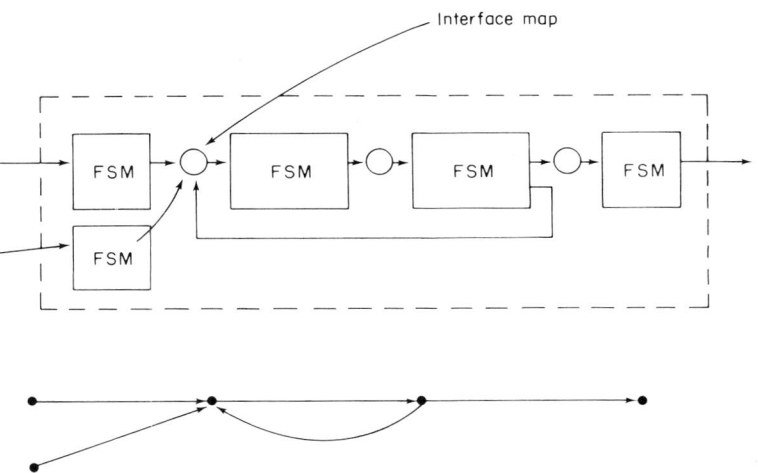

Fig. 2.3. An FSM network and its underlying diagraph.

influencers associates with each
element of D a subset of D (the FSMs
that immediately influence an FSM);
interface maps associates with each FSM
a mapping that translates the
composite[2] outputs of its influencers
into its input set.

A component FSM with no influencers (empty influencer set) is said to an *initial* component. An initial component does not receive input from any of the components. Several such components may exist. The crossproduct of their input sets constitutes the input set of the resultant of the FSM Network.

The *resultant* of an FSM Network is an FSM with the following structure:

inputs: the crossproduct of input sets of the initial components
states: the crossproduct of the component state sets
outputs: the crossproduct of the component output sets
transition function: obtained by applying to each component the input which is the translation of the composite output of its influencers
output function: the composite of the component output functions.

---

[2]"Composite outputs" here means the crossproduct of the output sets.

FSM Hierarchy (See Figure 2.4)
  Parameters: Tree, T
        set of FSMs indexed by nodes of T
        set of FSM Networks indexed by interior nodes of T

  Constraints: At each interior node, the FSM at (indexed by) the node is the resultant of the FSM Network at the same node.

  The component FSMs of the FSM Network at a node are those indexes by the nodes of its successors.

Thus, an FSM Hierarchy specifies a tree structure which forms the backbone for describing successively more detailed FSM networks. The FSM at the root node is the resultant of an FSM Network, whose component FSMs are associated with the next level nodes of the tree. Each of these FSMs is in turn, the resultant of an FSM Network of components at the next lower level, and so on. The leaf nodes represent FSMs which are undecomposed. We shall return to the subject of hierarchical decomposition in greater depth quite frequently throughout the book.

2.2. Mapping of Objects: Association, Abstraction and Specification

Classes of objects are often related in ways which can be formalized as mappings from one class to another. Of interest to us here are three types of mappings called *abstraction*, *association* and *specification*. The essential

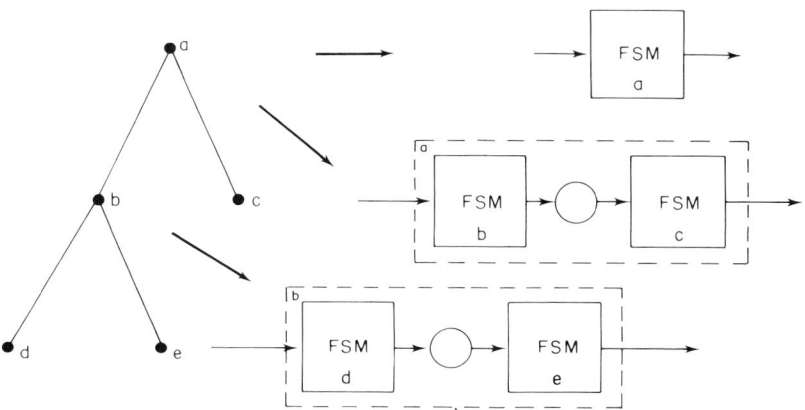

Fig. 2.4. FSM hierarchy. Tree at left indexes the FSMs at right (for example, node b indexes the FSM b and the FSM network consisting of FSMs d and e).

characteristic of the first two types is their many-one, non-invertable nature. In contrast, the third is distinguished by its one to one mapping of a smaller, more specific class into a larger, more general one.

Abstraction

Abstraction is the process by which details of an object are left unspecified in favor of a less cluttered appreciation of its structure. Abstraction is thus characterizable as a mapping of objects of one class into a second, presumably less complex class. The source class will be called "concrete" relative to the target or "abstract" class. One important effect of abstraction is that many source objects will be represented by the same target abstraction, or what is the same thing: given the abstract representation we can not uniquely recover the concrete object it represents.

In order to measure the complexity of a formalism let us select a *diversity measure* which counts the number of parameters in its structure specification and a *depth* measure which measures the number of levels of recursion in this specification. The depth of a formalism is 1 plus the maximum of the depths of its parameters; the depth of a parameter is the depth of its defining formalism. The depth of an "atomic" formalism (which does not employ parameters from other formalisms) is 1. Note that these are complexity measures, this time applied to the formalism itself rather than its objects.

Examples of Abstraction Mappings:
Abstracting the Influence Structure of a FSM Network

    Source Class: FSM Network with Diversity = 4, Depth = 2
    Target Class: DIG with Diversity = 2, Depth = 1
    Mapping: Assign to an FSM network a Digraph whose vertex
        set is the index set of the network and edges
        (A,B) where A is an influencer of B in the coupling.

Abstracting the Hierarchic Structure of an FSM Hierarchy

    Source Class: FSM Hierarchy with Diversity = 3, Depth = 3
    Target Class: Tree with Diversity = 1, Depth = 1
    Mapping: Project FSM Hierarchy onto its Tree parameter.

Association

The second type of mapping is that involved in dropping down from one level to the next in a hierarchy of system specifications, i.e. in going in the direction of "structure" to "behavior". This level dropping is formalized by considering each level as a formalism class and the mapping represents the *association* of a lower level object with a higher level one. Since the same behavior may be exhibited by many distinct structures, a many-one mapping also characterizes this association.

Examples of Association Mappings:

Associating the resultant of an FSM Network

>Source Class: FSM Network
>Target Class: FSM
>Mapping: Associate to each FSM network its
>   resultant FSM

Associating the top level network with an FSM Hierarchy

>Source Class: FSM Hierarchy
>Target Class: FSM Network
>Mapping: Associate with each FSM Hierarchy the FSM
>   Network
>   at the root of its defining Tree.

Associating the resultant FSM of a FSM Hierarchy

>Source Class: FSM Hierarchy
>Target Class: FSM
>Mapping: Associate to each FSM Hierarchy its root FSM

The FSM Hierarchy, FSM Network, and FSM formalisms constitute successively lower levels in a hierarchy of system specifications, soon to be discussed in more generality (Chapter 3). The first and second of the above association mappings provide means for descending this hierarchy one level at a time. The third is the composition of the first two providing the two step transition from FSM Hierarchy to FSM levels.

The inverse, set valued, mapping, of an association mapping is called a *realization* (or *implementation*) mapping. Elements in the inverse image of an object are called its *realizations*. Referring to the above mappings we have:

The FSM Network Realizations of an FSM = the FSM Networks that map into the FSM under the Network-to-FSM association.

The FSM Hierarchy Realizations of an FSM = the FSM Hierarchies that map into the FSM under the Hierarchy-to-FSM association.

In a design or synthesis problem, a lower level object is given and the problem is to find an "acceptable" higher level realization. For example, an acceptable Network realization might be sought for a given FSM. Complexity measures may be components in the "acceptability" criterion as we shall see (Section 2.3).

Specification

A subclass of a class of objects can often be specified most conveniently by introducing a new formalism for the subclass. Objects in the subclass can then be manipulated within the new formalism. However, in order to relate these objects to their brother objects in the wider class, a mapping must be provided which translates the special formalism into the more general one. With such a translation available, all the concepts and manipulations applicable to objects specified in the general formalism are applicable to objects specified in the special one as well.

To illustrate this idea, let us consider the subclass of finite state machines called finite memory machines. A finite memory machine is a finite state machine for which the state set takes on a very special form reflecting the fact that such a machine stores a finite number of its past inputs and outputs to compute its next output. This is a special case of memory retention since in general, the output produced by a finite state machine is not uniquely determined once one knows a finite part of its recent past input and output. A finite memory machine [1] can be specified as follows:

Finite Memory Machine (FMM)

> parameters:
>   input set, X
>   output set, Y
>   number of past inputs, n
>   number of past outputs, m
>   output function, f
>
> constraints:
> X, Y are finite sets
> n, m are non-negative integers
> f maps the crossproduct of X^n and Y^m into Y

The finite state machine which is specified by an FMM is constructed as follows:

Translation from FMM to FSM

    source: FMM
    target: FSM
    mapping:
      input set, X (given by FMM)
      output set, Y (given by FMM)
      state set, $X^n \times Y^m$
      output function, f
      transition function:
        $\delta((x_1,\ldots,x_n, y_1,\ldots, y_m),x) = (x,x_1,\ldots,x_{n-1},y,y_1,\ldots,y_{n-1})$
    where $y = f(x_1,\ldots,x_n, y_1,\ldots,y_m)$.

Note how the transition function handles the computation of the next output y and the insertion of this value and the input x into their respective pipelines and the shifting of the saved values over one place (the oldest values are lost in the process). The important point being illustrated is that the general form of this transition function is known once one states that the machine is of the finite memory subclass and supplies the special parameter values needed to distinguish between members of this class.

2.3. Induced Complexity Measures

New complexity measures arise from mappings between classes of objects. Let M be mapping from class C to class C'. Then a complexity may be attributed to an object in C by assigning it the complexity of its image object in C'. Thus any complexity measure of C' can be "pulled back" to apply to C. Examples of such *attributed* measures are:

Interactive complexity of FSM Networks: any measure of complexity of the associated Digraph such as maximum fan-in, number of strong components, etc. can be attributed to the FSM coupling and measures aspects of the interactive complexity of the network.

Hierarchical Complexity: any complexity measure of the defining tree of a FSM Hierarchy such as maximum number of successors, number of levels, etc., can be attributed to the FSM Hierarchy and measures its hierarchical complexity.

Such attributed measures are usually appropriate for abstraction mappings. In this case an object inherits the complexity of its abstraction.

In the case of association mappings the induction of complexity measures usually goes in the other direction: any complexity measure of the source class induces a complexity measure on the target class by the formula:

> complexity of object in target = minimum of complexities its realizations in the source class

The minimum operator is the one most often used to summarize the complexities of the inverse image. Representing the best realization of the image, it is most amenable to theoretical studies. However, in practice, other summaries such as the average, or other distribution statistic may be more helpful in estimating the typical complexity of a realization.

Induction of complexity measures can be compounded so that for example one can define:

> Interactive complexity of a FSM = the minimum of the interactive complexities of its FSM Network realizations

where as above, interactive complexity is one of the measures attributed to the Network by its Digraph.

> Hierarchical complexity of a FSM = the minimum of hierarchical complexities of its FSM Hierarchy realizations.

2.4. Computational Complexity

The study of computational complexity is a central theoretical area of computer science. It is useful to understand the distinction between it and the preceding concepts of structural complexity. The basic paradigm can be stated as follows:

> Given an object, and a question concerning it, what is the space or time required by a device or processor to answer to the question.

It is presumed that the computing device "knows" the formalism in which the object is specified, i.e. is able to properly interpret the object structure. Actually, the object is usually parameterized by some measure of its size, and the space/time complexity measure is studied as a function of this size. (This size measure is usually a formal complexity measure of the object, see later discussions.) By considering a class of computing models one can discuss summary measures as discussed above for the complexity of a question relative to this class. By considering a universal class, such as Turing Machines, and abstracting only the form of growth of complexity as size gets large, results can be obtained which are model independent, representing the inherent computational complexity of the question.

An example of this approach is the following:

Class: DIG
Question: Given digraph, and a pair of vertices,
is the second accessible from the first?
Size Measure: number of vertices

Class: DIG
Question: Does given digraph contain a
Hamilton Cycle?
Size Measure: number of vertices

In the case of the first problem, the time required to obtain an answer is known to be a polynomial function of the number of vertices. The second problem, however, represents a large class of problems called NP (Non-deterministic Polynomial) Complete of which it is not known whether the minimum solution time is polynomial or exponential [2].

As noted previously with respect to realization complexity, growth orders obtained as above provide a *lower bound* on the space/time function that characterizes a particular class of computing models. In practice other measures, such as the average, or most probable, space/time cost may be more significant.

The basic paradigm applies to the study of simulation complexity as well as we shall see.

3. Useful Abstractions

As Minsky ([3]) observed a model is not simply a model, it is a model which can answer certain questions about a certain object for a certain questioner. Abstractions are created in order to serve as models in this sense. We have already mentioned that they should provide a structurally less complex representation of an object. But abstraction serves a purpose only if it is informative about some aspect of the object of interest to the abstraction builder or manipulator. For example, the Digraph abstracted from a FSM Network is a useful abstraction for questions concerning the propagation of influence in the Network (e.g. what components will be immediately affected by a modification in a particular component). It is of no use however for questions concerning the structure of the Network components.

We are now in a position to formulate more closely what makes an abstraction useful. Let *Abs* and *Obj* be objects such that *Abs* is an abstraction of *Obj* according to the previous definition. Let *Q?* be a question about *Obj* and *Proc* a processor of questions. Consider the following definition.

Abs is *useful* for *Q?* about *Obj* if:

(i) *Q?* is applicable to *Abs*
(ii) *Abs* is valid for *Q?* wrt *Obj*
(iii) space/time(*Q?,Abs,Proc*) < space/time(*Q?,Obj,Proc*)

where space/time(*Q?,Obj,Proc*) is a measure of the computational complexity involved in computing the answer to *Q?* addressed to *Obj* on *Proc*.

Thus an abstraction is useful for a given question if it makes sense to address the question to it (i), it gives the correct answer, i.e. the same answer that querying the original object would (ii), and it is computationally easier to use the abstraction (iii). To summarize: the utility of an abstraction depends on its applicability, validity, and computational advantage with respect to questions concerning the object it models. Establishing that each of these requirements holds constitutes one component of the overhead that accompanies the use of abstraction as a key concept in managing complexity [4].

The concepts we have just stated apply, in particular, to the kinds of abstractions, namely system models, which are of central interest in this book. Indeed, we shall elaborate quite fully on the concepts of applicability, validity, and computational advantage in the context of multifacetted simulation modelling methodology in the sequel.

References

[1] Gill, A. (1965). *Introduction to The Theory of Finite State Machines*, McGraw Hill, NY.
[2] Hopcroft, J. E. and J. D. Ullman (1979). *Introduction to Automata Theory, Languages and Computation*, Addison Wesley, Reading, MA.
[3] Minsky, M. (1965). "Models, Minds, Machines". *Proc. IFIPS Conf.*, AFIPS Press, Montvale, NJ. pp. 45–49.
[4] Zeigler, B. P. and R. Rada (1983). "Abstraction in Methodology: A Framework for Computer Support". *Information Processing and Management* (in press).

Other Relevant Literature
Stanat, D. F. and D. F. McAllister (1977). *Discrete Mathematics in Computer Science*, Prentice Hall, NJ.

Chapter 3

SYSTEM SPECIFICATIONS AND MORPHISMS

This chapter continues laying the formal foundation begun in the preceding chapter. We conceive of models as means of specifying systems and develop a two-dimensional approach to such system specification: levels of system specification along one dimension, and formalisms for system specification along the other.[3]

A system is a coupling of interacting component systems. From this innocent looking definition much follows. Obeying the recursion implicit in the definition, we find that a system is a coupling of systems, each in turn being a coupling of systems,..., and so on ad infinitum. To be useful, such a decomposition process must be allowed to stop. We must, therefore, have a concept of system which allows us to treat it as atomic when we wish to stop decomposing, but also allows us to continue decomposing when we wish to do so. Set theory is just the right tool for this since what we require is a definition of a system as an abstract structure with the property that it can always be concretized by replacing it with a coupling of structures of the same kind.

Let us explore the implications of the last thought. Since a system may be a component of a larger system, we must provide it with an *interface*, i.e. a means to interact with other systems. The interface should represent the potential events which can occur at the system boundary. What goes on at this boundary is determined both by the system, within the boundary, and its environment, outside the boundary. The system's role in this arises by constraints it imposes on the interface, i.e. by what we call its *internal structure*. It is this internal structure which we must be allowed either to leave abstractly described (to stop decomposition) or to concretize (to continue decomposition) as desired.

To faithfully concretize internal structure, we must be able to *reconstitute* it

---

[3]The original formulation of this approach appeared in [1], Chapters 9 and 10, where many examples may be found to illustrate the concepts. The current formulation also appears in [2].

by combining the separate constraints that each component system's internal structure places on their mutual interfaces. A system *submits* to a (particular) decomposition if such a reconstitution is possible with this decomposition.

In the following, we review a hierarchy of system specifications which formalizes the foregoing notions of interface, internal structure and recursive decomposability at various levels, each level introducing more concreteness into the description of internal structure.

Exposition of the stratification begins by describing each of the levels of system specification starting with the lowest level (Figure 3.1). Then we provide so-called association mappings which take a specification at one level and produce its counterpart at the next lower level. This downward motion, in the structure-to-behavior direction, formally represents the process by which a simulator generates the behavior of a model. Each level also has a morphism concept that enables similarity comparisons between systems specified at that level. We show how the downward association of specifications is accompanied by a parallel association of morphisms.

While the downward association of specifications and morphisms is straightforward, the upward association is much less so. This is because many structures may exhibit the same behavior, so that recovering a unique structure from a given behavior is not possible except in special circumstances, called justifying conditions. Climbing up the hierarchy, the direction of structural inference, is the final topic in the first section.

Modelling formalisms are the topic of the second section. Each formalism

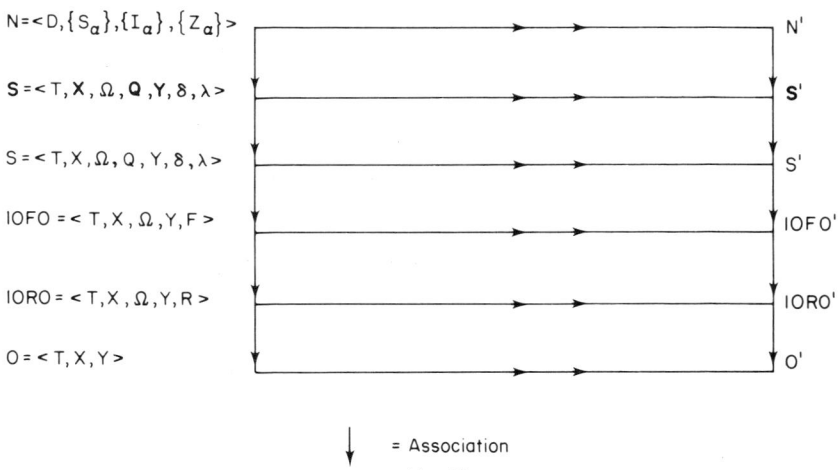

Fig. 3.1. Hierarchy of system specifications and morphisms.

assumes certain definite constraints on the class of systems that it can describe. The section sets the basis for the discrete event formalism that is the topic of the next chapter.

1. Levels of System Specification

We begin with the concept of observation frame which sets up the basic abstractions for observing a system as a black box. This concept is elaborated later into the concept of experimental frame (Chapter 5) which can characterize in more detail the observation and control constraints which may govern experimentation.

Level 0. Observation Frame

An *Observation Frame* is a structure

$$O = <T,X,Y>$$

where

T is a set, the *time base*
X is a set, the *input value* set
Y is a set, the *output value* set

subject to the constraint:

T is the R (the reals) or I (integers)

Time set

T is a set which represents clock time and serves to order events. Common choices for T are I, the integers, or R, the reals; S is said to be a *discrete time*, respectively *continuous time*, system accordingly. The essential properties required of the time base are that it be a linearly ordered abelian group under addition. Thus the rationals would also serve as a time base and in fact would better represent the time keeping ability of the digital computer in simulation.

Input value set

X is the set representing that part of the interface through which the environment affects the system, for example, through the inflow of information or material. Thus, the system is viewed as being subjected to elements from the inputs set X, at any time, over which it has no direct control. A common choice for X is R^n, for $n = 1,2,3...$, representing n real valued input variables. A second common choice is the discrete event set (Chapter 12), $X \cup \{\phi\}$ where X represents a set of external events and ϕ represents the non-event (i.e. the absence of an event).

Output value set

Y is a set representing that part of the interface through which the system affects the environment. The interpretation is the same as that for the input value set except for direction. Embedded in a larger system, the input (output) of a system component is the output (input) of its environment.

At this level we, the observers, are able to set up experiments on the system by stimulating it over an interval of the time base T with values from X and observing its response of values from Y.

An input segment describes a chronological pattern of inputs to the system over some time period. When embedded in a larger system, such a pattern is determined by the system's environment. When we isolate the system, by cutting its input wires so to speak, these wires are free, and we replace the environment by a set of input segments Ω (for reconstitutability Ω should include all the patterns receivable by S as a component in the larger system).

Thus, an *input segment* is a map of the form $\omega: <t_{in}, t_f> \to X$. Here $<t_{in}, t_f>$ is an interval of the time base between t_{in} (the initial instant) and t_f (the final instant) (t_{in} and t_f are arbitrary except that $t_{in} \leq t_f$). The set of all such input segments is called (X,T). The set of all output segments is called (Y,T).

Level 1. I/O Relation Observation

An *Input/Output Relation Observation (IORO)* is a structure:

$$IORO = <T, X, \Omega, Y, R>$$

where

$<T, X, Y>$ is an observation frame
Ω is a set, the *input segment* set
R is a relation, the *I/O Relation*

subject to the constraints:

$$\Omega \subset (X, T)$$
$$R \subset \Omega \times (Y, T)$$

such that if $(\omega, \rho) \in R$ then $\text{dom}(\omega) = \text{dom}(\rho)$.

(ω, ρ) is called an *input/output segment pair*. The restriction $\text{dom}(\omega) = \text{dom}(\rho)$ means that both segments of an I/O pair are observed over the same time interval (if $\omega: <t_{in}, t_f> \to X$, then $\text{dom}(\omega) = <t_{in}, t_f>$). A choice of Ω represents an expression of interest in only a subset, Ω, of all possible input segments. An I/O relation consists of output responses paired with input segments from Ω.

A common choice for Ω is the set of *piecewise continuous segments*. In this case $T = R$ and $X = R^n$, an n-dimensional real vector space. A second choice is the set of *discrete event segments* over X (a set of external events) and $T = R$. Such a discrete event segment is a mapping $\omega: <t_{in},t_f> \to X \cup \{\phi\}$ such that $\omega(t) = \phi$ except possibly for a finite set of event times $\{t_1,...,t_n\} \subset <t_{in},t_f>$. Finally, when $T = I$, Ω is in effect a set of finite sequences.

Level 2. I/O Function Observation

An *Input/Output Function Observation (IOFO)* is a structure:

$$IOFO = <T,X,\Omega,Y,F>$$

where

$<T,X,Y>$ is an observation frame
Ω is a set, the input segment set
F is a set, the set of *I/O Functions*

subject to the constraints

$f \in F$ implies $f \subset \Omega \times (Y,T)$
f is function such that $\mathrm{dom}(f(\omega)) = \mathrm{dom}(\omega)$

At this level we have knowledge of the "initial state" of the system in the sense that if it is in a state represented by f then the input segment ω will yield a unique response $f(\omega)$.

Level 3. I/O system

An *I/O System* is a structure

$$S = <T,X,\Omega,Y,\delta,\lambda>$$

where

T,X,Ω,Y are as in Level 2
δ is a function, the *transition function*
λ is a function, the *output function*

Internal state set

The set Q of internal states represents the memory of the system, i.e. the residue of its past history which will affect its future response. This is the heart of the modelling of internal structure which was referred to earlier.

State transition function

The state transition function is a map $\delta: Q \times \Omega \to Q$. Its interpretation is that if the system is in state q at time t_i and an input segment $\omega: <t_i,t_f>: \to X$

is applied then $\delta(q,\omega)$ is the state of the system at time t_f. Thus the internal state at any time and the input segment from that time on uniquely determine the state of the end of the segment.

That the state set Q must summarize all relevant information about the past is enforced by requiring the transition function to satisfy the *composition property*:

For every $q \in Q, \omega \in \Omega$ and t in the domain of ω (the interval on which it is defined), we have:

$$\delta(q,\omega) = \delta(\delta(q,\omega_{t>}),\omega_{<t})$$

where $\omega_{t>}$ is the part of ω between t_{in} and t and $\omega_{<t}$ is the part between t and t_f. This requires that the state $q_t = \delta(q,\omega_{t>})$ pertaining at any time t summarize all previous history so that continuing the experiment from this state will result in the same final state as would be the case if no interruption were made.

Finding the proper state space (i.e. one permitting the composition property) of a system is not a trivial matter. But once done it enables us to replace the past by an abstract quantity in the present. This formulation of internal structure greatly simplifies our ability to deal with decomposition (concretization of internal structure) and simulation (relation to other systems).

It should be clear that the state set is a pure modelling concept. Nothing in the real system need directly correspond to it. A modeller hypothesizes that the state set of his model captures the information necessary to predict the system's behavior, but the state set need not be identifiable with directly observable features of the system.

Output function

The output function is a map $\lambda: Q \to Y$. When the system is in stage q the value of the output is $\lambda(q)$. Thus λ relates the hypothetical internal state of the system to the effect of the system on its environment. In other words, $\lambda(q)$ is what can be sensed by the environment (or measured by the observer) when the system is supposed to be in state q. Usually, λ is a many-one mapping so that the state cannot be directly inferred from the output.

Level 4. Structured System

Briefly put, a system specification at this level is the same as that at Level 3 except that each of the sets and functions are *structured*, i.e. made more concrete by being represented as crossproducts of more elementary sets and functions.

First let us review the concept of a *structured set*. This is a structure:

$$\mathbf{A} = <A, D, \{A_\mathbf{a} | \mathbf{a} \in D\}, i>$$

where

A is a set (the set to be structured)
D is an ordered set $\mathbf{a}_1, \mathbf{a}_2, \ldots$,
(the *coordinates*)
$A_\mathbf{a}$ is a set (the range set of $\mathbf{a} \in D$)
i is a function (the assignment function)

subject to the constraint:

$$i: A \to X_{\mathbf{a} \in D} A_\mathbf{a}$$

is a one-one map.

Structured sets are the system theoretic mechanism for representing the use of variables in modelling and simulation practice. As we shall see later, the concept of variable relates intimately to measurement and other pragmatic considerations (Chapter 10). For now, we consider a *variable* to play the role of a co-ordinate in a structured set, i.e. it has a name (e.g. **a**) and a range set ($A_\mathbf{a}$).

A *structured function* maps one structured set to a second. Thus if *A* is structured by **A** and *B* is structured by **B** then $f: A \to B$ is said to be structured. In this case, *f* is built out of a family of coordinate functions, $f_\mathbf{b}$, one for each coordinate **b** of **B**. Moreover, each $f_\mathbf{b}$ need not depend on all of the coordinates of **A**. So we may associate with each **b**, a subset of coordinates $l_\mathbf{b}$ on which it depends. We shall not pursue this possibility further here (see [1], pages 247–360).

Level 5. Coupling of Systems

We have already discussed the basic concepts involved in coupling systems together in our discussion of FSM Networks (Section 2.1) which were couplings of finite state machines. Similar concepts apply when considering arbitrary systems but many variations are possible depending on the particular class of systems being coupled. Our objective at this point is to provide a conceptually straightforward definition which conveys the basic concepts.

A *Coupling of Systems* (System Network, Multicomponent Model) is a structure:

$$N = <D, \{S_\mathbf{a}\}, \{I_\mathbf{a}\}, \{Z_\mathbf{a}\}>$$

where

D is a set of *component* names

for each $\mathbf{a} \in D$,

$S_\mathbf{a}$ is a system, component $\mathbf{a}$
$I_\mathbf{a}$ is a set, the set of *influencers* of $\mathbf{a}$
$Z_\mathbf{a}$ is a function, the *interface map* of $\mathbf{a}$

subject to the constraints:

$$S_\mathbf{a} = <T, X_\mathbf{a}, \Omega_\mathbf{a}, Q_\mathbf{a}, Y_\mathbf{a}, \delta_\mathbf{a}, \lambda_\mathbf{a}>$$
$$I_\mathbf{a} \subset D$$
$$Z_\mathbf{a}: X_{\mathbf{b} \in I_\mathbf{a}} Y_\mathbf{b} \to X$$

Note that the specification designates a set of components $<D, \{S_\mathbf{a}\}>$ and a *coupling scheme* $<\{I_\mathbf{a}\}, \{Z_\mathbf{a}\}>$. An *initial* component is defined as one without influencers ($\mathbf{a}$ is initial if $I_\mathbf{a}$ is empty). We consider the initial components as the only ones which will receive the external input to the multicomponent system. Any other component receives input which is instantaneously determined by the outputs of its influencers as translated via its interface map. This is more restricted than one would like in practice where a component might be allowed to receive both internal and external input.

We shall return to a further discussion of the coupled system level in Section 15.3.

2. Association Mappings for Moving Down the Hierarchy

We now describe for each system specification level $i > 1$ the association mapping which associates a system at level $i-1$ with a system at level i. By composing these association mappings we can associate a system at level 1 with a level i system.

Level Transition $5 \to 4$

Given a system specified as a coupling of systems at level 5 we wish to associate with it a system at level 4 when this is possible. Let N be a specification at level 5 and let S_N be an object associated with it to be defined as follows (this object may or may not be a system):

$$S_N = <T,\mathbf{X},\mathbf{\Omega},\mathbf{Q},Y,\delta,\lambda>$$

where

T is the common time base
$\mathbf{X}$ = composite($\mathbf{X_a}|\mathbf{a}$ is initial)
$\mathbf{\Omega}$ = composite($\mathbf{\Omega_a}|\mathbf{a}$ is initial)
$\mathbf{Q}$ = composite($Q_\mathbf{a}|\mathbf{a} \in D$)
Y = composite($Y_\mathbf{a}|\mathbf{a} \in D$)
δ = to be considered in a moment
λ = composite($\lambda_\mathbf{a}$)

The object S_N fails to be a system when the transition function is not well defined. When systemhood is known to hold we call S_N a *composite* system.

The use of the composite operator in the above contexts is explained as follows: For structured sets **A** and **B**, composite(**A**,**B**) is a structuring of the cross product $A \times B$ obtained by employing the structurings for A and B independently, i.e., composite(**A**,**B**) structures the set $A \times B$ with coordinates obtained as the disjoint union of those of A and B; the range sets being the union of the collections $\{A_\mathbf{a}\}$ and $\{B_\mathbf{b}\}$; the assignment function taking (a,b) into $(i_A(a),i_B(b))$. Thus input set X is the composite of the input sets of the components.

For segment sets Ω_1 and Ω_2, composite (Ω_1,Ω_2) is the set of segments $\omega: <t_0,t_1> \to X_1 \times X_2$ with $\omega = (\omega_1,\omega_2)$ such that $\omega_1: <t_0,t_1> \to X_1 \in \Omega_1$ and $\omega_2: <t_0,t_1> \to X_2 \in \Omega_2$. Thus the composite consists of all segments that can be constructed by applying equal domain segments from the two sources in parallel.

For functions, $f_i: A_i \to B_i$, $i=1,2$, the composite function is obtained by applying each of the component functions to its domain, i.e. composite(f_1,f_2) maps composite$(\mathbf{A_1,A_2})$ to composite$(\mathbf{B_1,B_2})$ such that (a_1,a_2) is mapped to $(f_1(a_1),f_2(a_2))$. Thus the output function is the composite of the output functions of the components.

At this level of generality the transition function can only be defined indirectly. The situation is similar to that which holds when defining the system specified by a differential equation. Only in the case that the equation has unique solutions can a (well defined) system be associated with it. Likewise, the coupling of systems sets up a system of recursive equations for which solutions may or may not exist. For specific classes of systems such as discrete time systems, discrete event systems, and differential equation systems, the reasons why a network may not define a system are well understood. We summarize this information in Table 3.1.

Table 3.1. Coupling of systems in various formulations.

Formalism	Sufficient conditions for existence of resultant	Simulation method
Differential equation (Ordinary)	*Lipschitz conditions	*Discrete time approximation based on Taylor and rational functional expansions
	*No Algebraic cycles	*Discrete event approximation
Discrete event	*Legitimacy	*Discrete event simulation strategies
Discrete time		
Moore components	*Always exists	*Each component advanced each time step
Mealy components	*No algebraic cycles	*Discrete event realization

Let $q \in Q$ and $\omega \in \Omega$. A segment Φ in (Q,T) is associated with q and ω if it is defined on the same interval as ω (domain(Φ) = domain(ω) = $<t_0,t_1>$) and
(1) Φ starts at q: $\Phi(t_0) = q$
(2) Φ satisfies the following equations (one for each $\mathbf{a} \in D$)

$$\delta_\mathbf{a}(q_\mathbf{a}, \omega_{\mathbf{a},t>}) = \Phi_{\mathbf{a},t>}$$

for all $t \in <t_0,t_1>$.

These equations require that a potential state trajectory Φ simultaneously satisfy the input/output requirements and internal constraints of each of the component systems recall the discussion of reconstitutability at the beginning of this Chapter). $\omega_\mathbf{a}$ is the input segment seen by system $S_\mathbf{a}$: if $\mathbf{a}$ is an initial component this is just the $\mathbf{a}$th component of the external input ω; if $\mathbf{a}$ is not initial, then $\omega_\mathbf{a}$ is the result of applying the interface map $Z_\mathbf{a}$ to outputs of the influence of $\mathbf{a}$, i.e.

$$\omega_\mathbf{a}(t) = Z_\mathbf{a}(\lambda_{\beta 1}(\Phi_{\beta 1}(t)), \ldots, \lambda_{\beta n}(\Phi_{\beta n}(t)))$$

Clearly a prerequisite for the equation to be satisfied is that $\omega_\mathbf{a}$ belong to $\Omega_\mathbf{a}$, i.e. be one of the input segments recognized in the definition of $S_\mathbf{a}$. Assuming this is true, the equation goes on to require that the state trajectory generated by component $\mathbf{a}$ when started in state $q_\mathbf{a}$ upon receiving $\omega_\mathbf{a}$ is just the $\mathbf{a}$th component, $F_\mathbf{a}$.

We say that the network, N has *unique solutions* if for each $q \in Q$, and $\omega \in \Omega$, there is exactly one solution, call it $\Phi_{q,\omega}$ to the equation set (1,2). In this case, we can define the transition function δ of S_N as follows:

$$\delta(q,\omega) = \Phi_{q,\omega}(t_1)$$

where t_1 is the endpoint of the domain of ω. This states that the state reached from q after an input ω is just where the state trajectory $\Phi_{q,\omega}$ takes it to at the last instant of the interval.

Note that δ is presented in structured form since it is defined for each component **a** using the **a**th component of $\Phi_{a,\omega}$. Thus is δ is well defined we associate the structured system S_N with the coupling of systems N.

Because of its implicit form, the general coupling of systems' framework cannot guarantee that solutions exist at all or are unique. Even if solutions exist and are unique, there is no guarantee that that any constructive method exists to compute the state trajectories given an initial state and input segment. It is therefore fortunate that many special classes are known for which solution methods exist (Table 3.1). Such solution methods are at the heart of the simulation concept (Chapter 5).

Level Transition 4 → 3

Given a system in structured form we may readily associate with it a system at level 3 by "forgetting" the structuring of the sets and functions at level 4. Another way of saying this is that we do not pay attention to any of the extra structure that we have at level 4 when operating at level 3. For example, let us list the tuples of each structured set in some order and then assign to each tuple its integer rank in the order; now using the structured (transition and output) functions we fill in tables which explicitly realize these functions. The new sets are all subsets of the integers and the functions are mappings of integers to integers.[4] The point is that the new system that we obtain cannot be distinguished from the original one at level 3. A formal statement of this indistinguishability is that the two systems are isomorphic at level 3, a concept to which we return in Section 3.3.

We formalize the "forgetting" of structural detail as follows: Given a structured system $\mathbf{S} = <T,\mathbf{X},\Omega,\mathbf{Q},\mathbf{Y},\delta,\lambda>$ we associate with it a system at level 3, $S = <T,X,\Omega,Q,Y,\delta,\lambda>$ where X,Q,Y are the sets structured by $\mathbf{X},\mathbf{Q},\mathbf{Y}$ respectively.

Level Transition 3 → 2

Given a system at level 3, $S = <T,X,\Omega,Q,Y,\delta,\lambda>$ we wish to associate with it an I/O Function Observation, $\text{IOFO} = <T,X,\Omega,Y,F>$ where T,X,Ω,Y are the same as in S, and F is defined as follows:

---

[4] Assuming that the original sets are all countable.

With every state $q \in Q$, and input segment $\omega: <t_0,t_1> \to X$ in Ω, there is associated a unique *state trajectory*

$STRAJ_{q,\omega}; <t_0,t_1> \to Q$ such that $STRAJ_{q,\omega}(t_0)=q$ and $STRAJ_{q,\omega}(t)=\delta(q,\omega_{t>})$ for $t \in <t_0,t_1>$

In case the system at level 3 was derived from one at level 5, this definition of state trajectory will just reproduce the state trajectory used in the definition of the system transition function. The observable trace of this trajectory is the *output trajectory* associated with $q \in Q$ and $\omega \in \Omega$,

$$OTRAJ_{q,\omega}: <t_0,t_1> \to Y$$

where

$$OTRAJ_{q,\omega}(t) = \lambda(STRAJ_{q,\omega}(t))$$

for $t \in <t_0,t_1>$.

The I/O function associated with state $q \in Q$,

$$f_q:\Omega \to (Y,T)$$

where

$$f_q(\omega) = OTRAJ_{q,\omega}$$

Thus f_q maps an input segment ω into the output segment produced in response to ω, when starting the system in state q.

From this we see how the functions f_q represent knowledge of the initial state of the system.

Finally, the set of functions specified by the IOFO is $F = \{f_q | q \in Q\}$.

Level Transition $2 \to 1$

Given an IOFO $<T,X,\Omega,Y,F>$ we associate with it a IORO $<T,X,\Omega,Y,R>$ where R is the union of all the functions $f \in F$.

In other words, by throwing away the knowledge of which initial state was responsible for producing a response to an input, we obtain the I/O relation in which many output segments may be paired with the same input segment.

Level Transition $1 \to 0$

Given the IORO $<T,X,\Omega,Y,R>$, we merely extract the frame $<T,X,Y>$ on which it is based.

3. Hierarchy of System Specification Morphisms

Systems specified at each level of the hierarchy may be compared for structural resemblance making use of the features available at that level. A mapping between two systems which preserves the features at level i is called a *morphism* at level i. Such a morphism will always run from a "larger" system to a "smaller" one in the sense that the mapping will always be an onto one. Thus the "larger" system will have at least as many elements (e.g. states) as the "smaller" one.

Applied to the simulation relation (Chapter 5), the simulator will play the role of the "larger" system, and the model will be the "smaller" one. To prove that the simulator correctly implements the model one turns to an appropriate level of specification and a morphism to establish the meaning of "correct" and proceeds to try to prove that the morphism does indeed hold (see Chapter 17).

Likewise, the modelling relation (Chapter 5) can be formulated as involving a morphism from the real system as the "larger" system and the model as the "smaller" one. In this case, the appropriate level to begin with is that of the I/O relation, since the real system is to be regarded as a source of data. To make stronger statements about the relation between real system and model requires justification of the kind to be discussed in Section 6.8.

The morphisms can also be restricted so that the mapping is one-one as well as onto, i.e. constitutes a one-one correspondence. In this case, the morphism is called an *isomorphism* and asserts that the systems have the same structural features recognizable at the level at which the morphism holds.

We proceed to review representative morphisms at each of the levels avoiding a more complete exposition as in [1], Chapter 10 in the interest of a simpler presentation. However, in practice the greater flexibility in establishing correspondences that is expressed in the general formulation is often required to prove a desired simulation or modelling relation.

Level 0. Observation Frame Morphism

There is a *morphism* from $<T,X,Y>$ to $<T',X',Y'>$ if $T=T'$, $X' \subset X$, and Y' inc Y.

$<T,X,Y>$ is *isomorphic* to $<T',X',Y'>$ if $T=T'$, $X=X'$, and $Y=Y'$.

If there is a morphism from frame O to frame O', we say that O' is derivable from O. If two systems are isomorphic at the frame level they are said to be *compatible*.

Level 1. I/O Relation Morphism

There is a *morphism* from $<T,X,\Omega,Y,R>$ to $<T',X',\Omega',Y',R'>$ if observation frame $<T',X',Y'>$ is derivable from $<T,X,Y>$ and $R' \subset R$, i.e. each I/O segment pair (ω,ρ) observed in the "smaller" system is observable also in the "larger" system.

Two I/O relation observations are *isomorphic* if their observation frames are compatible and the two I/O relations are equal (equivalently, all sets and relations are equal).

Two systems that are isomorphic at the I/O relation level are also said to be *relationally equivalent* or *indistinguishable*. An observer or other system which has access to these systems only through the input/output behavior cannot distinguish between them in any way.

Level 2. I/O Function Morphism

There is a *morphism* from $<T,X,\Omega,Y,F>$ to $<T',X',\Omega',Y',F'>$ if the observation frame $<T',X',Y'>$ is derivable from $<T,X,Y>$ and $F' \subset F$, i.e. each I/O function in the "smaller" system is also realized in the "larger" system in the sense that for each $f' \in F'$, there is an $f \in F$, such that $f' \subset f$.

Two I/O Function Observations are *isomorphic* if their observation frames are compatible and the two I/O function sets are equal (equivalently, the two systems realize the same set of I/O functions).

Systems that are isomorphic at the I/O function level are also said to be *behaviorally equivalent*.

Level 3. I/O System Morphism

Let $S_i = <T_i,X_i,\Omega_i,Y_i,\delta_i,\lambda_i>$ $i=1,2$ be I/O systems. There is a *morphism* from S_1 to S_2 at the I/O system level if the observation frame $<T_1,X_1,Y_1>$ of S_1 is derivable from that of S_2, $\Omega_2 \subset \Omega_1$, and there is a homomorphism from S_1 to S_2 as follows: A mapping h from a subset Q_1' of Q_1 onto Q_2 is a *homomorphism* from S_1 to S_2 if the following commutativity conditions hold:

(i) preservation of transition function:

$$h(\delta_1(q,\omega)) = \delta_2(h(q),\omega)$$

(ii) preservation of output function:

$$\lambda_1(q) = \lambda_2(h(q)) \text{ for all } q \in Q_1', \text{ and } \omega \in \Omega_2$$

The subset Q_1' is called the set of *representing* states. For a representing state $q \in Q_1'$ of the "larger" system, the state $q' = h(q)$ in the "smaller" system is the state it represents; q and q' are called *corresponding* states. The preservation conditions say that every state of the "smaller" system has a corresponding state in the "larger" system that represents it (h is onto) and that corresponding states always transit to corresponding states under the same input segment, and produce the same observable output.

It can be shown that the I/O functions of corresponding states are equal, i.e. $f_q = f_{h(q)}$ when restricted to Ω_2. States (of the same or different systems) which have the same I/O function are said to be *behaviorally equivalent*. Such states cannot be distinguished by any experiment involving input/output observations since they produce the same response segment to every input segment. Thus, states which correspond under a homomorphism are behaviorally equivalent when segments are restricted to Ω_2.

S_1 and S_2 are *isomorphic* if their observation frames are compatible, $\Omega_1 = \Omega_2$, and there is a homomorphism from one to the other in which the state map (h) is a one-one correspondence (the set of representing states is required to be the whole set).

Level 4. Structured System Morphism

Let S_1 and S_2 be systems specified at level 4. There is a *structured system morphism* from S_1 to S_2 if there is a morphism from S_1 to S_2, the I/O systems associated with the two systems respectively where we require the homomorphism h to take a special form:

Let D_1 and D_2 be the co-ordinate sets of the structures $\mathbf{Q}_1$ and $\mathbf{Q}_2$ (of the state sets Q_1 and Q_2) respectively. Let d be a mapping from D_1 onto D_2 called the *co-ordinate* map. For each $\mathbf{a} \in D_2$, $d^{-1}(\mathbf{a})$ is the *block* of co-ordinates which represent $\mathbf{a}$ in D_1. Let $h_\mathbf{a}$ be a map from the states of this block onto the range set $Q_\mathbf{a}$, called a *local map*. We require that h be constructed as a composite of the $h_\mathbf{a}$, $\mathbf{a} \in D_2$.

In other words, there is a structured system morphism from S_1 to S_2 if the observation frame of S_1 is derivable from that of S_2, Ω_1 is a subset of Ω_2 and there is a homomorphism h from S_1 to S_2 which is constructed as a composite from local co-ordinate maps.[5]

Two structured systems are *isomorphic* if there is a structured system morphism from one to the other in which the homomorphism h is an isomorphism and is constructed as a composite of local maps where the co-ordinate map d is a one-one correspondence.

Level 5. System Coupling Morphism

Let $N_i = <D_i, \{S_\mathbf{a}\}_i, \{I_\mathbf{a}\}_i, \{Z_\mathbf{a}\}_i>$, $i = 1,2$ be two System Couplings Specifications. To understand the morphism concepts appropriate at this level, we consider first the case where the number of components in both systems is the same. Then we show how the more general situation reduces to this one.

First consider the case where the same number of components are to be coupled together in both specifications. We may then set $D_1 = D_2 = D$. For a

---

[5]This is called a weak structured morphism in [1]; a strong morphism being a weak one that also preserves influence relations.

morphism between two couplings of systems, we require that corresponding system components be morphically related, that the influencer relations and interface maps be preserved. We say that there is a *simple morphism* from N_1 to N_2 if:

For each $\mathbf{a} \in D$,
    (a) there is a structured system morphism
        $S_{\mathbf{a},1}$ to $S_{\mathbf{a},2}$ with homomorphism $h_{\mathbf{a}}$
    (b) $I_{\mathbf{a},1} = I_{\mathbf{a},2}$
    (c) $Z_{\mathbf{a},2} = Z_{\mathbf{a},1}$.

Requirements (b) and (c) can be considerably relaxed so that true simplification of the influencer and interface maps may also be expressed ([1], Chapter 10).

However, as it stands the morphism does reflect the conceptual framework of *top-down* design. In this paradigm, we develop a decomposition of a system by successive refinement of component systems leaving the influencer and interface relations fixed once they have been defined. Thus starting with $N_1 = <D,\{S_{\mathbf{a}}\},\{I_{\mathbf{a}}\},\{Z_{\mathbf{a}}\}>$, we proceed to replace each $S_{\mathbf{a},1}$ by a "larger" $S_{\mathbf{a},2}$. Requiring there be a morphism from $S_{\mathbf{a},2}$ to $S_{\mathbf{a},1}$ means in particular that the input and output sets of the "larger" system must include those respectively of the original. Thus we can safely plug in the former in place of the latter without violating the interface specifications. Furthermore the morphism requires that the internal transition function and the output map of the replacement preserve the operation of the original. Thus we should expect that the result of making such a replacement will not change the behavior of the overall system. Indeed, this invariance will be established soon.

Now consider the case where N_2 has a smaller number of components than N_1. Let d be a map from D_1 onto D_2. For each $\mathbf{a} \in D_2$, consider the block of components $d^{-1}(\mathbf{a})$ in D_1 that represent $\mathbf{a}$. This block of components can be coupled together using the influencer and interface maps given for the network of which it is a part. We shall not go into the details, but one can construct a new version of N_1 by carrying out the coupling internal to each of the blocks to obtain a set of structured systems, and then coupling these systems using the influencer and interface maps that run between blocks. Let N_1/d refer to this blocked version of N_1.

We say that a *morphism* runs from N_1 to N_2 if there is a component map $d: D_1 \to D_2$ (onto) and a simple morphism from the blocked version N_1/d to N_2 (Figure 3.2).

Two system coupling specifications are *isomorphic* if there is one-one correspondence between their components such that corresponding components are isomorphic (at level 4) and the corresponding influencers and interface maps are identical.

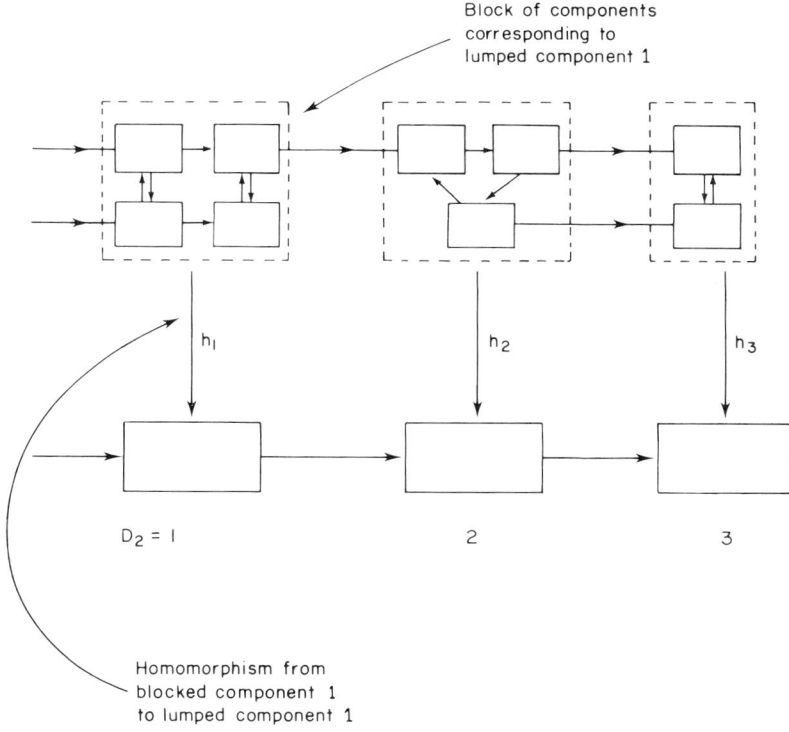

Fig. 3.2. Illustrating morphism at the coupled system level.

Summary

We note that the morphisms at each level are defined so as to be reflexive and transitive. Moreover, isomorphisms are defined so that if systems are isomorphic at some level, then there are morphisms running in both directions between them.

4. Moving Down the Morphism Hierarchy

Along with the descent through the hierarchy of specifications is a parallel descent through the levels of specification morphisms. Thus if there is a morphism at some level between two specifications at that level, then there are systems associated with these specifications at all lower levels. We now show that the pairs of specifications at the each level are related as well by a morphism at that level. Specifically, we need only show how to a morphism at level i induces a morphism at level $i-1$, since by iteration, the existence of morphisms at all lower levels will follow. The converse problem of inferring the existence of morphisms at higher levels than that of the morphism one has in hand is much harder (Section 3.5). The downward direction of this section is especially applicable to proving correctness or other equivalences between given system specifications. Here one tries to make use of as much of the structural features of the objects as possible, i.e. prove the existence of a morphism at a high level of structure. Having shown this, we are allowed to infer from the downward association of morphisms, that the two systems so related structurally are also related behaviorally. For example, one may use the system coupling morphism to show that a simulator correctly implements a model, i.e. reproduces its state and I/O behavior.

Transition $5 \to 4$

Let $N_i = <D_i, \{S_a\}_i, \{I_a\}_i, \{Z_a\}_i>$, $i=1,2$ be two System Couplings Specificaations with a morphism running from N_1 to N_2. Let S_1 and S_2 be the structured systems associated with N_1 and N_2 respectively. We must show that there is a morphism at level 4 from S_1 to S_2. Consider first the case of the simple morphism. In this case, one constructs a state map h as a composite of the local maps $\{h_a\}$, each h_a being given as a homomorphism from component **a** of N_1 to the corresponding one in N_2. The map h is then shown to be the homomorphism from S_1 to S_2 required to constitute the morphism at level 4.

In the general case, let S_1' be the structured system associated with the blocked version N_1' (induced by the given component map d). One first shows that S_1' and S_1 are isomorphic at the structured system level. Also notice that the simple morphism from N_1' to N_2 induces a structured system morphism from S_1' to S_2, as we have just shown. Then by transitivity of morphisms, we have a structured system morphism from S_1 to S_2.

Transition $4 \to 3$

By the definition of morphism at the structured system level, it follows immediately that such a morphism induces a morphism at the I/O system level.

Transition $3 \to 2$

Let there be a morphism at the I/O system level run from S_1 to S_2 with underlying homomorphism $h:Q_1' \to Q_2$. We have indicated that the I/O functions f_q and $f_{h(q)}$ are equal (restricted to Ω_2) for all such corresponding states. Since h is onto Q_2, this implies that F_2 is a subset of F_1. Thus there is a morphism at the I/O function level from the IOFO associated with S_1 to that associated with S_2.

Transition $2 \to 1$

A morphism at the I/O function level from $IOFO_2$ to $IOFO_1$ means that the set of I/O functions F_1 is included in that of F_2. Thus the union of $f \in F_1$ is included in the union $f \in F_2$. It follows that there is a morphism from $IORO_1$ to $IORO_2$, the I/O relation observations associated with $IOFO_1$ and $IOFO_2$ respectively.

Transition $1 \to 0$

Follows from the definitions.

5. Climbing up the Hierarchy

The problem can be stated as follows: given a pair of systems specified at level $i+1$ whose associated specifications at level i are related by a level i morphism, under what conditions can we infer that a level $i+1$ morphism connects the level $i+1$ specifications?

These conditions are called *justifying conditions*. They should be shown to be necessary for making the level jump in the sense that deleting any one of the clauses making up the justifying condition will invalidate it, i.e. a counter example to its claim can be adduced. Such a counter example also demonstrates the distinctness of levels, since it shows that two specifications that are not morphically related at level $i+1$ associate to specifications that are morphically related at lower level i.

Here we shall present only the justifying conditions for level jumps $1 \to 2$ and $2 \to 3$. A fuller treatment is available in [1], Chapter 13.

Level Transition $1 \to 2$

Let S and S' be compatible systems specified at level 3 (i.e. S and S' share the same observation frame $<T,X,Y>$). Let IORO and IORO', the I/O Relation Observations associated with S and S' respectively, be indistinguishable at level 1, i.e. $R = R'$. Now let IOFO and IOFO' be the I/O Function Observations associated with S and S' respectively at level 2. We can infer that there is a morphism from IOFO to IOFO' at level 2, i.e. the systems are behaviorally related, under the condition that system S' is *identifiable*. S' is *identifiable* if for every state q' in Q', there is an I/O pair (ω',ρ') which identifies q' in the sense that q' is the only state possible for S' to be in at the end of any experiment in which ω is applied as input and ρ is observed as output.

The proof shows that for any $q' \in Q'$, since q' is identifiable by some (ω',ρ'), all pairs $(\omega'\omega, \rho'\rho)$ in R' are of the form $\rho = f_{q'}'(\omega)$ and since $R = R'$, the function $f_{q'}'$ is also an I/O function of S. Thus $F' \subset F$ as required.

Level Transition 2 → 3

Let S and S' be systems at level 3 as above. For the associated I/O Function Observations at level 2, let there be a morphism from IOFO to IOFO', i.e. $F' \subset F$. The justifying condition for the inference of a system morphism at level 3 from S to S' is that S' is *reduced*. S' is *reduced* if distinct states of S' have distinct I/O functions, i.e. for every distinct pair of states $r,s \in Q'$, there is an input segment $\omega \in \Omega'$, such that $f_r(\omega) \sim = f_s(\omega)$ (so every pair of distinct states can be distinguished by some experiment).

The proof sets up a correspondence between Q and Q' such that equivalent states are paired. Since $F' \subset F$, this correspondence is onto Q'. Also, since S' is reduced, this correspondence is a mapping. Moreover, since equivalent states are sent to equivalent states under all input segments, the mapping is a homomorphism.

The paradigm to which the upward climb of the morphism hierarchy applies is that of structural inference. Such a situation prevails when one has a model of a real system which has been validated at the level of the I/O Relation Observation. The justifying conditions enable one to make stronger inferences concerning the predictive and structural validity of the model (Section 5.7).

6. Modelling Formalisms

While the hierarchy of system specifications just described is a necessary framework for multifacetted system modelling, it is not sufficient by itself. True, any formal model description should ultimately end up as a system. However, to write such a description directly is not a practical task for any but the simplest of models. Instead models are represented in various special formalisms such as those of differential equations, automata or discrete event. Each such formalism can be viewed as selecting a special class from the set of all systems. Once such a formalism is laid down so is the information common to the subclass of systems being referred to. Thus, to express a model in such a formalism we need only give the information necessary to distinguish this model from the others in the class. In this way, we can regard a special model formalism as providing a shorthand way of specifying a subclass of systems. And a model expressed in the formalism is a *system specification*, i.e. it indirectly selects a particular system from the set of all systems in the subclass.

As illustrated in Figure 3.3, each formalism can be regarded as a means for specifying a system at some level of the specification hierarchy. We shall take I/O System as the basic level at which to discuss the role of modelling formalisms. Thus, a formalism is to be a shorthand means for specifying a class of I/O systems. Actually, practical model construction proceeds at the higher levels (Structured System and Coupling of Systems) and we shall be concerned with these levels in later chapters. Any such system specification must eventually uniquely specify the sets and functions of some system

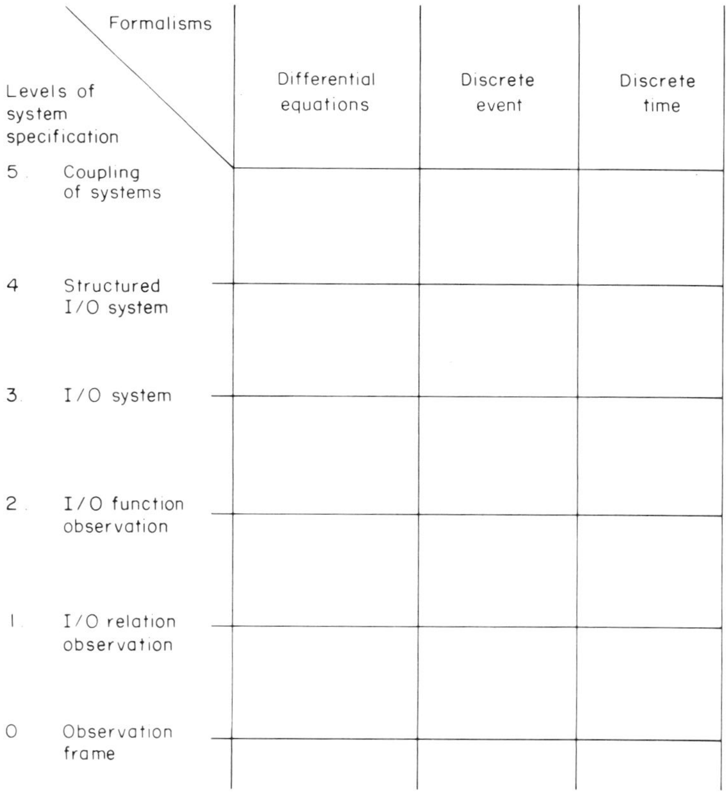

Levels of system specification	Formalisms	Differential equations	Discrete event	Discrete time
5	Coupling of systems			
4	Structured I/O system			
3	I/O system			
2	I/O function observation			
1	I/O relation observation			
0	Observation frame			

Fig. 3.3. System specifications and formalisms.

description. We refer to the sets X (input), Q (states) and Y (outputs) and the output function λ as constituting the *static structures* of the system.

The remainder, namely T (time base), Ω (input segment set) and δ (state transition function) constitute the *dynamic structure*. The distinction is a formal one in that the first group does not contain any reference to the time base while the second group most certainly does. The static structure provides a framework for taking "snapshots" of the system while the dynamic structure provides the framework for the changes such "snapshots" would record over time.

Now each special formalism puts its own restrictions on the possible static and dynamic structures it wants to encompass. In the subsequent chapters we shall consider such structure specification in detail. Meanwhile, let us see how the most common formalisms restrict the class of systems.

6.1. Subclasses of Systems

The following table characterizes the subclasses of systems corresponding to the major modelling formalisms by characterizing the restrictions they imply on the static and dynamic elements of the system description.

	differential equation	discrete event	discrete time
time base T	continuous reals	continuous reals	discrete integers
basic sets X, Q, Y	real vector space	arbitrary	arbitrary
input segments	piecewise continuous segments	discrete event segments	sequences
state and output trajectories	continuous segments	piecewise constants segments	sequences

Notice that the distinguishing mark of discrete event systems is that state changes occur discretely (as opposed to continuously in differential equation systems) and may be non-uniformly distributed in time. These state changes are called (endogenous or internal) *events* which along with externally occurring (exogenous) events form the basis for the specification of model structure and behavior. It is these events that are handled by the scheduling mechanisms of discrete event simulators. We shall see soon (Chapter 4) how they can be abstracted from the scheduling mechanisms to constitute the dynamic structure of discrete event models.

In contrast to the discrete event formalism, the discrete time formalism assumes that changes in state can occur at each time step. This leads to a relatively simple shorthand for model specification — a basic form of which is the sequential machine or automaton — because scheduling need not be specified. However, the assumption that any component can change state at any time step also entails significant disadvantages from the model expression and simulation efficiency point of view. Comparison of formalisms along these lines is discussed in Section 6.4.

The formalisms Q are all taken to specify time invariant systems. A system $S = <T,X,\Omega,Y,\delta,\lambda>$ is *time invariant* if the following hold:

(a) Ω is closed under translation, i.e.

$$\text{if } \omega \in \Omega, \text{ then so is TRANS}_\tau(\omega)$$

where $\text{TRANS}_\tau(\omega)$ is movement of ω forward on the time base by τ units.

(b) δ is time invariant:

$$\delta(q,\omega) = \delta(q,\text{TRANS}_\tau(\omega))$$

for all $q \in Q$, $\omega \in \Omega$, and $\tau \in T$.

A time invariant system formulation disallows the explicit appearance of the time at which an input is applied in the transition function. Such a model treats the rules of interaction as law-like in the sense of being the same no matter when applied.

The relation between a modelling formalism and the I/O system formalism is one of specification, in the sense of Section 2.2. Each formalism requires a translation map that maps its objects into I/O systems and thereby provides an interpretation of its parameters.

We shall briefly review the discrete time and differential equation formalisms and their associated translation maps. For a more complete treatment see [1,3].

Differential equation formalism

In contrast to the other major formalisms, a differential equation does not prescribe next states directly but provides implicit constraints on how such changes are to occur. The way it does this is to specifiy how the derivative (rate of change) of a state trajectory at any time t depends on the state and the input value at t. To display a state trajectory thus requires in principle, *solving* rather than computing. In practice, interesting models are almost never tractable to analytic solution techniques so that the differential equations are converted into numerically computable form, most often as discrete time systems.

A *Differential Equation System Specification* is a structure

$$D = <X, Q, Y, f, \lambda>$$

where
    X is a set, the input value set
    Q is a set, the state set
    Y is a set, the output value set
    f is a function, the *rate of change* function
    λ is a function, the output function
subject to the following constraints:
    X, Q, Y are real finite dimensional vector spaces
    $f: Q \times X \to Q$
    $\lambda: Q \to Y$

The translation of a DESS D into a structured I/O system S_D follows the pattern followed when a structured system is associated to a coupling of systems (Section 3.2). The time base of S_D is taken to be the reals, and the input segment set Ω is the set of piecewise bounded continuous functions on finite intervals of the reals. A segment over (Q,R) is a solution associated with a state $q \in Q$ and input segment, $\omega \in \Omega$ if it starts at q and satisfies the differential equation at each instant in the domain of ω. If solutions exist and are unique for each (q,ω) pair, then the DESS specifies an I/O system S_D. The Lipschitz conditions are known to guarantee unique solutions. In this case, a solution is called a state trajectory and the transition function is defined for (q,ω) as the state reached at the end of the state trajectory associated with q and ω.

Although methods for computing state trajectories of DESS commonly are based on Taylor series expansions, another approach is suggested by the system formulation. First one chooses a set of generators ([1], Chapter 9) for the input segment set, i.e. a set of segments which when concatenated together in finite combinations yield all possible input segments. Then state trajectories are computed using classical means but only for the generators so that a tabular representation of the transition function is built up for a variety of initial states and generators. Having done this numerical integration once for each such (state, generator) combination, one is in a position to compute a state trajectory for generated segments by iteratively applying the tabular form of the transition function — this makes use of the fact that Q is a state set and the transition function satisfies the semigroup property (Section 3.1). The method can be modified so that if a state is reached which was not one of those initially selected, a reversion to numerical integration takes place, and the result is added to the table. A related system theoretic based approach maps the differential equation system into the discrete event formalism (Chapter 4).

Discrete time systems: Sequential machine formalism

The sequential machine or automaton formalism[6] can be interpreted as specifying a class of discrete time systems. Moreover, it can be shown that any discrete time system can be specified by a sequential machine although it may not always be helpful to do so. We proceed to briefly review this formalism and its translation into the systems formalism.

A Sequential Machine is a structure:

$$M = <X, Q, Y, \delta_M, \lambda>$$

where

X, Q, Y are sets of inputs, states, and outputs respectively

δ_M is a function, the single step transition function

and

λ is a function, the output function

subject to the constraints:

$$\delta_M : Q \times X \to Q$$

[δ_M dictates the state transition made in one time step of model operation; it is extended to the multistep transition function of the system specified by M] and

$$\lambda : Q \to Y$$

[this is the Moore form; in the Mealy form the output may depend on both state and input].

With a sequential machine M we associate an I/O system

$$S_M = <I, X, \Omega, Q, Y, \delta, \lambda>$$

which shows the time base to be the integers, the input value set, state set, and output sets of S_M to be those specified by M, and the output function λ to be that of M.

The set of input segments $\Omega = (X, I)$ the set of all segments on an integer time base. A segment ω on $<t_{in}, t_f>$ can be specified as a sequence $\omega(t_{in})$, $\omega(t_{in}+1), \ldots, \omega(t_f)$ beginning at t_{in}. Thus Ω is isomorphic to the set $X \times I$, the set of pairs (x, t) where x is a sequence over X and t is the time at which its application is initiated. Actually, the behavior of S_M will not depend on the initial time t as it will be time invariant. We define the *extended* transition function

$$\delta' = Q \times X^* \to Q$$

---

[6]This is the same formalism as that of the FSM (Finite State Machine) introduced in Chapter 2, except that the state set is not restricted to being finite.

where
$$\delta'(q,\Lambda) = q \quad (\Lambda \text{ is the empty sequence})$$
and
$$\delta'(q,sx) = \delta'(\delta(q,s),x)$$
with $s \in X$, and $x \in X^*$.

This recursive definition reflects an iterative procedure that characterizes step by step digital simulation; δ' may also be called the *multiple step transition function* as $\delta'(q,x)$ is the state arrived at after n steps, where n is the length of x.

The system transition function is defined in terms of the extended transition function:
$$\delta(q,\omega) = \delta'(q,x)$$
where x is the sequence underlying ω.

7. Closure of Formalisms under Coupling

A formalism is said to be *closed under composition* if any composite system obtained by coupling components specified by the formalism is itself specified by the formalism. The differential equation, and sequential machine formalisms are known to be closed under composition [1, Chapter 9]. The significance of such closure is that it facilitates hierarchical construction of models by recursive applications of the coupling procedure.

More specifically, consider a coupling of systems (level 5) specification
$$N = <D, \{S_a\}, \{I_a\}, \{Z_a\}>$$

The association mapping N into a composite system S_N at level 4 (when the latter exists) demonstrates the closure of the systems formalism under composition. Closure of a formalism under composition requires that if each S_a is specified in the formalism then so is S_N. Such closure sets up the possibility for hierarchical model specification as first discussed in Section 2.1. We shall return to this topic in 15.3.

8. Summary

This chapter has presented a set of levels for system specification organized into a hierarchy ascending from behavioral input-output to increasingly structural specification. The organization is such that system specifications can be related by use of morphisms at each level. A morphism encodes the concept of structural or behavioral equivalence at a level. Association mappings provide the means to ascribe a lower level specification to a higher one. In this way, descending one step at a time, the input/output behavior (IORO at the lowest non-frame level) may be computed for a coupling of systems (a highest level specification). This association mapping also induces an association of morphisms, in that if two specifications are morphically related at level i, then their associated specifications at lower levels are related by morphisms at these levels as well.

Formalisms are shorthand means for specifying subclasses of systems. The closure of a formalism under composition is required in order for the formalism to lend itself to hierarchical model specification.

References

[1] Zeigler, B. P. (1976). *Theory of Modelling and Simulation*, Wiley, NY.
[2] Zeigler, B. P. (1983). "System Theoretic Foundations of Modelling and Simulation", Chapter 7. In *Simulation and Model-Based Methodologies: An Integrative View* (eds, T. I. Oren, M. S. Elzas and B. P. Zeigler), Springer Verlag, NY.
[3] Vansteenkiste, G. C. and J. Spriet (1982). *Computer Aided Modelling and Simulation*, Academic Press, London.

Other Relevant Literature
The following are source literature in general systems concepts:
Padulo, L. and M. A. Arbib (1974). *System Theory*, Saunders, Philadelphia.
Wymore, A. W. (1967) *A Mathematical Theory of Systems Engineering: The Elements*, Wiley, NY.
Klir, G. J. (1979), "General Systems Problem Solving Methodology." In *Methodology in Systems Modelling and Simulation* (ed, B. P. Zeigler *et al.*), North Holland, Amsterdam.
Mesarovic, M. and I. Takahara. (1975). *Theory of Hierarchical Control*, Academic Press, NY.
Pichler, F. (1983). "Symbolic Manipulation of System Models". In *Simulation and Model-Based Methodology: An Integrative View*, (eds, T. I. Oren *et al.*), Springer-Verlag, NY. Develops the hierarchy of systems specification as a basis for model manipulation.
Kindler, E. (1979). "Dynamic Systems and Theory of Simulation" *Kybernetica*, **15**, 2. Employs systems concepts to develop a theory of simulation language semantics.
Oren, T. I. (1979), "Concepts for Advanced Computer Assisted Modelling". In *Methodology in Systems Modelling and Simulation* (eds, Zeigler *et al.*) North Holland, Amsterdam; Provides a taxonomony of formalisms for modelling.
Nance, R. E. (1981). 'The Time and State Relationships in Simulation Modelling. *C.A.C.M.* **24**, 173–189. Discusses the importance of formalism for discrete event simulation practice.

Chapter 4

FORMALISM FOR DISCRETE EVENT SYSTEMS

As suggested by the title of this book, development of the theory of multifacetted modelling will be done in the context of discrete event simulation. This chapter begins the discussion of discrete event system specification (DEVS), a formalism introduced by [12] to provide a formal basis for specifying the models expressable within discrete event simulation languages (SIMSCRIPT, SIMULA, GPSS, etc.). We shall see the three primary "world views": next event, activity scanning, and process interaction can be expressed within the formalism. Several examples of models specified in the formalism have appeared in the literature [3,4,6,10]. The formalism has also served as a basis for model and simulation program development [1], advanced simulation language design [5,10] and verification of simulation programs [2]. After introducing the DEVS formalism we show how it specifies the subclass of discrete event systems in the hierarchy of system specifications (Section 3.1). Examples and interpretations of the formalism are then presented which will be useful later on.

1. The DEVS Formalism

We start with a presentation of the DEVS formalism focussing on the transition system, omitting the output portion for the moment.

A *discrete event system specification* (DEVS) is a structure

$$M = <X,S,\delta,ta>$$

where

X is a set, the names of *external event* types
S is a set, the *sequential states*
δ is a function, the *transition specification*[7]
ta is a function, the *time advance function*

---

[7] δ is a specification of the system transition function *not* the latter function itself.

with the following constraints:
(a) *ta* is a mapping from S to the non-negative reals with infinity:

$$ta: S \to R_{0,\infty}^+$$

[$ta(s)$ is interpreted as the time the system is allowed to stay in state s if no external events occur].
(b) The *total state* set of the system specified by M is

$$Q = \{(s,e) | s \in S, 0 \le e \le ta(s)\}$$

[both the sequential state s and the elapsed time e spent in this state are significant state variables].
(c) The transition specification δ consists of two parts:
  (1) The *internal* transition function

$$\delta_\phi: S \to S$$

[if no external events arrive, the system will transition from sequential state s to $\delta_\phi(s)$ after $ta(s)$ time units. Simultaneously the elapsed time component e is reset to zero].
  (2) The *external* transition function:

$$\delta_{ex}: Q \times X \to S$$

[if an event $x \in X$ arrives, and the system has been in state s for an elapsed time e, it transitions immediately to $\delta_{ex}(s,e,x)$. Simultaneously, the elapsed time component is reset to zero].

A full explication of the semantics of the DEVS is given by [12], Chapter 9. We shall give some examples of DEVS specification after showing how the DEVS formalism is just what is logically required to specify the operation of a discrete event simulator.

2. The System Specified by a DEVS

Chronology of events is represented in system theoretic formalism by the concept of time segment, a mapping from a time interval to some descriptor set. In order to distinguish between event occurrence and non-occurrence we introduce the artifact of a non-event symbols. The *non-event closure* X^ϕ of a set X is obtained as follows: If X does not contain a non-event symbol then ϕ is adjoined to it for this purpose; otherwise $X^\phi = X$ and ϕ is identified with the existing non-event symbol.
A *DEVS segment* over X is a mapping

$$\omega: <t_i, t_f> \to X^\phi$$

where $<t_i,t_f>$ denotes a time interval in the reals and ω satisfies the condition:

$$\omega(t_j) \sim = \phi$$

for at most a finite subset $\{t_j\}$ (possibly empty) of $<t_i,t_f>$. The instants $\{t_j\}$ are the event times in the observation interval $<t_i,t_f>$.

If X is a set of external event names for a DEVS M then the segments over X represent possible *input segments* to which it responds. For example, X may be a set of job identifiers and M a processor. An input segment then denotes a chronological sequence of job arrivals to M.

With a DEVS M we wish to associate a system S_M which represents its response to inputs segments. However, in order to be able to do so, the DEVS must possess the property called *legitimacy*. Roughly legitimacy prevents a DEVS from getting into an infinite sequence of states in which the time clock of its simulator would not advance beyond a certain point. For more information on this and what follows the reader is referred to Chapter 9 of [12].

We associate a system S_M with a legitimate DEVS M as follows:

$$S_M = <R,X^\phi,\Omega,Q,Y,\delta,\lambda>$$

which indicates that the time base T is continuous ($=R$) and the input set is the non-event closure of the external event set X. The input segment set Ω is the set of discrete event segments over X^ϕ and the state set Q is a set of total state pairs $\{(s,e)|s \in S, 0 \le e \le ta(s)\}$. The transition function δ is defined as follows:

Let $\omega: <t_{in},t_f> \to X^\phi$ with event times $\tau_1,...,\tau_n$ (where we shall set $\tau_1 = t_f$ if the event set is empty). We shall show how to compute the state that S_M will be in at time τ_1 given that it started in state $q = (s,e)$ at time t_{in} and received input segment ω. Let $t_1,...,t_m$ be the sequence of instants in the interval $<t_{in},\tau_1>$ and $s_1,...,s_m$ be the corresponding sequence of sequential states that the model traverses in this interval. Noting that no external events occur in the interval, we may generate the sequences of event times and sequential states starting from an instant t_O preceding t_{in} which represents the point at which the system entered state s. Thus we have:

$$t_O = t_{in} - e$$
$$t_{i+1} = t_i + ta(s_i)$$
$$s_O = s$$
$$s_{i+1} = \delta_\phi(s_i)$$

and m is the smallest i for which $t_{i+1} > = \tau_1$.

Reviewing, the sequence $t_1,...,t_m$ is the set of *internal* event times at which internally scheduled events occur before the incidence of the first external event. The sequence $s_1,...,s_m$ is the corresponding sequence of sequential states that the system moves through during this time. The total state that the system is in at time τ_1 is $q' = (s_m, \tau_1 - t_m)$.

The total state pertaining just after the external event x_1 arrives at τ_1 is

$$q_1 = (\delta_{ex}(q', x_1), 0)$$

Using the same method we can compute the total state pertaining just before and just after the next external event time τ_2. This is iterated until τ_n is reached. Finally having the state just after τ_n, the same method is used to compute the state at t_f, $\delta(q, \omega)$.

Exercise. Show that the transition function as defined has the composition property (Section 3.1). Note the essential role that the elapsed time component plays in guaranteeing that this property holds.

We have yet to mention the output set Y and the output function $\lambda: Q \to Y$ which must be supplied in order to complete the system specification. These interface concepts do not become important until one considers linking the system with other systems or the observer. Examples will appear in Section 4.5 and Chapter 8.

Having the state transition and output functions we can compute the state and output trajectories according to the formulae of Section 3.3.

3. Interpreting the DEVS

A state $s \in S$ for which $ta(s) = \infty$ is said to be *passive*. The system will stay in such a state indefinitely, only possibly changing state under the influence of an external event. Of course, the internal transition function δ_ϕ need not be defined for a passive state. A state which is not passive is said to be *active*; in particular, a state s for which $ta(s) = 0$ is said to be transitory. Sequences of transitory states provide the means to express intermediate model computations which do not consume model time — the clock of a simulator is not advanced in the execution of such states.

Let $x \in X$ be an external event and (s, s') be a pair of sequential states such that $\delta(s, e, x) = s'$, for some elapsed time $e \in [0, ta(s)]$. If s is passive and s' is active then x is said to *activate* the model; in particular if s' is transitory, then x is said to activate the model *immediately*. Conversely, if s is active and s' is passive then x is said to *passivate* the model.

The same terminology applies to transitions caused by internal events. A model passivates itself by entering into a passive state from an active state. However, by definition it is not possible for a model to activate itself in the sense of going from a passive state to an active state without the intervention of an external event.

Activation and passivation are important roles that external events can play. However, more subtle is the case where the states pertaining before and after the incidence of the event are both active. To discuss this case, we must explicitly represent the time left component of the model state.

A DEVS is in *explicit* form if the sequential state takes the form of a pair (s,σ) such that $ta(s,\sigma)=\sigma$. Thus the time left component σ is explicitly represented.

Let $x \in X$ arrive when the model has been in sequential state (s,σ) for elapsed time e and cause the transition to (s',σ') where $(s',\sigma')=\delta((s,\sigma),e,x)$. The event x is said to be *ignored* if $s'=s$ and $\sigma'=\sigma-e$, i.e. the only result has been to update the time left component to account for the passing of elapsed time e; the model remains scheduled to undergo a transition from the same state at the same time that it was before. An event which is not ignored is said to cause an *interrupt*. Such an event causes a non-trivial state change and/or a rescheduling of the model's next internal transition.

4. Example: A Generalized Queue

To illustrate the DEVS formalism we shall develop a class of models for single server queueing systems. The class is quite flexible in that it includes the simplest first come, first served discipline as well as more sophisticated job scheduling algorithms such as prioritized preemption and round robin. To construct this class we shall use the approach of developing a special formalism and mapping it into the DEVS formalism (Section 2.2.2). We also employ the terminology introduced by Parnas [8]: a *V-function* or Value function returns a value when applied to an object but does not change its state; an *O-function* or operation function provides a means of performing such a change in state. Generality is achieved by specifying only the axioms that such functions must satisfy without committing oneself to their complete definitions.

More specifically, as illustrated in Figure 4.1, first, we define a Job Description Structure which enables us to stipulate what must be known about the jobs to be processed. Then we define a Queueing structure which stipulates the functions required to insert jobs in the queue and to select a job to be processed. Finally, we employ these structures to specify a class of DEVS which embodies the dynamics they set up.

A *Job Description Structure* (JDS) is an object

$$J = <X, id, tl, \text{update}>$$

$J = <X,id,tl,\text{update}>$
Job Description Structure --->
    JDS

$QU(J) = <S,\text{first},\text{insert},ts>$
Queue based on JDS, J ---->

$M_{QU(J)} = <X,S,\delta_\phi,\delta_{ex},ta>$

DEVS

Fig. 4.1. Specification of generalized queue.

where
  X is a set, the set of job descriptors
  id is a V-function, the identifier of a job
  tl is a V-function, a job's processing time left
  update is an O-function, for updating job status
subject to the constraints:

$$id: X \to I^+$$
$$tl: X \to R_0^+$$
$$\text{update}: X \times R_0^+ \to X$$

such that if update$(x,t) = x'$ then

$$id(x') = id(x)$$

and

$$tl(x') = tl(x) - t$$

[update(x,t) reduces the processing time left of job x by the elapsed time t; the updating affects only the time left component of the job description]

We define a predicate to indicate whether a job is finished its processing:

$$\text{done}(x) \equiv [tl(x) = 0]$$

A *Queue based on a JDS J* is a structure

$$QU(J) = <S,\text{first},\text{insert},ts>$$

where

  S is a set, the set of line configurations
  first is a V-function, selecting the first in line
  insert is an O-function, for inserting a job
  ts is a V-function, for determining the time slice

2. Fundamentals and formalism

subject to the constraints:
   S consists of the set of all finite subsets of X satisfying:
      the empty set $\phi \in S$
      if $s \in S$, and $x, x' \in s$, then

$$x \sim = x' \text{ implies } id(x) \sim = id(x')$$

[a line cannot contain two job descriptors with the same id]
          first:$S \to X$ such that first$(s) \in s$ for $s \sim = \phi$
          insert:$X \times S \to S$ such that if insert $(x,s) = s'$

then

$$x \text{ is\_\_in } s'$$

and

$$y \text{ is\_\_in } s'/x, \text{ if, and only if, } y \text{ is\_\_in } s$$

where
   x is__in $s \equiv$ [there is an $x' \in s$ with

$$id(x') = id(x) \text{ and } tl(x') = tl(x)]$$

and

$$s'/x = s' - \{x'\}$$

where
   $id(x') = id(x)$ and $tl(x') = tl(x)$
[we allow the insert operation to change any aspect of a job except its id and time left components; FIFO (first in, first out) and line configurations based on priority schemes can therefore be accommodated]

$$ts: S \to R_0^+$$

[selects a next time slice based upon the current line configuration, includes the case of constant time slice]

We define the following:

$$\text{empty}(s) \equiv [s = \phi]$$

$$\text{rest}(s) = \begin{cases} \phi & \text{if empty}(s) \\ s/\text{first}(s) & \text{otherwise} \end{cases}$$

4.1. Specification of the DEVS Class of Queues

We are now ready to provide the translation which takes a Job Description Structure and a Queue based on it and produces the DEVS which portrays how such a queue operates. Such a translation could be programmed so as to produce a simulation program which would implement the queueing system specified by the JDS and the Queue parameters.

A Queue Structure $QU(J)$ based on JDS J specifies a DEVS

$$M_{QU(J)} = <X, S, \delta_\phi, \delta_{ex}, ta>$$

where
  X is the set of job descriptors given by J
  S is the set line configurations given by QU(J)
  $\delta_\phi: S \to S$ is given by

$$\delta_\phi(s) = \begin{cases} \text{rest}(s) & \text{if done(update.first}(s)) \\ \text{insert(update.first}(s)), \text{rest}(s)) & \text{otherwise} \end{cases}$$

where update.first(s) = update(first$(s), ta(s)$).
[the time slice allotted to the job being processed (namely, the one first in line) has just ended; if the job has finished then remove it from the line, otherwise reduce its time left by the time slice it received (namely, $ta(s)$) and replace into the line]

$$\delta_{ex}: Q \times X \to S \text{ is given by}$$

$$\delta_{ex}(s, e, x) = \begin{cases} \text{insert}(x, s) & \text{if empty}(s) \\ \text{insert}(x, \text{insert(update(first}(s), e), \text{rest}(s))) & \text{otherwise} \end{cases}$$

[if a job arrives to a non-empty line then the job currently being processed has its time left updated and then replaced into the line, after which, the incoming job is inserted into the line (if the insert operation realizes a preemption scheme this may result in its becoming the first in line)]

$$ta: S \to R_0^+ \text{ is given by}$$

$$ta(s) = \begin{cases} \infty & \text{if empty}(s) \\ \text{minimum}\{tl(\text{first}(s)), ts(s)\} & \text{otherwise} \end{cases}$$

[if the line is not empty then allot a processing time to the first job in line of at most the allowed time slice]

Example: Classical FIFO discipline

Let us express a queue in which jobs wait in line in order of arrival, and are given the processing time required for completion. Let X be a set of triples (i,σ,p) such that $id(i,\sigma,p)=i, tl(i,\sigma,p)=\sigma$ and p is a non-negative integer which will indicate the place of the job in line. Thus (i,σ,p) represents a job with identifier i, a processing time left of σ (initially this is the processing time required for completion, and a position p in the line (in this representation p functions like the numerical ticket some stores provide customers upon entry to the store).

Let update$((i,\sigma,p),t) = (i,\sigma - t,p)$

For $QU(J)$ let $ts = \infty$ and if $s = \{(i_1,\sigma_1,p_1),...,(i_n,\sigma_n,p_n)\}$ then first$(s) = (i_{min},\sigma_{min},p_{min})$ where $p_{min} = $ minimum$\{p_i\}$, and insert$((i,\sigma,p),s) = s \; U \; \{i,\sigma,p_{max} + 1)\}$ where $p_{max} = $ maximum$\{p_i\}$.

Thus a new arrival is explicitly placed at the end of the line by giving it a higher number than those of the jobs already in line.

Exercise. Make the above substitutions in the definition of $M_{QU(J)}$ and simplify as far as possible.

Example: Prioritized queue with pre-emption

To express a queue in which jobs wait in order of priority, interpret the p of the previous job triple as a priority. Change the "insert" operation so that an arriving job is added without change into the line. Let the "first" operation select the job in line with highest priority (break priority ties using the identify attribute, there are of course other ways). Pre-emption will occur when an incoming job has higher priority than that of the job in process — it will automatically displace the latter as the job being processed.

Exercise. Define the Job Description and Queue structures which realize the following types of queues:
(1) Round Robin with constant time slice (if job in process is not finished at the end of the time slice, it is replaced in line according to the FIFO discipline).
(2) Shortest Remaining Time First (the job selected by first has the smallest processing time left).
(3) Priority with no pre-emption (a job selected by using criteria in the following order: it has already received some processing, it has highest priority, it has arrived first).
(4) Starvation prevention (a job which has waited longer than a given threshold is selected by first regardless of all other criteria; this requires a modification of δ_ϕ).

4.2. Illustrating the Terminology of Section 3

Let us view the terminology of Section 3 in the light of the generalized queueing example. The empty line configuration ϕ constitutes the only passive state. All other states are active. The transitory states are those in which a job with zero processing time left has been selected by first. An empty queue is activated by every job arrival. The model cannot be passivated by an arrival, however, it does passivate itself when the line becomes empty.

The definition of insert guarantees that the arrival of a job is never ignored.[8] An arrival causes an interrupt in the case of pre-emption (when the arriving job displaces the job in process) otherwise the time left of the job in process is appropriately updated.

5. Special Classes of DEVS Models

Special forms of discrete event specification are useful in realizing experimental frames in the simulation of DEVS models. Specifically, the concepts of generator, acceptor, and transducer can be readily generalized to the DEVS context from their automata theoretic origins. The concepts of active and passive DEVS are useful in this generalization. Chapters 13 and 16 will put these concepts to use.

A DEVS is *passive* if its time advance function is identically infinite. Such a DEVS is not capable of initiating activity on its own since the time to its next internal transition is always infinity. It can however respond to external events by changing its state and producing an output.

An *active* DEVS is one which is not passive. Such a DEVS has at least one active state in which it can initiate activity of its own choosing by means of its self scheduling mechanism. Indeed, at one extreme, one may visualize an active DEVS as always engaged in autonomous state change activity, only now and then being interrupted by external input events.

A *generator* is an active DEVS which is input-free. Such a DEVS is capable of scheduling itself for a state transition, producing an output as a function of this state, and rescheduling itself for the next such iteration. The output function is defined in such a way that an output segment produced by such a generator when started in an initial state satisfies the constraints of a DEVS segment. Generators may be used to implement arrival processes (in the way that blocks of the same name function in GPSS) among other classes of input segments to models.

---

[8] We have not considered the case of an erroneous job arrival, i.e. where the job's identifier is already possessed by some job in line. In this case the transition function could be defined so that the event was in fact ignored.

2. Fundamentals and formalism

Formally, a *generator* is a DEVS

$$G = <S, \delta_\phi, ta, Y, \lambda>$$

where the symbols have the meanings standard in the DEVS context. If the output function is given the special constraint:

$$\lambda(s,e) = \phi \qquad \text{except when } e = ta(s)$$

then output events will occur only at the internally generated event times. An example is shown in Figure 4.2.

The set of all output segments generated by G, called its *generated set* is given by

$$\Omega_G = \{OTRAJ_q | q \in Q\}$$

Exercise. Show how to compute the output segment produced by a generator when started in an initial state.

An *acceptor* is a passive DEVS with state set partitioned into two sets, the accepting and non-accepting states. In addition the acceptor designates a state in which the system is always initialized. An input segment is accepted by such a system, if it causes the acceptor to reach an accepting state at the end of its application.

An *acceptor* is a DEVS

$$A = <X, S, \delta_{ex}, q_0, F>$$

where X, S, and δ_{ex} have the usual meanings and

q_0, the initial state

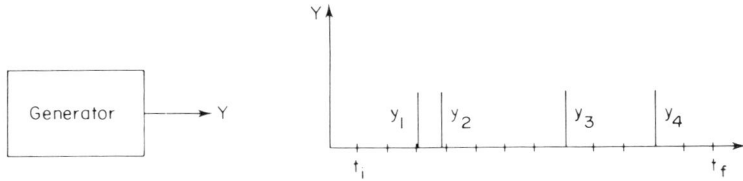

$G = (S, \delta_\phi, t, Y, \lambda)$

Example: Generator of a job arrival process
$\quad S = [0,1] \times [0,1]$ (random number seeds)
$\quad \delta_\phi(r_1, r_2) = (\Gamma(r_1), \Gamma(r_2))$ (Γ = random generator)
$\quad ta(r_1, r_2) = $ interarrival (r_1)
$\quad \lambda(r_1, r_2) = $ job identifier (r_2)
$\qquad\qquad\qquad$ (both random variables)

Fig. 4.2. Example of DEVS generator.

and

F, the set of accepting states

are subject to the constraints:

$$q_0 \in Q$$

and

F is a subset of Q (where Q is the total state set).

Note that the time advance function need not be included in the specification since it is identically infinite.

The set of segments accepted by A is given by

$$\Omega_A = \{\omega | \delta(q_0, \omega) \in F\}$$

In analogy with automata theory, Ω_A may be termed the language accepted by M. Ω is said to be accepted by A if $\Omega = \Omega_A$.

Exercise. Show that the languages accepted by DEVSs are closed under complementation, in particular, if Ω is accepted by A, then its complement is accepted by an acceptor which is the same as A except that its set of accepting states is $Q - F$.

Language acceptance by DEVSs constitutes a second form of specification of DEVS segments. Acceptors may be used to check that model trajectories satisfy specified constraints. For example, this may involve checking for proper initialization, return to initial state (used in regenerative sampling), or attainment of equilibrium. Since acceptors may be coupled together with boolean operators, acceptors for checking complex conditions may be built up from more simple ones. An example is given in Figure 4.3.

Exercise. Show that the DEVS languages are closed under union and intersection.

A *transducer* is a passive DEVS object with a designated initial state. When started in such a state, the transducer maps its input segments into output segments. A transducer is called a *final value* transducer, if all output is suppressed until the end of the input segment interval. Transducers may be employed to gather statistics about model trajectories in a manner similar to the ACCUMULATE mechanism in SIMSCRIPT 11.5 [9] or the abstract data types defined by Landwehr [7].

A *transducer* is a DEVS

$$T = <X, S, \delta_{ex}, Y, \lambda, q_0>$$

where q_0 is the initial state.

The (final value) *function* computed by T is the mapping

$$f_T : \Omega \to Y$$

defined by

$$f_T(\omega) = \lambda(\delta(q_0, \omega))$$

where Ω is the set of DEVS segments over X and δ is the transition function of the system specified by T. An example is shown in Figure 4.4.

5.1. Finite Memory DEVS

An important means of constructing DEVS generators, acceptors, or transducers is to save a finite number of past inputs and outputs for use in computing the next output. Known more generally as a finite memory system, the sequential state set of such a model takes on a special form: if n past inputs and m past outputs are saved then a state is a pair of vectors, one of size n of values of X and the other size m of values of Y. As was done in Section 2.2 a finite memory DEVS can be specified by supplying only the information required to uniquely identify a particular finite memory DEVS. Thus the following specification is appropriate:

$$F = <X, Y, n, m, f>$$

where
    X is a set, the input set
    Y is a set, the output set
    n is a non-negative integer, the
        input order
    m is a non-negative integer, the
        output order
and
    f is a function, the output
        function
such that

$$f : X^n \times Y^m \to Y$$

A FM-DEVS, F translates into a passive DEVS

$$M = <X, S, Y, \delta_{ex}, \lambda>$$

where
    $S = X^n \times Y^m$
and
    $\delta_{ex} : Q \times X \to S$

is defined by

$$\delta_{ex}((x_1,\ldots,x_n,y_1,\ldots,y_m),e,x) = (x,x_1,\ldots,x_{n-1},y,y_1,\ldots,y_{m-1})$$

where
$$y = f(x_1,\ldots,x_n,y_1,\ldots,y_m,e)$$
and
$$\lambda = f$$

By providing for more specialized means of specifying the basic forms of DEVS system, a very flexible means for user specification of experimental frame components may be constructed (Chapter 13). Tables 4.1 and 4.2 display the kinds of acceptors and transducers which are commonly employed in simulation studies.

Exercise. Using the finite memory DEVS and the examples in Figures 4.2–4.4 provide system specifications for the devices in Tables 4.1 and 4.2.

Table 4.1. Acceptors representing common experimental frame components.

(1) initial (subset): accepts an input segment if its initial value lies in the specified subset
(2) continuation (subset): accepts an input segment if it never enters the specified subset
(3) repeat (subset, n): accepts an input segment if it returns to the specified subset exactly n times
(4) constant: accepts an input segment if its event values are all identical
(5) reached constant (n): accepts an input segment if its last n event values are identical
(6), (7) same as (4), (5) except that values are required to be close within given metric and tolerance

Table 4.2. Transducers representing common experimental frame components.

(1) count: counts the number of events in the input segment
(2) elapsed time: accumulates the total elapsed time
(3) sum: accumulates the sum of the input values (assumed numerical)
(4) integrate: accumulates the time weighted integral of the input segment
(5) average: computes the average of the input values
(6) time-average: computes the time weighted average of the input segment
(7) maximum: computes the maximum of the input values
(8) minimum: computes the minimum of the input values
(9) median: computes the median of the input values

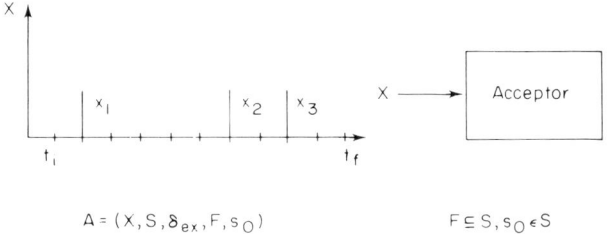

$A = (X, S, \delta_{ex}, F, s_0) \qquad F \subseteq S, s_0 \in S$

Note: $ta(s) \equiv \infty$ so does not have to be given

Example: Acceptor of segments of constant input

X = arbitrary
$S = \{ s_x | x \in X \} \{ \cup \ 0, 1 \}$
$F = \{ s_x | x \in X \}$
$s_0 = 0$
$\delta_{ex}(0, e, x) = s_x$
$\delta_{ex}(s_x, e, y) = \begin{cases} s_x & \text{if } y = x \\ 1 & \text{otherwise} \end{cases}$
$\delta_{ex}(1, e, x) = 1$

δ_{ex}:

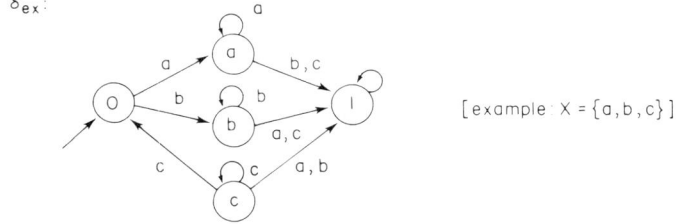

[example: $X = \{a, b, c\}$]

Fig. 4.3. Example of DEVS acceptor.

6. Summary

This chapter laid the basis for employing the discrete event formalism in multifacetted modelling methodology. Discrete event systems specification (DEVS) was discussed at the I/O systems level and examples were given of such systems. Coming chapters will extend the formalism to higher levels of the hierarchy of system specifications.

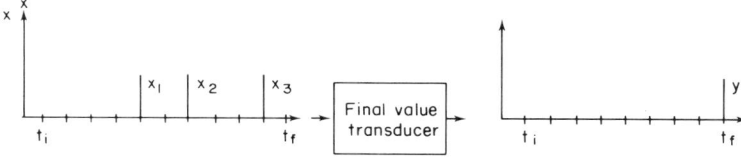

$T = (X, S, \delta_{ex}, Y, \lambda, q_0) \quad q_0 \in Q$

Example: Time averager

$X = R$

$S = \{(r,n) | r \in R, n \mid N\}$

$\delta_{ex}(r,n,e,x) = (r \cdot e + x, n+1)$

$Y = R$

$\lambda(r,n,e) = \dfrac{r \cdot e}{n}$

$q_0 = (0,0,0)$

Final value at end of input segment is time averaged value of input.

Fig. 4.4. Example of DEVS transducer.

References

[1] Aggarwal, S. (1983). "Discrete Event Formalism and Simulation Model Development". In *Simulation and Model-based Methodologies: An Integrative View*, (eds, T. I. Oren et al., Springer-Verlag, NY.

[2] Cutler, M. M. (1980). *A Formal Program for Discrete Event Simulation and its Use in the Verification and Validation of System Models and Implementations.* UCLA, Doctoral Dissertation.

[3] Evenczyk, D. and Zeigler, B. P. (1977). "Formalization and confirmation of the Boyd-Epley operating system model". *Computers and Mathematics with Applications* **3**, 1–13.

[4] Goldstein, M. and Zeigler, B. P. (1975), "Dynamic models for variance reduction in particle transport simulation". *Int. J. Systems Science* **6**, 765–786.

[5] Hooper, J. W. & K. D. Reilly (1982). "Analyzing Simulation Strategies". *Int. Jnl. Com. Inf. Sci.* (submitted).

[6] Hogeweg, P. H. (1980). "Locally Synchronised Developmental Systems", *Int. J. General Systems* **6**, 57–73.

[7] Landwehr, C. E. (1980). "An Abstract Type for Statistics Collection in SIMULA", *ACM Transactions on Programming Languages and Systems*, **2**, 544–563.

[8] Parnas, D. L. (1972). "On the Criteria to be used in Decomposing a System into Modules". *C.A.C.M.* **15**, No. 12.

[9] Kiviat, P. J., R. Villanueva, and H. M. Markowitz (1975), "SIMSCRIPT 11.5 Programming Language", (ed, E. C. Russell), CACI.
[10] Sklenar, J. (1981). "Apparatus for Formal Description of Discrete Systems. *Int. J. General Systems* **7**, 225–233.
[11] Subrahmanian, E. and R. L. Cannon (1981). "A Generator Program for Models of Discrete Event Systems". *Simulation*, March.
[12] Zeigler, B. P. (1976). *Theory of Modelling and Simulation*, Wiley, NY.

Chapter 5

A WORKING FRAMEWORK

We have by now acquired some familiarity with the multifacetted system approach, its role in decision making and the necessity to develop adequate methodologies to facilitate this approach. We have also learned some of the fundamental tenets of systems theory. In this chapter, we will lay out a preliminary framework for multifacetted system modelling employing the systems theory language. This framework will serve as a basis for further elaboration as we proceed to develop useful methodologies.

Basically, modelling and simulation involves three types of entities — *real system*, *model* and *simulator*. These entities cannot be properly understood in isolation but must be seen in their interrelation to one another, As shown in Figure 5.1a there are two fundamental kinds of relationships: *modelling* deals with the relationships between real systems and models and *simulation* refers to the relationships between models and computers.

As a first crack at definitions, we can think of the real system as a *source of data*, the model as *a set of instructions* for generating data and the simulator as *a device capable of carrying out model instructions*. The modelling relation concerns the validity of the model as a representation of the real system. Alternately this must come down to assessing the degree to which the model data agrees with the real system data. The process of ascertaining this degree of agreement is called *validation*. The *simulation relation* concerns the faithfulness with which the simulator executes the instructions intended by the model. Since the data that is compared with the real world is produced by the simulator, we must first establish the correctness of the simulator if our testing of the model is to make any sense. The process of assessing the correctness of the simulator, with respect to the model is called *verification*.

1. System Formulation

To further our understanding of these entities and relations and especially the distinctions they embody we shall employ the systems formalism of the

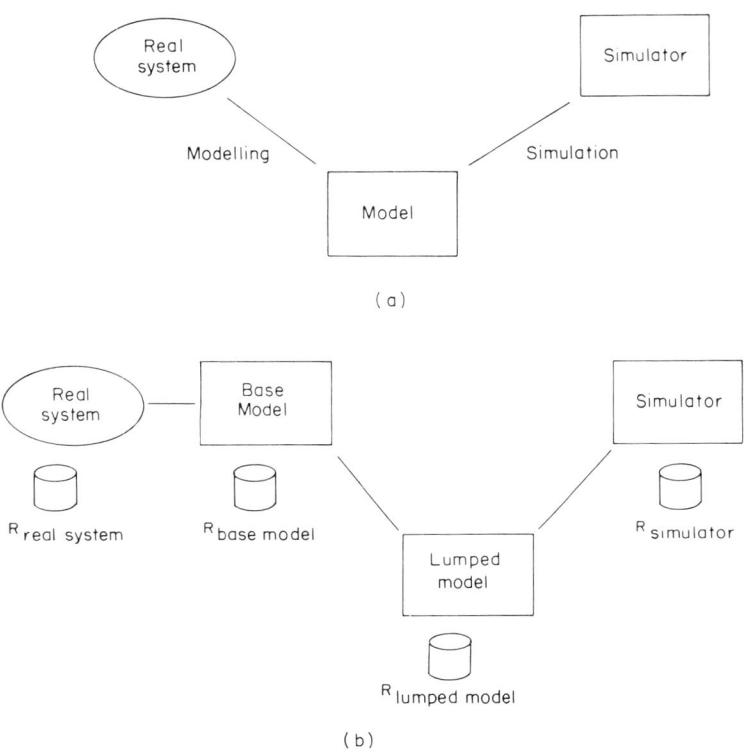

Fig. 5.1. Entities and relations of modelling and simulation.

preceding chapter. Let us regard each of the entities as *systems*, i.e. as formal objects of the type defined in Chapter 3. To make the distinction clear between the informal entity and its formal representative we shall use italics type. Thus *real system, model* and *simulator* are the formal systems representing a real system, model and simulator respectively. The *data elements* we have referred to informally can now be defined as *input-output segment pairs*.

The real system is a source of such data which we obtain by input-output observation. This means that we are concerned with $R_{real\ system}$, the I/O relation of the real system. As we do more and more experimentation with the real system we may expect to acquire more and more data elements. But at any time t, we will have available only a subset $R^t_{real\ system}$ of the potentially obtainable pairs, $R_{real\ system}$ (recall the partial data discussion of Section 1.2).

The model is also a source of such data. We obtain R_{model} by simulation, i.e. by experimentation with the simulator. Actually we do not obtain this data directly. What we do obtain from the simulator are elements of its own I/O relation $R_{simulator}$. To say that the simulator *correctly simulates* the model is just to say that $R_{simulator} = R_{model}$, i.e. that the model and simulator are behaviorally equivalent.

Likewise, to say the model data agrees with that of the real system is to say that $R_{model} = R_{real\ system}$. Such a concept of validity represents an ideal that realistically can never be achieved, but can be strived for. To this we return in a moment.

Ideally a completely valid and verified (real system, model, simulator) combination is represented by the equation:

$$R_{real\ system} = R_{model} = R_{simulator}$$

2. Knowledge of Internal Structure

All is not so straightforward as it appears however. We have not yet captured the modeller's relation to the entities in the framework. Since we have constructed the model we may assume that we know its system representation viz., *model*.

But all we can presume to be able to know about the real system is what can be observed about it, i.e. its I/O relation $R_{real\ system}$. In fact all we really do know about it at any time t is $R^t_{real\ system} \subset R_{real\ system}$. In other words, the real system should be regarded as a source of data and nothing more than that. Even to propose that there is a unique real system underlying our experimentation must be considered an hypothesis expressing belief in such ideas as the consistency of real world data and the universality of our systems formalism (i.e. its ability to express all of the models that could be valid representations of the real system).

Thus to be clear in our thinking we postulate the existence of a unique system specified at the highest level (Coupling of Systems) called the *base model*. This *base model* is distinguished from the *lumped model*, the model that we are constructing to meet our current objectives.

We now have the situation illustrated in Figure 5.1b. $R_{real\ system}$ represents the data potentially obtainable from the real system. By our postulation, there is a *base model* such that $R_{base\ model} = R_{real\ system}$. In contrast to the lumped model, we do not have the base model description. Indeed the heart of the system description is its internal structure, especially its transition function. So the difference between the base model and the lumped model is that we know the internal structure of the first but not of the second.

Finally, we shall assume that we know the system description of the simulator, viz, *simulator*. Although the simulator is also a real system we presume to know its internal structure. This in effect assumes we have a completely valid model for the real simulator. This is a good working assumption given for example, a reliable computer. (If it is in doubt, we can treat the computer as a real system and apply our modelling methodology to it!)

To summarize, we can assume we know the system descriptions of the model viz, *lumped model* and of the simulator but *not* that of the real system *base model*. This difference in status has profound implications for validation and verification methodologies.

3. Verification

In the verification situations we can hope to establish full correctness in a finite time. This is because we have the systems, *simulator* and *lumped model*, so if we can establish a morphic relation which places the known internal structures into correspondence, we can be guaranteed that $R_{simulator} = R_{lumped\ model}$ as desired. Thus access to the state spaces and defining functions of the model and simulator greatly facilitates the understanding and execution of the verification process.

4. Validation

The modelling relation is much more problematic than the simulation relation. Since we do not have the base model structure we cannot hope to establish a morphic relation with the lumped model by directly placing their internal structures into correspondence. We are limited in the final analysis to comparison of $R^t_{real\ system}$ with $R^t_{lumped\ model}$, the I/O segment pairs so far observed in the real system and the model.

This points out the essential difference in difficulty between the validation and verification process. But of course it only begins to suggest methodologies for their execution.

5. Experimental Frames

To proceed further with the modelling relation requires that we bring modelling objectives into the discussion. Recall from Section 1.1 how the objectives brought to the modeller by the decision maker oriented the modelling effort. We shall not deal with the objectives as such but only with their effect in limiting the real system data to a much smaller subset of relevance to the objectives. Such limitation is captured by the concept of experimental frame.

Basically an experimental frame specifies a limited set of circumstances under which a system (real system or model) is to be observed or subjected to experimentation. It thus establishes a data space — a framework into which only data relevant to a set of objectives is allowed.

To understand how such a frame can come about recall Figure 1.1 which illustrated how management, control or design were concerned with influencing the controllable part of a real system in order to achieve some objectives. Recall from Section 2.1 that the input set formalized that part of the system interface which was not under the system's control. Likewise, we define *system input variables* as variables that are presumed not to be under control of the system. These variables constitute the concrete form in which the system experiences the external influences of its environment or the experimenter. Indeed, the input set is recognizable as the crossproduct of the range sets of the input variables (the crossproduct was defined in Section 2.1, the range set of a variable is the set of values it can take on).

Recall that the output set summarized that part of the system interface by which the system makes itself felt externally. Thus we define *system output variables* as variables which are under the control of the system and which can potentially influence the system environment or can be measured by an observer. The crossproduct of the range sets of the output variables constitutes the output set.

In agreement with the point of view of the last section, we can attribute the input and output variables of a real system to the base model that represents it. This leads us to define *model input variables* and *model output variables* as the variables that structure the input and output sets, respectively of a model. Then for total precision, the system input/output variables are identified as those of the base model, while the model of current interest, viz., the lumped model, also has input/output variables.

Now we shall define an *experimental frame* as specifying two sets of variables, called the *frame input variables* and the *frame output variables*, respectively. (This is a preliminary definition, adequate for the present purposes. Later we shall allow the experimental frame to specify much more about the experimentation (Chapters 12, 13).) We designate a variable as a frame input or output variable to express our interest in *treating* it as an input or output variable of a real system or a model. A frame is *applicable* to a system or a model, if the frame input and output variables can be consistently treated as input and output variables, respectively, of the system or model. A direct test of applicability is that the input and output variables of the frame match exactly those of the system or model. We shall provide a more adequate and formal definition later (Section 13.1.10).

We can now see how an experimental frame can express the objectives with which a modelling project is undertaken. We may be interested in directly manipulating certain variables of a real system. In this case we would make these variables input variables of a frame for this objective. We may believe that other variables influence the system as well and thus we include them also in the input variables of the frame. Similarly, we may be interested in treating a variable of a real system or model as an output variable. For example, our objective may be to predict the performance of a system in its execution of a certain task. In this case, variables that can be employed to evaluate this performance would constitute the output variables of the frame representing this objective.

This constitutes our introductory exposition of the relation of the experimental frame to modelling objectives. This relation will be later elaborated as part of an objectives driven methodology for experimental frame construction (Chapter 12).

6. Experimental Frames and Validity

Let us return to the role played by experimental frames in our formal framework. We have not made (and will not make) any distinction in the type of experimentation, whether on the real system or on the model, specified by an experimental frame. Thus a frame E may be applicable to a *real system* or to a *model*. In the first case, the data potentially obtainable are denoted by $R_{real\ system}/E$ and in the second case by R_{model}/E. The slash(/) can be read: "as viewed in E" or "modulo E" or "within E."

We now have at our disposal a better notion of validity, one which takes account of objectives. A model is *valid for a real system in frame E* if:

$$R_{real\ system}/E = R_{model}/E$$

Thus validity is no longer absolute but relative to a particular frame. A model may be valid in one frame but not in another. In other words, a model may be able to answer questions about a real system relevant to one set of objectives but may not have this capability for some other set of objectives.

Returning to the base model concept, since by definition $R_{real\ system} = R_{base\ model}$, the base model is valid in every frame E. It represents the comprehensive model which, however desirable, is unfeasible to build.

However it is possible to imagine constructing lumped models which are valid in particular experimental frames. Considering a model in isolation, at any given time t, all we have on which to base a validation process would be a comparison of $R^t_{real\ system}/E$ and R^t_{model}/E. Such a comparison should result in an assessment of the confidence that we can have that the model is indeed valid in E. However if we take into account the existence of other models in the model base, their confidence levels and their relation to the given lumped model, we may be able to better assess its confidence level (recall the discussion of Section 2.5).

7. Varieties of Validity/Structural Inference

The validity we have been discussing so far may be termed *replicative*. It concerns ability of the model to replicate the input/output data of the real system. A stronger form is *predictive* validity for which we require that it be possible to identify the state that the model should be set into after some finite observation of the real system. Moreover, after such initialization, the model is capable of predicting the response of the real system to any input segment. A yet stronger form is *structural* validity in which we require that the model structure truly represent the internal operation of the real system.

These forms of validity may be readily formalized with the concepts so far developed.

A model is *replicatively valid* in frame E for a real system if

$$R_{model}/E = R_{base\ model}/E$$

The model is *predictively valid* in frame E for the real system if it is replicatively valid and in addition

$$F_{model}/E \subset F_{base\ model}/E$$

where F_S/E is the set of I/O Functions (Section 2.2) of system S as observed within experimental frame E.

The model is *structurally valid* in frame E for the real system if it is predictively valid and in addition there is a morphism from *base model* to *model* in frame E.

In view of the discussion in Section 5.2, these definitions may be reformulated as follows, all with respect to frame E: A model is replicatively valid if it is indistinguishable from the real system at the I/O relation level. It is predictively valid if in addition there is an I/O function morphism from the *base model* to the model. It is structurally valid if in addition there is a system morphism from *base model* to the model at the system level.

The justifying conditions in Section 3.5 enable us to infer higher level validity from the primary replicative validity. This is called *structural inference*: If a model is replicatively valid and identifiable, then it is also predictively valid. If in addition it is reduced, then it is also structurally valid.

Thus if a replicatively valid model is identifiable one can determine the state it should be initialized to after observing an input/output pair which identifies this state. If it is also reduced one can assert that there must be a morphic relation the hypothetical base model (even though the base model cannot be known).

There is a catch: replicative validity requires complete agreement of the two I/O relations. But only a finite portion of either relation will have been observed at any time. So even with a lumped model that is identifiable and reduced it is not possible to gain complete confidence in its predictive and structural validity, although more confidence can presumably be placed in such a model than would otherwise be justified for a given level of replicative agreement.

8. Valid Simplification

Having clarified the concept of validity for models, we should ask the question: why use lumped models in the first place? The answer, as we have already suggested, is that the base model is unfeasible to build. But what determines the feasibility of a model and why should a lumped model be more feasible than the base model?

The concepts of computational complexity and useful abstraction introduced in Sections 2.2 and 2.3 can be brought to bear here. Let us make the following identifications:

$$\text{model} = \text{computational object}$$
$$\text{model behavior} = \text{questions about object}$$
$$\text{simulator} = \text{processor}$$

Then we can see that the *simulation complexity* of a model can be defined in terms of measures of the space/time cost incurred by a simulator to generate its behavior. And just as with computational complexity in general, this space/time cost can be related to structural complexity measures of the model such as those mentioned in Section 2.1.

Further let us make the following identifications:

$$\text{base model} = \text{object}$$
$$\text{lumped model} = \text{abstraction}$$
$$\text{experimental frame} = \text{type of question}$$

Then the concept of useful abstraction tells us that a model M is useful in an experimental frame E with respect to a base model B
  (i) E is applicable to M
  (ii) M is valid in E with respect to B
  (iii) the simulation complexity of M is less than that of B (in frame E on the same processor).

Generally, though not always ([1], Chapter 16) the simulation complexity increases with model size (measured by structural complexity) so that lumped models may be much cheaper to simulate than base models. Indeed, a base model may not be feasible (in this sense) to simulate at all.

Further insight is gained by interpreting the processing to be performed on the model, not as behavior generation, but as analysis. For example, the modeller may be considered to be a processor of the model when trying to understand its structure. Then again the computational complexity (or here the psychological complexity) involved in understanding a large model may be much greater that that required for a small model. So a base model may be infeasible (in this sense) to comprehend. A third way in which a lumped model may have a computational advantage over a base model is in the parameter identification process (Sections 9.5 and 4.3).

We say that a lumped model is a *valid simplification* of a base model in frame E if it is useful in the above sense.

Formal definitions of experimental frame, validity, applicability of frames to models, and valid simplification will be given in Chapters 12 and 13.

9. Implication for Computer Assistance to Modelling and Simulation

Recall Figure 1.4 which depicted modelling and simulation as an ongoing process organized around the model and data bases. On the basis of our new concepts, let us draw some implications for what kinds of facilities should be provided by the computer to assist in this process.

As illustrated in Figure 5.2 we should, first of all, add a base of experimental frames to the model and data bases. We should then have assistance in constructing an experimental frame to represent the data space of a set of objectives as well as assistance in locating this frame within the base of frames.

The computer should be able to locate any models in the model base relevant to the new frames. Indeed there may already exist a model to which the frame is applicable, i.e. a model capable of generating the kind of data demanded by the frame. Such a model can be said to be capable of providing answers to any questions that can be posed within the frame, but of course, the answers need not agree with what the real system would say. In other words, applicability is a necessary, but not sufficient condition, for validity of a model in a frame.

In case there is no model to which the frame applies, there should be assistance for constructing such a model either from scratch or by simplifying, modifying and coupling the relevant existing models to form a composite model.

The data base should be organized according to the frames so that real

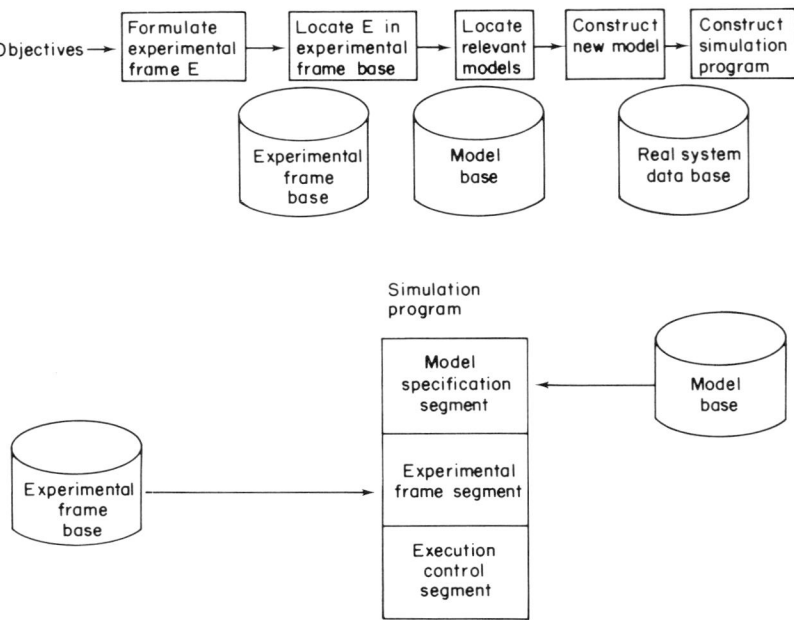

Fig. 5.2. Computer assistance to modelling.

system data elements belonging to a given frame can be conveniently stored and retrieved. The same goes for data elements of computationally expensive models. (The criterion for "computationally expensive" here is that it is cheaper to store simulation results rather than re-run the simulation.) The ability to access data by experimental frame keys would play a fundamental role in the validation process which involves comparison of data sets within the same experimental frame.

There should be assistance in simulation program construction and verification. Indeed our framework suggests how such programs should be constructed. As illustrated in Figure 5.2, there should be a *model specification segment* and an *experimental frame segment*. The latter only specifies the kind of experimentation to be done, not any particular experiment. Thus there must finally be an *execution control segment*, which selects an input segment and an initial state setting to perform a particular experiment. Since the model and experimental frame could originate from their respective bases there should be a test of applicability of the frame to the model before the segments are allowed to be combined.

This concludes our introductory discussion of the multifacetted system modelling framework. In coming chapters we shall define the concepts which were just introduced in greater depth. Current state-of-the-art simulation software cannot provide all the assistance in modelling which these concepts suggest. Nevertheless it is useful to examine such software from the perspective of the framework since we can then better understand what is available now and what is the potential for future development.

10. Summary

Modelling concerns the relation between a real system and a model; simulation concerns the relation between a simulator and a model. The system descriptions (and internal structures) of the model and the simulator are known but that of the real system is not. Verification, the process of establishing the behavioral equivalence of the model and simulator, can be established by proof of morphism relating simulator and model internal structures. A parallel approach to validation involving behavioral equivalence of real system and model is not possible, so validation is not a finite process.

An experimental frame captures the restrictions on experimentation and observation of real systems or models implicit in a given set of modelling objectives. The fundamental validity concept of multifacetted modelling is that of validity of a model with respect to a real system in a given frame.

The modelling and simulation process should be supported by an experimental frame base as well as model and data bases. Computer assistance should facilitate construction, manipulation, storage and retrieval of models and frames, as well as the construction of programs for simulating models within frames.

Succeeding chapters will define the above concepts in greater depth and will discuss some contemporary simulation software from the perspective they offer.

References

[1] Zeigler, B. P. (1976). *Theory of Modelling and Simulation*, Wiley, NY.

Other Relevant Literature
  In order to provide a clean framework for multifacetted modelling, we may have given an impression that validation and verification are easier to accomplish than they actually are. To correct this impression, we recommend reading the following:
Oren, T. I. (1981). "Concepts and Criteria to Assess Acceptability of Simulation Studies: A Frame of Reference". *C.A.C.M.* **24**, 180–189.

Sargent, R. (1983). "Validation of Simulation Models". In *Simulation and Model-Based Methodology: An Integrative View* (eds., T. I. Oren *et al.*), Springer-Verlag, NY.

Henize, J. (1983). "Critical Issues in Evaluating Socio-Economic Models". In *Simulation and Model-Based Methodology: An Integrative View* (eds, T. I. Oren *et al.*), Springer-Verlag, NY.

Measures for the complexity of comprehension of models and programs respectively, are discussed by the following authors:

Overstreet, C. M. (1982). *Model Specification and Analysis for Discrete Event Simulation*, Doctoral Dissertation, VPISU, Blacksburg, VA.

Halstead, M. H. (1977). *Elements of Software Science*, Elsevier-North Holland, NY.

PART 3
MULTICOMPONENT MODELS

Chapter 6

MULTICOMPONENT MODELS OF SPATIALLY DISTRIBUTED SYSTEMS

This chapter initiates the discussion of multicomponent models. It considers models of spatially distributed systems with the property of uniformity (or homogeneity). A classical prototype of such models is the cellular automaton which represents both space and time in discrete form. By retaining the discreteness of space but changing to discrete event time flow, we arrive at the concept of discrete event cellular models. The uniformity property enables us to characterize the model components with a single prototypic description rather having to specify each one separately. Aside from this simplification, the essentials of multicomponent discrete event system specification are evident in the cellular model definition.

The second part of the chapter considers the issue of formalism adequacy — how well can a formalism represent real world relationships and how do formalisms compare in this regard? Various sides of this issue are considered for the case of cellular models but the results and approach extend much more generally.[9]

1. The Cellular Automaton Formalism

A cellular automaton (also called a tessellation automaton or iterative array when external input is allowed) arises by placing copies of the same sequential machine at each of the lattice points of a two-dimensional array[10] and connecting them together in a uniform fashion. As all the sequential machines are supposed to undergo state transitions in parallel and simultaneously, this composition specifies a discrete time system. The uniformity in

---

[9]Some of the material in this chapter first appeared in [15].

[10]Actually, the dimension may be allowed to be specified as well. We shall freely employ examples of one or three dimensional models. Moreover, the space can be wrapped into finite torroidal form [1,2,12].

space, which will be apparent shortly, means that the system will possess spatial invariance (with respect to translation) as well as time invariance. While the cellular automaton underlies many models of discrete natural systems such as neuronal networks and tissue segregation models, it is more importantly a means of understanding many more common formalisms which can be viewed as extensions of the basic concept. Of these, we mention developmental system formalism [7], and partial differential equation [1] models, which when discretized for computer simulation can take the form of cellular automata. In this chapter we shall introduce another cellular automaton inspired formalism, the discrete event cellular space models, obtained by placing copies of the same DEVS component at lattice points of a planar array.

A cellular automaton is characterized by specifying three parameters: a set of states S, a neighborhood N, and a transition function T. The interpretation of the triple $<S,N,T>$ is as follows:

One imagines a checkerboard stretching out towards infinity in both north-south and east-west axes (Figure 6.1). At each square is located a cell with state set S. The neighbors of a cell located at square (i,j) (where as will be apparent "neighbors" is intended in an informational sense) are simply determined from the neighborhood N: in fact, N is a finite ordered set of integer pairs and the neighbors of this cell are located at the squares obtained by adding (i,j) to each pair of N (vector addition).

Now imagine a global state of this system pertaining at some time instant t, i.e. assign to each cell a state from the set S. Then the system will move to a

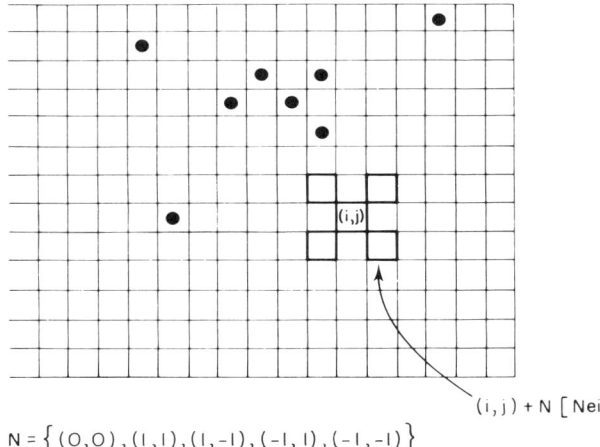

$N = \{(0,0),(1,1),(1,-1),(-1,1),(-1,-1)\}$

$(i,j) + N$ [Neighbors of (i,j)]

Fig. 6.1. Portrayal of cellular automation.

succeeding global state at time instant $t+1$ which is determined as follows: simultaneously, for each cell apply the transition function T to the states of its neighbors and let the result be the state of the cell at $t+1$.

Note that the cellular automaton, abbreviated CA, specifies a system at the structured system level. That is to say, the sequential machines at the lattices do not appear in modular form with their input and output ports identified — the connection of the wires has already taken place and is represented by the neighborhood concept. The neighbors of a cell are those whose present states influence its next state, viz., its *influencers*. This contrasts with the discrete event specification in which the neighborhood is given a different interpretation.

A CA specifies an autonomous (input free)[11] discrete time system. To see this it is enough to show how to associate a sequential machine with the CA which will represent the single step transition of the system. This process, depicted in Figure 6.2, is accomplished as follows:

Let CA $= <S,N,T>$. The automaton specified by CA is denoted M_{CA} and is defined by:

$$M_{CA} = <Q,\delta>$$

where

$$Q = \{q \mid q : I^2 \to S\}$$

and

$$\delta : Q \to Q$$

is defined by:

$$\delta(q)(i,j) = T(q \mid N + (i,j))$$

for each $q \in Q$, and $(i,j) \in I^2$.

CA $= <S,N,T>$

Cellular Automaton ‑‑‑>

  $M_{CA} = <Q,\delta>$
  sequential machine
  specified by CA ‑‑‑‑>

    System specified
    by M_{CA}

Fig. 6.2. Translation of CA specification into system specified at structured system level.

---

[11] The cellular automaton, and cellular space concepts in general, can be readily extended to allow external input to the model.

Note that the set of states of M_{CA} is the set of global states of the system. A *global* state q is an assignment to each of the cells of an element of S, which in contrast is called a *local* state. Likewise, the function, δ of M_{CA} is called the *global transition function* in contrast to the local function T.

Having the sequential machine M_{CA} representing the single step transition of the system, we obtain the multi-step transition function for the system S specified by M_{CA} in the usual way (Section 3.6). Thus the transition function δ^+ of S is defined by

$$\delta^+(q,t) = \begin{cases} q & \text{if } t=0 \\ \delta(\delta^+(q,t-1)) & \text{otherwise} \end{cases}$$

In other words, the iterative application to a global state q of the single step transition function generates a chronological sequence of global states, the tth of which is $\delta^+(q,t)$.

Example. A simple cellular automaton is derived from discretizing the partial differential equation model for diffusion. It has the following description

$CA_{\text{diffusion}}$:

$$S = R_0^+$$
$$N = \{(0,0)\}, (1,0), (0,1), (0,-1), (-1,0)$$
$$T: S^5 : \to S$$

where

$$T(s_1, \ldots, s_5) = \text{average}(s_1, \ldots, s_5)$$

i.e. the next state of a cell is the arithmetic average of the present states of its neighbors (including itself).

A cell in state r signifies the presence of r units of a diffusable quantity such as heat at its location. A global state represents a spatial distribution of this quantity.

Exercise. Let the initial global state q_o be such that all cells have 0 units of the diffusable quantity except the cell at the origin (0,0) which has 1 unit. Compute and display the first few global states in the state trajectory beginning with q_o. Note the pattern in which the diffusion occurs. How does it relate to the form of the neighborhood N?

As given, the CA specification does not place a constraint on its initial global state. However, it is necessary to do so if the resulting system is to be within the capability of any conceivable computer to simulate. One way to do this is to require that there be a local state, called a quiescent state, such that at most a finite set of cells do *not* find themselves in this state initially. The idea is that most of the space will be in the quiescent state and will remain in it, only a finite number of cells being capable of leaving this state at any time step.

A local state s is a *quiescent state* if when the neighbors of a cell are in s at time t, then the cell will be in s at time $t+1$. Succinctly, s is quiescent if $s = T(s,s,\ldots,s)$. Every local state of the diffusion model example satisfies this criterion since the average of a constant is the constant itself. If initially all of the cells are in the same quiescent state then this global state will persist through time, i.e. such a global state is an equilibrium state. More interesting are global states in which a finite set of cells are non-quiescent. Due to finiteness of the neighborhood, such global states always evolve to global states with the same property. Moreover, if the quiescent state is known, then most of the cells which are in this state at any time do not have to be examined by a simulator. This renders the simulation *computable*, i.e. performable by conceivable computers.

Exercise. Show that a global state q consisting only of quiescent states is an equilibrium state, i.e. $\delta(q)=q$. Let a global state have *finite support* if at most a finite set of cells are not in the quiescent state. Show that the set of global states with finite support Q_f is closed under transition, i.e.

$$q \in Q_f \text{ implies } \delta(q) \in Q_f$$

How can the simulator know which cells may leave the quiescent state? Examine the consequences of the following observation.

Lemma. If the neighbors of a cell do not change state from $t-1$ to t then the cell can not change state from t to $t+1$. (Here change of state means that the state and its successor are not the same.)

Exercise. Prove the Lemma.

Thus a cell may (but need not) change state from time t to $t+1$ only if there is a cell in its neighborhood which has changed state from $t-1$ to t. The simulator thus has to know at each time step which cells changed state to determine those cells which can change state in the next time step in these circumstances.

The *inverse neighborhood* $-N$ is the set of mirror images of the elements in N, i.e. $(i,j) \in -N$ if, and only if, $(-i,-j) \in N$.

Lemma. A cell (i,j) is in the neighborhood of a cell (k,l) if, and only if, (k,l) is in the inverse neighborhood of (i,j). A cell may change state from t to $t+1$ if, and only if, it is in the inverse neighborhood of a cell that changed state from $t-1$ to t.

Exercise. Prove the Lemma. When are N and $-N$ equal? Thus the set of cells that may change state from time t to $t+1$ is just the union of the inverse neighbors of the set that changed state from $t-1$ to t.

After introducing discrete event cell space models, we shall return to examine the relative abilities of the cellular automata and discrete event cell space models to represent distributed systems.

2. Next Event Cell Space Models

Discrete event cell space models preserve the spatial discreteness of cellular automata as well as their space and time invariance. However, the time base is no longer discrete but is continuous, that is to say, there is no intrinsic time step for such models. Events, i.e. cellular state transitions, may occur at irregularly spaced intervals, not necessarily synchronized to the beat of a clock as in the cellular automaton case. These events are all scheduled to occur as a consequence of the actions of cells: a cell may schedule events to occur to itself as well as to its neighbors, and these events may in turn schedule other events, and so on.

A *Next Event Cell Space* (NEVS) is specified by a quadruple $<S,N,T,\text{SELECT}>$, where the first three parameters bear a resemblance to those of the cellular automaton, but their interpretation is quite different. As before, one imagines an infinite checkerboard, with each square containing a cell. However, the state of any cell is now a pair (s,σ) where s is an element of S and σ is a non-negative real number (it will be useful to allow σ the value of infinity ∞ as well). A cell in state (s,σ) will remain in "sequential" state s for a time σ before undergoing a self induced transition. However, in the meantime its state may be altered as a result of some other cell's action. Thus s and σ are called the *sequential state* and *time left* components of the total state (s,σ). If cell c is in state (s,σ) at time t, this can be expressed as:

(∗) Cell c is in sequential state s and is scheduled for a transition in time σ (or at time $t+\sigma$).

As before the neighborhood of a cell is computed by adding its coordinates to the prototype neighborhood N (a finite set of pairs as before). When a transition event occurs to a cell, it and its neighbors undergo an immediate state change as dictated by the transition function T. Thus, in contrast to the automaton case, T takes a list of pairs (s,σ) — the total states of the cell and its neighbors just before the event — and produces a list of such pairs, giving the total states of these cells just after the transition.

In contrast to the cellular automaton, the updating of an NEVS model takes place at irregularly spaced computation instants. Let t_i be such an instant, then its successor t_{i+1} is computed as follows:

Imagine the global state at time t_i to be a list of pairs (s,σ) containing the total states of all the cells at this time. Let σ^* be the minimum of all the σ's on the list. Then $t_{i+1} = t_i + \sigma^*$.

The global state at time t_{i+1} is computed as follows:

Call the cells whose σs at t_i were equal to σ^* the IMMINENT cells. These cells are all scheduled for a transition event at t_{i+1}. If there are more than one, apply the SELECT function to choose one of the IMMINENT cells call it c^*. (Thus the SELECT function provides the desired tie breaking rules when more than one transition event is scheduled for activation at the same time.) The chosen cell c^* carries out the transition function as indicated above resulting in a new global state which is related to that pertaining at t_i as follows:

The total states of the neighbors of c^* are as dictated by T; the total state of every other cell is converted from (s,σ) to $(s, \sigma - \sigma^*)$.

2.1. Examples of Discrete Event Cell Space Models

Some examples of next event cell space models will be given to illustrate their operation and versatility. They will also serve to point out the differences in expressive capability relative to cellular automata.

Example 1: motion of a single body

Let us begin simply by expressing the motion of a single object in a NEVS. The object is to move to the right, at a speed of one square every MOVE-TIME seconds. An appropriate NEVS is defined as follows:

Every cell has two states, one to indicate the presence of the object and the other its absence: thus $S = \{0,1\}$. The absence indicator 0 is a *passive* state, i.e. a cell in this state will never become active of its own accord. This is represented by the total state pair $(0,\infty)$ which indicates that the time left for a cell in sequential state 0 to change state is ∞. On the other hand, the presence indicator 1 is an *active* state (opposite of a passive state); indeed, it is the source of motion in this example: when a cell is in total state $(1,\sigma)$, it is scheduled to pass on the object in σ time units.

To get the object to move we must arrange for a cell in state 1 to cause its right adjacent (physical) neighbor to enter state 1 after staying in this state for MOVE_TIME seconds. Thus for the neighborhood N we require only the two integer pairs $\{(0,0), (1,0)\}$ — adding (i,j) to each we obtain $\{(i,j), (i+1,j)\}$ the co-ordinates of the cell at (i,j) and its right neighbor.

In the light of the above comments, the transition function can be expressed as follows:

> (*) set your own sequential state to 0
> and passivate (set your own time-left to ∞)
> set right neighbor state of 1 and
> schedule a transition event there in MOVE__TIME
> (set neighbor's time__left to MOVE__TIME)

The SELECT function is arbitrary in this example since at most one event is scheduled at any time.

Starting from an initial global state in which all but one cell are in state $(0,\infty)$, the exception being in state (1,MOVE__TIME), successive global states are generated every MOVE__TIME seconds in which a 1 is seen to travel horizontally to the right.

Example 2: traffic congestion

Note that the above model does not specify what happens in the case of collisions — when two or more objects want to move into the same square. The present example illustrates this phenomenon. Consider cars which move in one direction at a constant speed when possible (travelling in an east bound line for example). When the car immediately ahead obstructs progress a car seeks to pass preferably to the right, but also to the left if the first is not possible. The patterns of traffic congestion which develop can be studied with NEVS models of which the following is prototypic:

The state set $S = \{0,1\}$ represents absence and presence of a car as in Example 1. We put in the neighborhood N all cells immediately above and below the center cell to allow for the possibility that a car may move in these directions if direct progress is blocked. Let N be ordered in the order of preferred motion as illustrated in Figure 6.3. Then the transition function T can be expressed as follows:

> (*) Find the first unoccupied neighbor cell
> (scanning in the preference order)
>
> set this cell to state 1 and schedule a
> transition event there in MOVE__TIME
>
> set your state to 0 and passivate
> leave all other cells unaffected
> (let their sequential states
> remain unchanged and subtract your
> time__left ($\sigma*$) from theirs)[12]

---

[12]This is to update the time__left values appropriately as with all non-neighboring cells in the space.

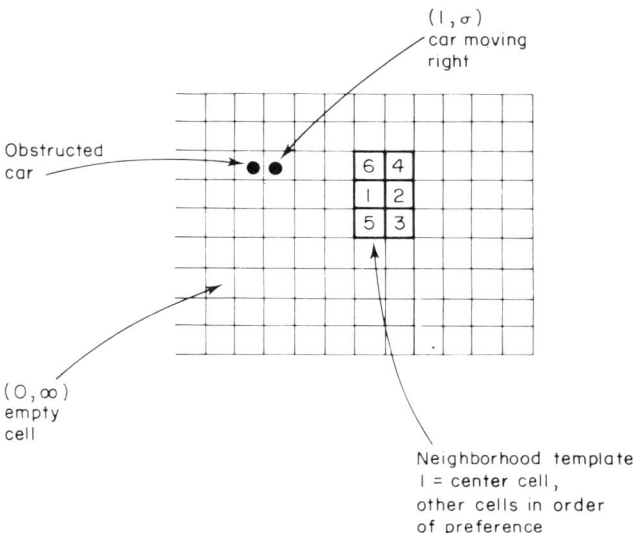

Fig. 6.3. Illustrating traffic next event cell space model.

        if no unoccupied neighbor exists
            reschedule yourself in time CHECK__AGAIN__TIME
            leave all other cells unaffected.

The SELECT will determine which of two or more cars contending for the same square at the same time will move into it. Thus, the order in which IMMINENT cells are selected does make a difference in this example. SELECTION, based on "right of way" principles for example, therefore incorporates essential model hypotheses in this case.

An initial global state in this model consists of putting a finite set of cells in sequential state 1 with a time__left σ between 0 and MOVE__TIME; all other cells are set to the passive $(0,\infty)$ state. The pattern then generally moves right with rearrangements as cars take diversionary maneuvers to bypass obstructions.

Note that to make the simulation more realistic, MOVE__TIME and CHECK__AGAIN__TIME can be made to be random variables so that cars move at randomly assigned rather than constant speeds. Note that in this model cars must continually check their neighborhood for the opening of a square. This may realistically represent the driver's reaction time but it is neither elegant nor efficient when this reaction time is negligible compared to the speed of travel. Such conditional waiting can be nicely expressed by extending the NEVS to include activity scanning and process interaction mechanisms (see Section 6.2.3).

Example 3: plant growth: gravity signal

This example shows how next event models can handle "action at a distance" in a very natural manner. Consider embedding a plant in a one dimensional cell space as illustrated in Figure 6.4. Growth is made to occur at the tip of the plant by sprouting a new cell every GROWTH_TIME hours. The extra weight is transmitted to the stem and base cells by a gravity signal which travels instantaneously downward. Thus at the same computation instant at which the tip extends itself the weights on the other cells are simultaneously increased. This happens no matter how long the plant becomes. The following NEVS will exhibit this behavior:

Let the state set S be $\{0,1,2,3,...\}$ where 0 is a passive environmental state and $i > 0$ is a plant cell supporting a total weight of i units (itself plus $i - 1$ cells

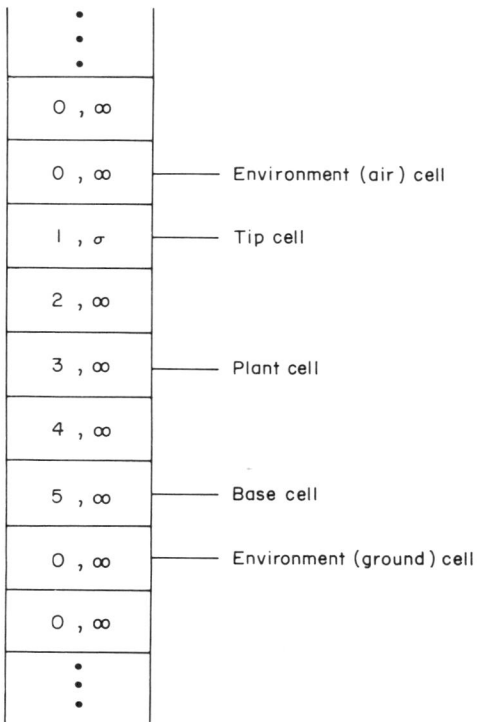

Fig. 6.4. Representing a plant as a one-dimensional next event cell space.

above it). The neighborhood N consists of the center cell and the cells above and below it. The transition function T is expressed as follows:

> (∗) If you are the tip (center cell sequential state = 1)
> set the upper neighbor to state 1 and
> schedule a transition event to
> occur there in GROWTH_TIME
> If the lower neighbor is a plant cell (state = $i > 0$)
> add 1 to its state and activate it now
> (schedule a transition event
> to occur there in 0 time)
> set yourself to state 2 and passivate
>
> Otherwise (you are a plant cell)
> if the lower neighbor is a plant cell
> add 1 to its state and activate it now

Note that if the activated cell is a tip then it carries out three actions: sprout a new tip upward, start a gravity signal downward and transform itself to a passive plant cell; if the activated cell is a plant cell it passes on the gravity signal (unless it is a base cell known by the fact that its lower neighbor is an environmental cell).

When started in a global state such as illustrated in Figure 6.4, this NEVS will display plant extension with weight accumulation in "real time" (see later comparison with cellular automata).

Studies of plant growth using discrete event models may be found in [6]. Actually, it is interesting that a further extension of the cell space formalism is necessary to conveniently express growth in the interior of a plant — the neighborhood must be allowed to change ([5,6,7] and Problem 3).

2.2. Formal specification of next event cell models

Just as the cellular automaton specifies a discrete time system so does a discrete event cell model specify a discrete event system. As illustrated in Figure 6.5 to do so first associates with a NEVS a DEVS, which represents the single transition description of the system. The DEVS is translated into a system in a standard way and represents the multi-transition behavior of the system in the same way the extended transition function does in the discrete event case.

A NEVS (Next Event Cell Space is a structure)

$$CN = <S,N,T,SELECT>$$

where
S is a set of sequential states

Spatially distributed systems   103

CN = <S,N,T,SELECT>
Next Event Cell Space   --->

$M_{CN} = <S,\delta,ta>$
DEVS specified by CN   --->

System specified by M_{CN}

Fig. 6.5. Translation of next event cell space specification into system specified at the structured system level.

N is the neighborhood template
T is the local transition function
SELECT is the tie breaking function
subject to the following constraints:
    N is a subset of I^2, the influencees of the cell at the origin
    The total state at a cell is given by a pair (s,σ); accordingly T operates on $|N|$-tuples of such pairs:

$$T: (S \times R_{0,\infty}^+)^{|N|} \to (S \times R_{0,\infty}^+)^{|N|}$$

The SELECT function is a choice mapping

$$\text{SELECT: (subsets of } I^2) \to I^2$$

such that SELECT $(subset) \in subset$ for all non-empty finite subsets.
    We associate a DEVS M_{CN} with the NEVS CN:

$$M_{CN} = <\mathbf{S},\delta_\phi,ta>$$

where the global sequential state set is

$$\mathbf{S} = \{s|s:I^2 \to S \times R_{0,\infty}^+\}$$

and for $\mathbf{s} \in \mathbf{S}$, $\mathbf{s}(i,j) = (s(i,j),\sigma(i,j))$ the global time advance function is given by:

$$ta(\mathbf{s}) = \text{minimum}\{\sigma(i,j)|(i,j) \in I^2\}$$

[the next event will occur when the smallest time left component of the global state has expired]
let

$$IMMINENT(\mathbf{s}) = \{(i,j)|\sigma(i,j) = ta(\mathbf{s})\}$$

and

$$c(\mathbf{s}) = \text{SELECT}(IMMINENT(\mathbf{s}))$$

[IMMINENT(s) are the cells with the smallest time left and $c(\mathbf{s})$ is the cell selected to be activated by applying SELECT] then the global internal transition function is defined by:

1. $\delta_\phi(\mathbf{s})|(N+c(\mathbf{s})) = T(\mathbf{s}|(N+c(\mathbf{s})))$

and otherwise

2. $\delta_\phi(\mathbf{s})(i,j) = (s(i,j), \sigma(i,j) - ta(\mathbf{s}))$

Line 1 specifies the effect of the activated cell on its neighborhood, while line 2 updates all of the other cells' time_left values.

As in the cellular automaton case, the initial global states must be restricted for computability. Here the role of the quiescent state is played by the passive state. A cell is *passive* if its time_left σ equals ∞. An initial global state can consist of at most a finite set of non-passive cells.

Exercise. Explain how this enables a simulator to restrict its attention to a finite number of cells at any stage in the simulation.

Having the DEVS M_{CN}, it is now routine to obtain the system that it specifies and the state trajectories it generates.

2.3. Extension to Activity and Process Formalisms

The next event cell space can be readily extended to the most general discrete event form by including the capability for activity scanning. To do this we allow the time_left component of the local state to take on negative values as well as the non-negative values already in its range. If a cell has a negative σ the cell is past due, i.e. its event time has already been past but it has not been able to satisfy the condition enabling it to proceed to carry out its activity; in brief, it is blocked. In the pure activity scanning formalism, there is only one (condition,action) pair associated with a cell. In the process formalism, there may be several such pairs strung together in a sequence. The state of the cell then contains a phase variable which indicates which pair is currently in control. The traffic example (2), modified to reflect the blocking of cars, is given next to demonstrate this formalism.

Example 4: traffic model in extended formalism

The neighborhood N is the same as that of the original of Example 2 as is the non-phase component of the sequential state. The transition function is given as follows:

(∗) 1. waituntil(there is an unoccupied cell in your neighborhood)
   find the first cell in the preference order
   set it into state 1
      activate it in MOVE_TIME units from phase 1
   set your state to 0
2. passivate

In this language a *waituntil* indicates the presence of a condition to be tested when a process is due. A phase label marks such a condition and there follow statements for the action to be performed when, and if, the condition finally becomes satisfied.

Note that the waituntil statement permits a process to proceed as soon as the blocking condition permits it to do so. The next event formalism can not directly express this form of immediate activation, (but see Section 8.6). Thus the NEVS of Example 2 enables a car to become aware of movement possibilities only periodically. Moreover, we shall see how implementation of the waituntil construct can be done in an efficient manner so that a blocking condition is tested only when it could possibly change truth value.

Example 5: reaction systems: ecological, chemical, etc.

This example illustrates how collisions which result in changes of state are naturally expressed in the discrete event formalism. Such collisions occur for example, when two molecules collide and combine chemically. The example however, concerns ecological interactions of this nature. Each cell is to represent a "patch", a hospitable area isolated from other patches by inhospitable terrain. Prey species colonize such patches and are predated upon by predators. Both species tend to remain localized to a patch until forced by shortage of resources to migrate. How far migrants can get is species dependent (depending on such factors as speed of travel, ability to survive without food, etc.). The following prototype is drawn from the literature [13,8].

The phases are designated as follows: EMPTY, PREY (patch colonized by prey, population not yet at equilibrium), PREY' (prey population at equilibrium, sends out migrants), PRED (predators have immigrated to cell in PREY or PREY' state), and PRED' (predator population has reached maximum, sends out migrants, and dies out).

The neighborhood is the union of N_{PREY} and N_{PRED}, cells reachable from the center cell by prey and predator migrants respectively.

The transition function is expressed as follows:

(∗)   PREY.   hold(PREY__GROWTH TIME)
       PREY'.   hold(INTER__MIGRATION TIME)
              for each cell in N_{PREY}:
                 if the cell is EMPTY, then
                    activate it from PREY
              go to PREY'
       PRED.   hold(PRED__GROWTH__TIME)
              for each cell in N_{PRED},
                 if the cell is in PREY or PREY', then

```
                    reschedule it from PRED now
                    hold(PRED__EXTINCTION__TIME)
        EMPTY. passivate
```

We employ the phrase $hold(\tau)$ as the equivalent of "set σ to τ", i.e. reschedule the next event in τ units. The phase segments in the above example are all of the unconditional kind. Phase labels serve as convenient markers for segmenting activities: when the process in a cell is reactivated control passes to the phase in which suspension occurred (if suspension was caused internally) or to the phase specified by an externally caused interruption. An example of the latter is given above by the external interruption caused by the arrival of a predator in a patch colonized by prey. A cell in phase PRED may interrupt a cell in phases PREY or PREY' and set it into phase PRED to be activated immediately (σ is set to 0).

In actual studies, colonization and migration are made stochastic and the space is limited to a finite region. From an initial global state consisting of EMPTY, PREY and PRED cells, the evolution of the system is followed to see if global persistence of predators is maintained despite guaranteed local extinction (see Section 15.1.1).

3. Formalism for General Discrete Event Cell Space Models

The formalism of the Next Event model is extended to a general formalism which can be specialized to the three primary "world view" formalisms as follows.

A *Discrete Event Cell Space Model* (DEVCS) is a structure:

$$DE = <S, N, T, SELECT>$$

with the same meaning as before but with the following constraints:

$$S = P \times M$$

where $P = \{p_i | i = 1, ..., n\}$ denotes a phase set and M a memory set

$$N = N_{influencees} \cup N_{influencers}$$

the latter being neighborhood sets
T is a finite set of labelled pairs:

$$p_1.(c_1, f_1), ..., p_n.(c_n, f_n)$$

where for each $i = 1, ..., n$:

c_i is a condition predicate on $N_{influencers}$

and

f_i is an action function mapping $(S \times R_{0,\infty})^{|N_{\text{influencees}}|}$ to itself

Associating a DEVS M_{DE} with a discrete event cell space DE is an extension of the procedure given in the NEVS case. The major difference is that given a global state one must first separate out the due cells, i.e. those with negative or zero σ. The subset of the due cells whose condition (in the current phase) is satisfied is called the *activateable* set. If it is not empty, one of the activateable cells is activated using the SELECT function. Otherwise SELECT is applied to the IMMINENT cells (those with minimum *positive* σs) and the chosen cell is activated. The time advance is 0 in the first case and the minimum of the positive σ's in the second.

Exercise. Write the formal definition of M_{DE}.

Example 6: networks of computing devices

Von Neumann originally used the cellular automaton framework to embed networks of logical elements with the purpose of exhibiting systems with self reproductive ability [11]. Subsequently, this framework under the rubric "tessellation or iterative arrays" has been extensively employed for hardware design. Extending the same principles to the discrete event cell space, one can readily model networks of asynchronous elements for questions of logical correctness, coordination control and reliability. Such networks may model computer systems at all levels: whether the components be electronic elements, hardware devices, or full scale computers connected by communications links (see Problems at end of Chapter 12).

On the natural computing side, cellular automata have been used to model networks of neurons and the extension to discrete event models is natural (see example in Section 8.5).

4. Formalism Adequacy: Representing Real World Relationships

It should be apparent that different model formalisms carry along differing "world views" that determine the ease of expressing real world relationships within the constraints of the formalism. We cannot hope to fully consider the adequacy of a formalism to represent reality, but we can explore some of its aspects. One of these is a formal one: the relative expressive power of a formalism. A second concerns "intrinsic" simulation efficiency. We shall first comment upon these "hard" criteria as a backdrop for consideration of the "softer" questions such as convenience, versatility, naturalness, etc.

4.1. "Hard" Adequacy Criteria: Expressive Power

By the *expressive power* of a formalism is meant the class of systems it can specify. We can compare powers of two formalisms by asking whether every system specified by a model in one can also be specified by a model in the other. In this form, automata and discrete event systems are incomparable since the time bases are distinct. Thus we are led to the more sophisticated formulation: can every system specified in one be "simulated" by a system specified in the other? Now the formal concept of "simulation" is expressed in a hierarchy of system preservation relations (recall Section 3.3). At the lowest level, we require only that one system reproduce the external behavior of another in order to grant it simulation capability; at the higher levels, we require that the first represent more and more of the internal structure of the second to be accorded such status. Thus as we ascend the hierarchy, simulation related systems will be "closer" together in structure and behavior. Conversely, as we descend, systems so related may be more and more different in their detailed operation. A consequence of this fact is that as we loosen our definition of simulation, we increase the expressive power of formalism. However, because the two system structures are increasingly remote we place more of the burden on the observer to interpret the behavior of the simulator as revealing that of the simulatee.

With this as prologue, let us compare the powers of the cellular automata and discrete event cell space formalisms. We shall restrict our attention to computable models, i.e. models whose behavior can be computed by a (idealized) digital computer.[13] Then it is easily seen that the cellular automata can simulate any DEVCS in the sense of computing its behavior. The argument is simple: to be computable is more particularly, to be computable by a Turing machine. Now any Turing machine can be embedded in a cellular automaton space (using Von Neumann's construction [11] or the very straight forward approach of [9]). Composing the two simulations we get that any discrete event cell space model can be simulated by some cellular automaton.

Of course, the concept of simulation employed in this result is a rather weak one (lies at the lower level of the hierarchy) and states little more than that a cellular automaton can be used as a computer to generate the behavior of a DEVCS. It leaves the burden of writing and interpreting this simulation to the user and so says nothing of the ability of the automaton formalism to directly express NEVS-like models.

---

[13] In addition to restricting the global states to those with finite support, we also limit the real numbers employed in the DEVS formalism to the rationals. This is to make the comparison fair — we should not be taking advantage of the non-computable reals to demonstrate a non-achievable capability.

In fact, this ability is demonstrably limited when the simulation relation is tightened (hierarchical level increased) by requiring the simulation relation be at the structured system level and in "real time" (Section 3.3). Thus we require that there be a one-one correspondence between the cells of the simulator and simulatee, and a mapping from the local state set of the first on to that of the second. For correct simulation, we require that the mapping of global states set up in this way be a homomorphism of the systems specified by the two models. In other words, we require that the simulator reproduces the state behavior of the simulatee in a cell for cell fashion and in a fixed time relation to it.

Using the above simulation relation we shall show that every cellular automaton can be simulated by some DEVCS but that the converse does not hold. Thus the next event formalism is the more powerful under the real time structured system simulation relation. Moreover, as a consequence of this observation and the previous one, the unrestricted simulation capability of the two is the same, both being computation universal. Consider then the

Theorem (Real Time Structured System Simulation)

Given any cellular automaton CA, there is a DEVCS DE which simulates it in the sense that there is a real time structured system morphism from the system specified by DE to that of specified by CA. The converse is false.

First the positive part. Let $CA = <S,N,T>$ be a cellular automaton, We construct a DEVCS $DE = <S',N',T',SELECT>$ as follows:

S' contains two copies of S, one to hold a CA cell's present state and the other to hold the computed next state. It also contains a third boolean component, *changed*, which records whether a cell has truly changed state in the last CA global transition. Finally there are four phases: *compute, wait.1, update, wait.2*.

N' is the union of the neighborhood N and the inverse neighborhood $-N$.

The transition function T' is expressed as follows:

(∗) *compute.*   wait until (at least one cell in N has *changed* = true)
                apply T to the present states of the cells
                in N and store the result as the next
                state of the automaton cell

wait.1.   hold(0.5).

update.   if you truly changed state
                (next state not = present state)
                then set *changed* = true else false
                transfer the next state to the present state

```
wait.2     hold(0.5)
           if changed = true
           activate all cells in −N
           which are passive
           go to compute
```

The SELECT function is arbitrary: the result will not depend on the order in which scheduling ties are broken.

The simulation relation is as follows: Each cell in the DEVCS corresponds one-one with the CA cell with the same spatial co-ordinates. The local state set mapping merely projects out the present state component of the DEVCS and ignores the rest. The subset of representing global states of the DEVCs is the set of global states in which all cells are in the compute phase with either $\sigma = 0.5$ or are passive ($\sigma = \infty$). Thus if the DEVCS is in such a representing state, one should be able to read out the corresponding global CA state by inspecting the present state component of each cell.

The initial global state correspondence is established as follows: for each CA cell which is non-quiescent, its DEVCS counterpart is loaded with the CA cell's present state, its *changed* set to true and is started in the *compute* phase. The inverse neighbors of the non-quiescent cells are treated in precisely the same way (this is necessary since such cells may change state in the next time step). All other DEVCS cells are passivated. Thus all but a finite set of simulator cells have $\sigma = \infty$ initially as required for computability.

It can then be shown by induction on t that the global state of the DEVCS system will be a representing state at time t and will map to the global state of the CA system at time t. The proof makes use of the following points:

* In the *update* phase only those DEVCS cells having a neighbor cell with *changed* = true will not be passivated and compute a next state for $t+1$. That this is correct follows from the Lemma and the fact that *changed* is set true if, and only if, a cell indeed changed state on the last time step.

* A DEVCS cell computes the correct next state of its counterpart since it has access to the present states of each of the latter's neighbors. This is guaranteed by the 0.5 unit hold between the *compute* and *update* phases.

* A passive cell will be activated at time $t+1$ if, and only if, it has an active neighbor which changed state from $t-1$ to t. This activation was done by the neighbor at the end of its *update* phase.

Now for the *negative part*: By virtue of its finite neighborhood the cellular automaton eschews "action at a distance" [3]. By allowing instantaneously propagating signals, as in the plant growth example, the DEVCS formalism does not impose this restriction.[14] In particular the plant model can be shown not to be simulatable by any cellular automaton in the real time sense. To see this consider a plant configuration which has extended beyond the neighborhood of the base cell. To simulate the plant model in the required manner, the plant must be represented by a direct counterpart configuration in a CA space. For a correct simulation, the base cell (corresponding DEVCS local state) must increase by 1 at the next time step (for simplicity let GROWTH__TIME be unity). But now consider the same configuration in which the growth tip is removed. Then it can be readily checked that the DEVCS model will not change state (no gravity signal will be sent down to the base). But in the purported simulation the base cell readout must still increase by 1 since its neighborhood configuration has not been affected by the removal of the growth tip.

Note that it would not help to allow the CA to take a *fixed* finite number of transitions to compute one plant model global transition. If such a multistep simulation were correct then the DEVCS could also be simulated by a CA with a larger neighborhood on one time step.

Exercise. Verify that the global states produced every n times steps by a CA with neighborhood N can be reproduced by a second CA every step if it has a neighborhood of nN ($N + N + ... N$ (n times)). Does the same limitation apply in the case where the number of intermediate steps is not fixed?

In short, a cellular automaton realization of the plant model would require an infinite neighborhood to instantaneously pass the weight information generated at the arbitrarily extending tip to the base and stem cells.

4.2. "Hard" Adequacy Criteria: Efficiency of Simulation

The simulation efficiency of a formalism refers to the time and space required to simulate systems specified in this formalism. Since space and time are measured of a simulator, efficiency must be stated relative to a class of simulators, a simulation relation, and must aggregate over a set of simulation algorithms (recall Section 2.3).

The issue arises because the historical development of discrete event languages can be traced to the perception that efficiency in simulation could be achieved by skipping over time gaps in model behavior during which no significant activity occurs. The "world view" of the discrete event formalism is

---

[14]Holland has developed an extension of the cell space concept which removes the action-at-a-distance limitation as well [14]. His scheme lays down wires capable of passing instantaneous signals in such a way that no cycles or infinite paths are generated. In our scheme such conditions are possible and are the responsibility of the user to prevent (see Problem 1).

that only some components are likely to change state only some of the time. In contrast that of the discrete time formalism in general, and the cellular automaton in particular, is that every component is liable to change state at every time step. One might therefore expect, in models where activity[15] is relatively rare, that standard implementations of discrete event models would be more efficient in time and space than their discrete time counterparts. As the activity level rises, a crossover point should be reached where the advantage lies with the discrete time simulation [15]. Empirically, this has been found to be so.

Such studies however have not recognized the caveats stated above concerning the effect of simulation relation and class of simulation algorithms. As a case in point, the algorithm used by the DEVCS to simulate the CA in the last theorem provides an alternative approach to simulation of discrete time models. One first translates the given CA description into the DEVCS which represents it (this can be done algorithmically) and then simulates the latter using event set approaches, standard in the common discrete event simulation languages. Since in this simulation only cells that may possibly change state are considered by the simulator, one may expect to get a much more efficient simulation for low activity CA models.

Thus from the point of view of simulation efficiency, arguments against the use of discrete time formalism may lose their force considering the above results. Note however, that the discrete event formalism is in any case an essential intermediary in this approach. In the sequel we examine soft criteria for formalism adequacy and find more compelling arguments favoring the discrete event approach.

4.3. "Soft" Adequacy Criteria

The expressive power and simulation efficiency of a formalism throw light on, but do not close the question of its adequacy. We can see for example, that the two formalisms under discussion are computationally equivalent (a loose notion of simulation) but differ quite drastically when more direct model representation is required — the cellular automaton being incapable of exhibiting immediate signal propagation in real time. And although any cellular automaton can be directly represented in next event form, the representation $<S',N',T',\text{SELECT}>$ contains much additional apparatus not found in the original specification $<S,N,T>$. Thus if one had already developed a simple cell space model, one would have to complicate its

---

[15] Activity can be measured in terms of the number of non-trivial state changes per unit time interval — theoretical studies have shown that models exist at all levels of this complexity measure [10].

description in order to convert it to next event form. Moreover, the result would obscure the parameters of the original specification and would be more difficult to modify because of the interdependencies of the components of the new parameters (N' for example consists of N and $-N$: if a change is made in N a corresponding change must be made in $-N$). The same considerations hold, in even more striking form, when expressing simple NEVS models as cellular automata. To make this point we shall consider expressing the models given in Examples 1–4 as cellular automata.

Expression of motion

Consider Example 1 modelling the motion of a single body. To represent this in a cellular automaton, we require a neighborhood consisting of the center cell and its *left* adjacent cell (rather than the right neighbor as in the NEVS case). The reason is that motion consists of two parts: leaving one cell and entering the next. Since all state transitions are made independently and simultaneously by automaton cells, both cells affected by the motion must do some computation. The cell containing the object must release it (change from state 1 to 0) and the cell receiving the object must accept it — look to the *left* to see if the neighbor is a 1 and if so change itself to state 1. Thus what is a single action in the NEVS case, is redundantly expressed in two actions in the automaton case.

Handling of collisions

The redundancy of computation required is greatly compounded when collisions must be resolved as in the traffic congestion model in Example 2. Consider that to compute its next state, the automaton cell must independently decide whether any car it contains will move right and also whether any other car will move into it. To see whether its car can move, it must look not only for unoccupied cells in its own preference order but also check that an unoccupied cell is not "desired" by a car with higher priority (otherwise both cars will attempt to move into the same cell). To see whether a car will move into a cell, the cell must check whether any car which can move into it, will prefer to do so (a car below will move up only if all of its other possible moves are blocked). The reader can convince himself that this computation will require a neighborhood of approximately 20 cells (!) and is, of course, heavily redundant — the same checks being done independently by many cells.

Choice of time step

The actual value of the time step does not appear explicitly in the automaton formalism. However, when discrete time models are formulated for real systems, the choice of this value is always an issue. As discussed in [1] cellular automata can be viewed as providing the formal basis for the usual representation of partial differential equation models for computer simulation. Such models are based on continuous time and space variables which are discretized for numerical integration. The time step and spatial unit must be chosen carefully to conform to the rates of propagation expressed in the original model.

Similar considerations apply when the underlying model is a discrete event type. The most direct way to simulate an NEVS by a cellular automaton is to let the automaton cell state be a pair (s,σ) where now σ is a non-negative integer or ∞. The transition function sees to it that σ is treated as a time__left variable by decrementing it by 1 at every time step; when 0 is reached, a new total state pair is determined according to the transition function of the NEVS. Of course, the cell must also check at each time step for cells reaching $\sigma = 0$ for changes that they would cause to it.

The question is: what real value should the integer σ represent? Too small a discretization will result in much unnecessary recomputation of global states that essentially do not change; too large a value risks missing of events which would have occurred in the original. We can conceptualize this problem by limiting consideration to recurring cycles of a propagating nature (such as the motion of Examples 1 and 2 or the signals of Example 3) or of a local nature (such as that of Example 4). Then if the cycle durations are constant the best step size is given by their greatest common divisor. However, in even the simplest of models, these durations may not be constant due to being random variables or due to interactions which determine the continuation of the cycle (this is especially evident in Example 4 where the cycle must await successive prey and predator migrations in order to proceed).

In conclusion, no general procedure exists for selecting the step size underlying cellular automaton representation of NEVS models. Even if such a procedure were employed, the step size would have to be recalculated, and the time__left variables appropriately rescaled, for every change in the NEVS parameters. Thus a cellular automaton realization of models such as those of Examples 1-4 would be either inefficient or inflexible (or both).[16]

4.5. World View of Formalisms

Let us summarize the above findings in terms of the "world view" of the formalisms. The cellular automaton cell space is one in which each cell updates its state at each time step. The updating is done simultaneously and independently by the cells. The cell space thus expects a state change to occur

---

[16]It should be noted that the same difficulties *do not* apply to digital simulation of discrete event models (as they do to simulation of differential equation models). The reason is that the simulation strategies employed by such discrete event languages as GPSS, SIMSCRIPT and SIMULA are based on event-driven rather than fixed step time-advance (Chapter 17).

at every cell even though a true change may not in fact take place. To obtain the information necessary for computing its next state, a cell interrogates the cells of its neighborhood — which must be sufficiently extensive to provide this information. Since cells cannot immediately act on one another, a cell must predict what effect other cells would have had on it in the more interactive DEVCS, and this may require large neighborhoods and redundant computation. Since all cells use the information current at the same date (i.e. the previous global state) to compute their next state, the result is independent of any sequential order imposed by a sequential simulator.

The discrete event cell space on the other hand, considers cells to be capable of acting upon their neighborhood as well as themselves. Cells undertake their activity in one-at-a-time fashion so that the result is deterministic but also may be dependent on the order in which simultaneously scheduled cells are serialized by tie breaking rules. Updating of states occurs only via events which need not be uniformly distributed in time or space. The information required to schedule its transition event is contained within the cell state (in the time_left component). We shall return to the primary world views of the discrete event formalisms in Section 7.1.

5. Summary

This chapter has presented cellular space formalisms as important in themselves but also as a starting point for understanding more general multicomponent modelling formalisms. The cellular space formalisms specify systems at the structured system level, place components in a fixed geometric spatial lattice, and impose uniformity in the nature of the components and their interconnections. All of these restrictions may be lifted in more general formalisms while still retaining the fundamentals of the discrete time and discrete event dynamics. It is to the more general case that we now turn.

PROBLEMS

1. In the plant model (Example 3) activity caused by a gravity signal is limited to the cells in the stem of the plant configuration. However, in general it is not decidable whether activity in a DEVCS will be restricted to a given finite set of cells during a transitory state sequence (i.e. while the clock is not advanced). As a consequence it follows that it is not decidable whether an arbitrary DEVCS is legitimate or not (Section 4.2).

Show this by showing how to embed an arbitrary Turing Machine in a DEVCS in such a way that the clock is advanced only when the Turing Machine enters its halt state.

2. Design and implement a discrete event cell space simulator. Make it as general as you can, for example, placing as few restrictions on the local state set as possible. The user should be able to input a specification in DEVCS form, an initial global state, and a desired stopping condition. The simulator should then produce the global state trajectory of the model starting with the given global state and terminating upon meeting the stopping criteria.

3. Extend the formalism for cellular automata so that the neighborhood of a cell is a function of its state. Do the same for the discrete event cell space formalisms. Investigate the expressive power and other adequacy properties of the extended formalisms *vis a vis* the original ones.

References

[1] Barto, A. G. (1975). *Cellular Automata as Models of Natural Systems*, Doctoral Diss., CCS Dept., University of Michigan, Ann Arbor.
[2] Burks, A. W. (1970) (ed.). *Essays on Cellular Automata*, University of Illinois Press, Urbana, Ill.
[3] Burks, A. W. (1977). *Cause, Chance, and Reason*, University of Chicago Press, Chicago, Ill.
[4] Holland, J. H. (1970). "A Universal Computer Capable of Executing an Arbitrary Number of Subprograms Simultaneously". In *Essays on Cellular Automata* (ed., A. W. Burks). University of Illinois Press.
[5] Herman, G. T. and G. Rosenberg (1975). *Developmental Systems and Languages*, North Holland, Amsterdam.
[6] Hogeweg, P. H. (1980). "Locally Synchronised Developmental Systems", *Int. J. General Systems*, **6**, 57–73.
[7] Lindenmeyer, A. (1968). "Mathematical Models for Cellular Interactions in Development". *J. Theo. Biol.* **18**, 280–312.
[8] Sampson, J. R. and M. Dubreuil (1979). "Design of Interactive Simulation System for Biological Modelling". In *Methodology in Systems Modelling and Simulation* (eds, B. P. Zeigler *et al.*), North Holland Press, Amsterdam.
[9] Smith, A. R. (1972). "Simple Computation Universal Cellular Spaces. *J. Assoc. Computing Mach.* **18**, 339–353.
[10] Vollmar (1979). *Algorithmen in Zellularautomaten*, B. G. Teubner, Stuttgart.
[11] Von Neumann, J. (1966). *Theory of Self-Reproducing Automata*, (ed, A. W. Burks), University of Illinois Press.
[12] Zeigler, B. P. (1976). *Theory of Modelling and Simulation*, Wiley, NY.
[13] Zeigler, B. P. (1977). "Persistence and Patchiness of Preditor-Prey Systems Induced by Discrete Event Population Exchange Mechanisms". *J. Theo. Biol.* **67**, 687–714.
[14] Zeigler, B. P. (1977b). "Systems Theoretical Description: a vehicle reconciling diverse modelling concepts". In *Proc. Intl. Conf. Applied General Systems Research* (ed, G. J. Klir), von Nostrand.
[15] Zeigler, B. P. (1982). "Discrete Event Models for Cell Space Simulation". *Int. J. Theo. Phys.* **21**, 573–588.

Chapter 7

DEVS MODELS AT THE STRUCTURED SYSTEM LEVEL

This chapter presents a general formalism for specification of multicomponent DEVS at the structured system level and a user oriented language which can be employed to express the concepts of the formalism. The formalism can be specialized to each of the world views: next event, activity scanning, and process interaction. Since, the formalism takes the waituntil construct as a basic primitive we show how an efficient realization of this primitive can be realized in the SIMULA language. The chapter concludes with an introduction to hierarchically specified models.

1. Multicomponent DEVS in Structured System Form

A *multicomponent DEVS in structured system form* is a structure

$$DS = <D, \{S_i\}, \{I_i\}, \{T_i\}, \text{SELECT}>$$

D is a set, the *component names*;

for each i in D,

S_i is a set, the sequential states of i
I_i is a set, the *information neigbors* of i
T_i is a function, the transition function

and

SELECT is a function, the *tie-breaking selector*

subject to the constraints:

I_i is a subset of D
T_i maps $(S_i \times R_\infty)^n$ to itself

where $n = |I_i|$, the number of information neighbors

$$\text{SELECT:(subsets of } D) \to D$$

such that for any non-empty subset E, $\text{SELECT}(E) \in E$.

This specification generalizes the DEVS cellular space in that uniformity is no longer enforced: each component has its own individually specifiable sequential state set, information neighbors, and transition function. Aside from this non-uniformity in structure, the interpretation of the specification and the association of a DEVS with it remain the same. In particular, the specification may be specialized to each of the primary world view formalisms:

For *next event* specification, the range of the time left component associated with each component is restricted to $R_{0,\infty}^+$. The information neighbors of a component are called its *influencees*. When activated, the component maps the pairs (s_j,σ_j) of its influencees into new values using its transition function.

For *activity scanning* specification, I_i, the information neigbor set of component i is specified as a union of sets $I_{i,\text{ influencees}}$ and $I_{i,\text{ influencers}}$. The transition function is expressed as a pair (c_i,f_i) where c_i is a predicate on the influencers and f_i is a function operating on the influencees. When activated, the component i tests its activation condition c_i; if it is true, it goes on to apply its action function f_i which employs the pairs (s_j,σ_j) of its information neighbors to make new assignments to the influencees. If false, the component remains due for testing after every event.

Process interaction is specified in a manner similar to that of activity scanning except that each transition function is specified as a finite labelled set of condition, action pairs. The sequential state of a component is constructed from a phase variable with a range equal to the set of labels and (possibly) memory variables (see Section 2). When a component is activated, the condition,action pair labelled by the current phase is treated in the same manner as in the activity scanning case.

1.1. Example: Dynamics of Social Relations

Social science studies perceptions of humans with regard to their position within a group. Feelings of belonging to, and of attitude toward, the more esteemed subset of a group, the ingroup, may be surveyed at different points in time. Attempts are made to induce the dynamics of group interactions from such behavior data [1]. The simulation approach on the other hand postulates the existence of underlying processes and generates time series data which can be compared to that obtained by measurement. Such a model is now described as an example of a multicomponent DEVS at the structured system level.

The model identifies components with human individuals. Each individual may cycle in a process between two phases: being, and not being, a member of the ingroup. We postulate that each individual has a fixed set of others that he (used for either he or she) likes and a second disjoint set of dislikes (that such feelings are invariant is no doubt a simplification). An individual's attitude toward a group and its attitude toward him is expressed as an aggregate of these individual feelings. Conditions are right for an individual to join the ingroup when both attitudes are favorable. An individual remains in the ingroup so long as these positive feelings remain in effect.

Let D be a finite set of individual names. There are two fixed relations on D which describe person-to-person feelings: *likes* and *dislikes*. The relations are disjoint (a cannot both like b and dislike b) and irreflexive but not necessarily symmetric (a likes b does not imply b likes a). Thus with each individual i we associate the four sets: i__likes, i__dislikes, likes__i, dislikes__i. Individual i's relation to members of the ingroup at any time is determined by the intersections of the four sets with the ingroup. When an individual j in any one of these sets changes membership status (joins, or quits, the ingroup) repercussions may be felt by i. Otherwise put, the influencers of i's conditions for joining or for quitting are the individuals in the union of the four sets. respectively, the individuals that i likes and dislikes, and that like and dislike i. To summarize the model structure is given by:

D = finite set of names
S_i = {member, non-member}
$I_i = i$__likes $\cup$ i__dislikes $\cup$ likes__i $\cup$ dislikes__i

and the transition function of component i is given in process interaction form as follows:

(*) non-member.   waituntil(conditions for joining)
                      hold(reaction time)
    member.   waituntil(conditions for quitting)
                      hold(reaction time)
                      go to non-member

where:
   conditions for quitting $\equiv [\sim$ conditions for joining]
   conditions for joining $\equiv [i$__favors group and group favors__$i]$
   i__favors group $\equiv [|\{j:i$ likes j and S = member$\}| >$
                       $|\{j:i$ dislikes j and S_j = member$\}|]$
   group favors__$i \equiv [|\{j:j$ likes i and S = member$\}| >$
                       $|\{j:j$ dislikes i and S_j = member$\}|]$

The model is initialized by setting each component into a phase which corresponds to the individual being initially a member, or not a member of the ingroup. As time proceeds, considerable shifting of positions may occur until finally an equilibrium state may be reached in which there is no subsequent change in ingroup composition. Corresponding to actual survey measurements a contingency table may be constructed in which the state of the model initially is correlated with that at some later time.

Let us refer to assessing the status and attitude of an individual as conducting an interview. Then an individual who has been interviewed may be assigned to one of 4 cells: $\{++,+-,-+,--\}$ with the first place indicating membership in the ingroup and the second, attitude toward it, e.g. $++$ indicates membership in the group and a positive attitude towards it ($i\_$favors group is true). An individual interviewed at two points in time may be assigned to one of 16 cells in the 4×4 contingency table shown in Figure 7.1, e.g. $(+-,--)$ represents the case of individuals who at the time of the first interview were members of the ingroup but had a negative attitude toward it; at the second interview, these individuals had left the ingroup and continued to have an unfavorable attitude toward it. Counts of individuals in each cell may be compared with real survey data. Note that to produce the information required for such contingency tables, *the model must keep track of individual identities* (Markov models often employed in stochastic model situations cannot suffice for handling such indefinite memory requirements).

1.2. Realization of the Formalism in SIMULA

SIMULA [2] is an elegant process interaction language which is perhaps closest in spirit to the DEVS multicomponent formalism at the structured system level. Standard SIMULA does not however contain the waituntil construct crucial to the latter formalism. We will demonstrate one approach to the extension of SIMULA process class which adds the waituntil

	2nd interview			
	+ +	+ −	− +	− −
+ +				
+ −				
− +				
− −				

1st interview

Fig. 7.1. Contingency table for social relations.

procedure in a manner oriented to efficient use of the influencer concept.

Our approach is to define a subclass of SIMULA class PROCESS called EXTENDED which adds two procedures *waituntil*(condition) and *wake*(list__of__processes) (Figure 7.2). A subclass of a class generates objects which contain all of the attributes and procedures of its superior but may define additional ones of its own (Section 10.2.3 elaborates on this theme).

Two procedures are made available by declaring a class to be a subclass of Class Extended: 1)Waituntil(*B*) will test predicate B and will passivate the process if it is false, setting a flag DUE which indicates that the process has been passivated in connection with a waituntil (as distinct from passivate statements appearing elsewhere in the process description); 2) Wake(*L*) will activate either a single process or a list of processes provided they have been passivated by a waituntil test.[17]

```
process class extended;
begin
boolean due;

procedure waituntil (B);
name B; boolean B;
begin
start:
if B then go to exit;
due: = true;
passivate;
due: = false;
go to start
exit:end of waituntil;

procedure wake(L);
ref(linkage)L;
begin
ref(process)p;
inspect L
   when process do
      if L qua extended.due then
      activate L after current
   when head do
      if L qua head.empty then go to exit
      for p:- L qua head.first,p.suc
            while p=/=none do
      if p qua extended.due then
      activate p after current
exit: end wake;
end extended;
```

Fig. 7.2. Extension of SIMULA to include waituntil procedures.

---

[17]Actually, this construction has to be further extended so as to allow a process to belong to more than one "wakeup" list simultaneously. This can be done with the help of a class of objects which point to extended processes.

A multicomponent DEVS in process form can then be implemented in SIMULA as follows. For each component $i \in D$, the transition function T_i is expressed as a subclass of Class Extended. (In the case of a DEVCS (Section 6.3) only one such class need be defined due to the uniformity of the structure.) An instance of the class defined for T_i is constructed (using the New statement) and a reference to it is given to a variable which uniquely represents the component. Each of these extended process objects also maintains a wakeup list of components. A reference to component i appears in the wakeup list of component j if, and only if, i is an influencer of $j (i \in I_j)$.[18] Wake statements are placed in the process description for i at points where at least one of the components in the wakeup list may have the truth value of its waituntil predicates changed by the change in state of component i. An example, from the implementation of the social relations model (Section 1.1) is shown in Figure 7.3.

It should be apparent that a translation from a multicomponent DEVS specification into a SIMULA implementation could be performed algorithmically. Were such a translator to be provided, the modeller would be able to state his model at the DEVS level of abstraction, and not have to worry about the many implementation details required to express the model in the SIMULA language. Of course, it might even be better to have a new language which provides a choice of modelling formalisms and handles the formalism

```
extended class persons;
begin
. . . . . . . . . . . . . .
. . . . . . . . . . . . . .
. . . . . . . . . . . . . .

member: waituntil (condition__for__quitting);
membership: = false;
wake(whom__you__like); wake(whom__you__dislike);
wake(who__likes__you); wake(who__dislikes__you);
hold (reaction__time);

nonmember: waituntil(condition__for__joining);
membership: = true;
wake(whom__you__like); wake(whom__you__dislike);
wake(who__likes__you); wake(who__dislikes__you);
hold(reaction__time);
go to member;
end of persons;
```

Fig. 7.3. Implementation of social relations model in SIMULA.

---

[18] A component j also appears in the wakeup list for component i if there is a passive component k which is an influencee of j and an influencer of i.

to code translation in a manner most suited to the formalism involved. The next section indicates how such a language might look.

2. User Oriented DEVS Language

The discrete event formalisms we have been introducing have reached the complexity where they can readily capture much complex real world behavior. As we have indicated in Chapter 3, formalism provides a concise, abstract means of characterizing a class of structured objects. Formalism is useful especially for communicating the constraints satisfied by the objects in question and the relation of the class to other classes. However, the abstract nature of a formalism is not well suited to describing particular objects with their multitude of idiosyncracies. For this a more "user-friendly" language equivalent to the formalism in question should be provided. We shall now provide such a language for the general multicomponent discrete event formalism. Actually, we have been using this language all along, but is now time to crystalize the statements we have introduced and to complete the presentation. Our approach will not be to formalize the language for this would inevitably subject its syntax and semantics to constraints arising from some envisioned computer implementation. It is therefore to be regarded as an aid in working with the precision of a set theoretic specification, not a substitute for it.

Consider then the statements that may be used to specify a multicomponent discrete event specification:

$$DN = <D, \{S_i\}, \{I_i\}, \{T_i\}, \text{SELECT}>$$

To specify D:
    Declare components: {list of names}
To specify S:
    Declare phases of i: {list of names}
    Declare variables of i: {list of name:range pairs}
To specify I_i:
    Declare influencees of i: {component names}
    Declare influencers of i: {component names}
[not applicable for the next event formalism]
To specify SELECT:
    Tie breaking order is: {list of component names}
[this is a commonly employed special form of SELECT]
To specify T_i:
    {finite list of *statements* of the following type}
    assignment statement:

variable name: = {expression evaluating to range of variable}
if...then...else statement:
    if {predicate} then *statement* [else *statement*]
plus the following (depending on the formalism):
Next Event Formalism:
    (re)schedule {component name/this component}
                in {non negative real} units
[set component's σ to {non-negative real}]
    activate {component name}
[set component's σ to 0]
    passivate {component name/this component}
[set component's σ to ∞]

Activity Scan Formalism:
    waituntil({predicate})
[must be first statement in list; exactly one waituntil may occur]
    plus all of Next Event statements

Process Formalism:
    {phase name}.waituntil({predicate})
    {phase name}.hold({non-negative real}) units
[set your own σ to {non-negative real}]
            go to {phase name}
            (re)schedule {component name}
                [from {phase name}]
                in {non-negative real} units
[set component's phase to {phase name} and its σ to {non-negative}]
            activate {component name} [from {phase}]
[set component's phase to {phase name} and its σ to 0]
            passivate {component name/(this component)}
                [in {phase name}]
[set component's phase to {phase name} and its σ to ∞]

Note that in the case of an assignment statement, we say that the component has *write* access to the variable on the left hand side; it has *read* access to any variable used in the expression on the right hand side. For example, in the assignment:

$$A := B$$

the component has write access to A and read access to B. Similarly a component must have read access to any variable appearing in a predicate of an if...then...else statement.

The declaration of *influencees* of a component i must include all components *j* whose variables are alterable by component *i*, or which are used to compute the transition function which does this alteration. In other words, *i* has read or write access to a variable of *j* only if *j* is declared as an influencee of *i*.

The declaration of *influencers* of component *i* must include all components *j* whose variables are referred to in the predicates of its waituntil statements. Component *i* needs read access to these variables in order to test whether the condition causing a wait has become true: changes in state of the influencers may enable the component to process with its activity or process.

A simple example of multicomponent specification will now be given employing the language just introduced.

2.1. Example: Coin Tossing

The most familiar of systems exhibiting chance or random behavior is perhaps the common coin toss. Once flipped, a coin may fall on the heads or tails face in a seemingly unpredictable manner. There may seem very little one can model about such a system except to assign a probability of its landing heads p_{HEAD}, hence also of tails since $p_{TAIL} = 1 - p_{HEAD}$. For a fair coin these probabilities are equal. Actually, coin tossing is a remarkably complex dynamic system for which some 15 models have been suggested [3]. It thus exemplifies the multifacetted modelling approach in a miniaturized form. One of these models is in discrete event form and is formulated within the language for multicomponent specification.

Coin tossing may be decomposed into two processes: a spin process in which the coin flips from heads to tails (as seen from above) and a translation process in which the coin is travelling through the air while spinning. When the coin lands the side of the coin facing up represents the result of the toss. A new toss is initiated from this state. The coin spends a fraction τ in heads (and $1 - \tau$ in tails) during one spin revolution. A fair coin has $\tau = 0.5$ by definition (Figure 7.4).

The coin tossing model is a multicomponent DEVS at the structured system level specified by:

Declare components: {spin process, translation process, toss}
Declare phases of toss: {WAIT}
Declare phases of spin process: {HEADS, TAILS}
Declare phases of translation process: {START, STOP}
Declare influencees of toss: {spin process, translation process}
Declare influencers of toss: {translation process}
Declare influencees of spin process: none
Declare influencees of translation process: {spin process}

The transition function of each component, specified in process formalism,

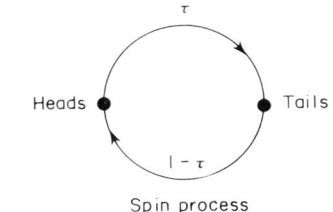

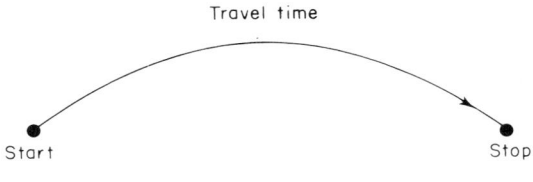

Fig. 7.4. DEVS coin tossing model.

is declared as follows:

spin process
    HEADS. hold(τ)
    TAILS.   hold($1-\tau$)
                go to HEADS

translation process
    START. hold(travel__time)
               passivate spin process
    STOP.   passivate

toss process
    WAIT.   waituntil(translation process phase = STOP)
                activate spin process
                activate translation process
                go to WAIT

We shall return to the coin tossing model to illustrate the specification of experimental frames in Section 12.5.1.

3. Hierarchical Multicomponent DEVS

Until now, our DEVS models have been constructed from components described by a single process. However, as indicated in Section 8.7, it is often more convenient to specify a component as an interaction of lower level components. For DEVS multicomponent models specified at the structured system level, this means that a component may be specified in terms of a number of processes.

3.1. Example: Subway Train

A subway train consists of a lead car (responsible for pulling the train), passenger cars, and a rear car (the last passenger car). Except for the lead car, each car is specified by three processes: control, loading and unloading. The hierarchical structure of the model is depicted in Figure 7.5. The symbolism employed will be fully explained in Section 10.2.2.

The lead car has only a control process which is responsible for initiating the motion of the train when, and only when, the loading and unloading of passengers in each of the cars have been completed. It must also signal the cars to begin passenger loading and unloading when, and only when the train has stopped at a station. The lead car does not communicate directly with each passenger car to perform its function. Instead, it signals only the car behind it that the train has arrived at a station; this signal is then passed from car to car until it reaches the rear car. The latter is responsible for initiating a ready signal which travels in the forward direction. The ready signal is delayed at a car until it has finished its passenger exchange processes. When it reaches the lead car, the latter can start moving the train to the next station.

The formal description as a multicomponent DEVS is given in Figure 7.6. We note that except for the lead and rear cars, which are necessarily somewhat different, each passenger car has the same sequential state set, (isomorphic) information neighborhood and transition function. Thus the passenger car portion of the train can be considered to be a one-dimension discrete event cellular space model (Section 6.3).

Although the model has a hierarchical structure in the sense that we can identify subprocesses of its components, it is readily apparent that the description of the subprocesses does not allow us to extract them as standalone DEVS models. To do this we shall have to develop the concept of modular specification as in the next chapter. To see the difference in the form of specification, the reader may wish to glance ahead to Figures 8.10 and 8.11, where a modular version of the same subway train model is given.

128   3. *Multicomponent models*

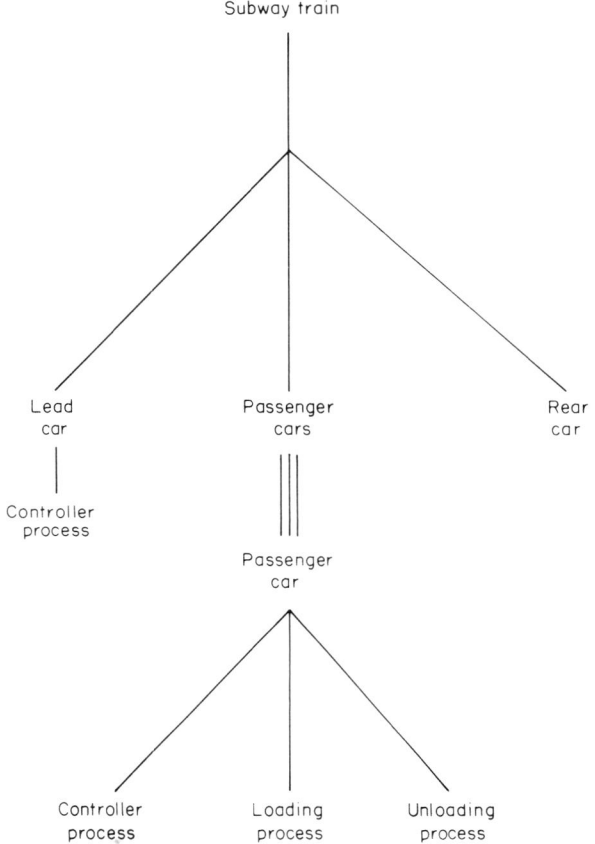

Fig. 7.5. Hierarchical structure of subway trains.

4. Summary

A formalism, and an associated user oriented language, for multicomponent discrete event model specification at the structured system level has been presented. In the next chapter, the modular form of multicomponent DEVS specification will be taken up and its relation to the non-modular form of the present chapter considered. Likewise, the theme of hierarchical model construction will be taken up in greater depth within the modular specification context.

DEVS models at the structured system level

Declare components:

$$D = \{0\} \text{ (lead car)},$$
$$1,\ldots,n-1 \text{ (passenger cars)}$$
$$n \text{ (rear car)}$$

Declare phases of lead car: {STOPPED, MOVING}

Declare variables of lead car:
    arrived: boolean

Declare phases of all other cars:
    the crossproduct of phase sets:

        {STOPPED,MOVING} (control process)
        {MOVING,WORKING} (unloading process)
        {MOVING,WORKING} (loading process)

Declare variables of all other cars:
    arrived, ready:boolean
                      (control process)
    front__door:range {open,closed}
                      (loading process)
    back__door:back__door:{open,closed}
                      (unloading process)
for $i=1,\ldots,n$ (except the rear car does not use *arrived*)

Declare influencers of lead car:

$$I_0 = \{1\}$$

[the influencer of the lead car's condition is the car behind; it has no influences other than itself]

Declare influencers of cars:

$$I_n = \{n-i, n\}$$

[the influencers of a passenger car are the the cars behind and ahead as well as itself (via the unloading and loading processes); it has no influencees other than itself]

    T_i is described in process form as follows:

lead car control process

STOPPED. waituntil (car__behind.ready)
              arrived := false
MOVING.   hold(travel__time)
              arrived := true
              go to STOPPED

passenger car control process

MOVING.   waituntil(car__ahead.arrived)
              arrived := true
              ready := false
              front__door := back__door := open

(Continued on p. 130)

STOPPED.   waituntil(front__door and back__door = closed
                        and car behind.ready)
           ready := true
           go to MOVING

rear car control process

MOVING.   waituntil(car__ahead.arrived)
           ready := false
           front__door := back__door := open
STOPPED.   waituntil(front__door and back__door = closed)
           ready := true
           go to MOVING

unloading process

MOVING.   waituntil(back__door = open)
WORKING.   hold(unloading__time)
           back__door := closed
           go to MOVING

loading process

MOVING.   waituntil (front__door = open)
WORKING.   hold(loading__time)
           front__door := closed
           go to MOVING

Fig. 7.6. Multicomponent discrete event specification for subway train model.

PROBLEMS

1. A farmer owns a flock of sheep which are supposed to remain in the barn all night. Now and then however a sheep leaves the barn. A wolf lies waiting close to the barn for this to happen. But it takes some time for the wolf to reach a wayward lamb and it makes some noise while it does so. This noise wakes the farmer and he runs out to shoot the wolf. Sometimes the farmer kills the wolf before it kills lamb, other times the wolf gets away and returns to his hiding place awaiting the next wayward lamb.
(a) Construct next event model, process interaction, and activity scanning models of what happens in the farm during the night. Use the user oriented language to construct the model. Categorize the components of each model as active or passive (recall Section 3 of Chapter 4).
(b) Which of the three models do you prefer as a description of the night time happenings? Indicate why from both the model expression and simulation efficiency points of view.

References

[1] Broekstra, G. (1979). "Constraint Analysis". In *Methodology in Modelling and Simulation* (eds, B. P. Zeigler *et al.*), North Holland, Amsterdam,
[2] Franta, W. R. (1977). *A Process View of Simulation*, North Holland, Amsterdam.
[3] Zeigler, B. P. (1979). "Structuring Principles for Multifacetted System Modelling". In *Methodology in Systems Modelling and Simulation* (eds, B. P. Zeigler *et al.*), North Holland, Amsterdam.

Chapter 8

MULTICOMPONENT DEVS: MODULARITY AND HIERARCHY

This chapter develops an approach to modular and hierarchical model construction that is to serve as a basis for exploring alternative architectures for multifacetted modelling and simulation support. System theoretic concepts developed earlier are marshalled (Chapter 3) to provide precise definitions of modular and hierarchical construction of discrete event specified systems. As we did for non-modular DEVS specification, we present a "user-friendly" language that mirrors the formal constructs but is also closer to a practical model specification tool.

While modular forms of model specification are necessary for flexible model assembly and disassembly, non-modular forms of model specification offer certain conceptual advantages. Transformations between the two forms of specification will be given so that the properties of each may be exploited.

The modularization of model and experimentation in simulation program design is an important application of modularity theory and is taken up in Chapter 16.

1. Modularity and Hierarchical Construction: Informal Concepts

Modularity and hierarchy have been recognized to be important properties in software system design, yet there is relatively little agreement on the essentials of these concepts among the members of the general software community [2-5]. So it is worth providing some informal definitions before proceeding. We shall understand a *module* to mean a program text that can function as a self-contained autonomous unit in the following sense: Interaction of such a module with other modules can occur only through predeclared input and output ports. Except for such interface variables, all other variables referenced in the module receive declarations local to it. The module may contain memory (saved) variables which retain their values between invocations.

Such a module is well characterized as an I/O system in the sense of Chapter 3. Recall that such a system is defined by an input/output interface which characterizes its interaction with other systems and a state object which represents the memory of the system. The notion of information hiding [3] is a natural feature of this systems concept since only those aspects of the system state which can be inferred from its input-output behavior are visible to, and can be manipulated by, the outside world.

Recall too, that systems may be coupled together to build composite systems. The systems so coupled are called component systems. A composite system may itself be employed as a component to be coupled together with other systems to form higher level systems. A *hierarchical* construction is a finite recursion of such couplings (See Section 15.3).

Thus a module is a programming construct that may be formally characterized as a system at the I/O system or structured system levels. A software system specified by interfacing such modules is then usefully characterized as a coupling of such systems. Note however that we should be careful to distinguish the characterization of a module in system theoretic formalism from its realization in programming language form.

These concepts apply to the discrete event world where the DEVS formalism is a system theoretic characterization of the programming constructs employed in discrete event simulation languages. By showing how to deal with DEVS systems at the coupled system level, we provide a basis for modular construction of discrete event simulation languages and simulators.

2. Multicomponent DEVS in Modular Form

Our first task is to provide the formalism for specifying discrete event systems at the coupled system level. In this formalism, each component system is specified as a DEVS and a means of coupling them is provided in the spirit of the requirement that all interaction must occur through the I/O interface. In contrast to the general system coupling (Section 3.1), in the DEVS case the coupling scheme is specified by telling how a component influence*s* others when it is activated rather than how it is influence*d*.

We shall show how coupling of DEVS component models results in a composite system which is itself described by a DEVS. An important consequence of this closure of the DEVS class under coupling is that we are guaranteed that models constructed from discrete event components in a hierarchical fashion are themselves discrete event in form. Later, we shall show how to design a simulator for a DEVS multicomponent model that is modular in form and that can be coupled together in a uniform manner with the simulators of other DEVS components to constitute a simulator for a composite DEVS model. These results provide a rigorous basis for the design of simulation languages and software which seek to support modular and hierarchical model construction (Chapter 18).

A *multicomponent DEVS in modular form* (also called a DEVS coupled system, or a DEVS network) is a structure

$$DN = <D, \{M_i\}, \{I_i\}, \{Z_{i,j}\}, \text{SELECT}>$$

where

D is a set, the *component names*;

for each i in D,

M_i is a component
I_i is a set, the *influencees* of i

and for each j in I_i,

$Z_{i,j}$ is a function, the *i-to-j output translation*

and

SELECT is a function, the *tie-breaking selector*.

The structure is subject to the constraints:

$M_i = <X_i, S_i, \delta_i, ta_i>$
I_i is a subset of D, i is not in I_i
$Z_{i,j}: S_i \to X_j$
$SELECT: subsets\ of\ D \to D$

such that for any non-empty subset E, $\text{SELECT}(E)$ is in E.

An example of such a multi-component system is shown in Figure 8.1. Note that in this approach we specify the coupling of the output of i to the input of j by designating j as an influencee of i and providing an output translation map $Z_{i,j}$ from the state of i to the input set of j. The interpretation will be that when an internal event occurs to component i, it sends a signal to component j that arrives at the same clock instant. The signal is computed on the basis of the state of i using the translation map $Z_{i,j}$.[19]

The result of M_{DEVN} of coupling DEVS components in the preceding manner is itself a DEVS specified as follows:

$$M_{\text{DEVN}} = <X, S, \delta, ta>$$

---

[19]Although the computation of a signal should be defined in terms of the output of a component, we choose not to do this here as a matter of formal convenience. By composing output and translation maps in such a formalism, we arrive at the translation map employed in the present form.

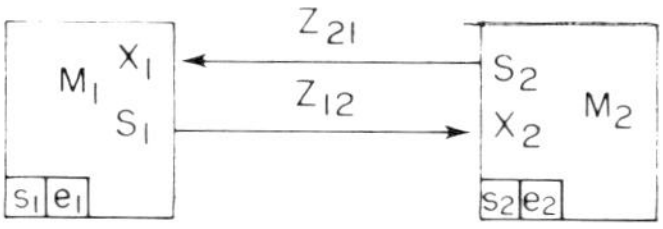

Multicomponent DEVS: Modularity and hierarchy

```
         Z₂₁
    ┌──X₁◄──────┬──S₂────┐
    │M₁        │   M₂   │
    │   S₁─────►X₂       │
    │    Z₁₂   │         │
    ├─┬─┐      ├─┬─┐     │
    │s₁│e₁│    │s₂│e₂│   │
    └─┴─┘      └─┴─┘
```

DEVN = $(\{1,2\}, \{M_1, M_2\}, \{I_1, I_2\}, \{Z_{12}, Z_{21}\}, \text{Select})$

$M_1 = \langle X_1, S_1, \delta_1, ta_1 \rangle \qquad M_2 = \langle X_2, S_2, \delta_2, ta_2 \rangle$

$I_1 = \{2\} \qquad\qquad I_2 = \{1\} \qquad\qquad \text{Select}(\{1,2\}) = 1$

$Z_{12}: S_1 \rightarrow X_2 \qquad\quad Z_{21}: S_2 \rightarrow X_1$

Fig. 8.1. Example of multicomponent DEVS in modular form.

where

$$X = \times X_i \text{ (crossproduct of all input sets)}$$
$$S = \times Q_i \text{ (crossproduct of all total state sets)}$$

[we have to know the total state pair (s,e) of each component to compute its internal and external transition behavior]
Let $s = ((s_1, e_1), \ldots, (s_i, e_i), \ldots)$ in S, then

$$ta: S \rightarrow R^+_{0,\infty}$$

is defined by

$$ta(s) = \text{minimum } \{\sigma_i | i \in D\}$$

where $\sigma_i = s_i - e_i$

[$ta(s)$ is the minimum of all time left values of the components; the system will remain in s during this time if no external events occur].
Let

$$\text{IMMINENT}(s) = \{i | \sigma_i = ta(s)\}$$

[these are the components which are scheduled to undergo internal transitions just after $ta(s)$ has elapsed]
Let

$$i^* = \text{SELECT}(\text{IMMINENT}(s))$$

[i^* is the component selected from those imminent as the one to undergo its internal state transition and to send outputs to its influencees]

Then
$$\delta_\phi : S \to S$$
is defined by
$$\delta_\phi(s) = s' = (\ldots,(s'_j,e'_j),\ldots)$$
where

$(s'_j, e'_j) = (\delta_{\phi, j}(s_j), 0)$ if $j = i^*$ ...1)
       $= (\delta_{ex, j}(s_j, e_j + ta(s), x^*_{i, j}), 0)$
                                                  for $j \in I_{i^*}$

where

$x^*_{i,j} = Z^*_{i,j}(s^*_i)$ ...2)
       $= (s_j, e_j + ta(s))$ otherwise   ...3)

[line 1 specifies that the internal transition of the imminent component i^* is as its own internal transition function dictates; line 2 specifies the effect of the signal sent by the imminent component on each of its influencees as determined by the external transition function of the influencee; line 3 acknowledges the effect of time advance on the elapsed time co-ordinates of all other components]

Also
$$\delta_{ex} : Q \times X \to S$$
is given by
$$\delta_{ex}(s,e,x) = s'$$
with
$$(s'_j, e'_j) = (\delta_{ex, j}(s_j, e_j + e, x_j), 0)$$
where $x = (\ldots, x_j, \ldots)$

[an external x arriving to the network an elapsed time e after the last (global) event, causes each component j to react appropriately to its projection x_j; of course, many of the components may ignore this input (Section 4.3).]

Having constructed a DEVS on the basis of the information provided by the multicomponent specification, our task is to show that it does indeed specify the resultant of the coupling of systems specified by the component DEVSs. To do this we must formally define the coupling scheme applicable in the general system formalism (Section 3.1) that represents the special form of coupling defined for in the DEVS case. With this definition in hand, we must then show that the resultant system at the structures system level, if it exists, is specified by M_{DEVN}. We spare the reader the details except to mention that the resultant will exist, if and only if, M_{DEVN} is legitimate (Section 4.2 of Chapter 4).

This argument shows that the resultant of coupling components, each of which is specified by a DEVS is itself specified by a DEVS. It establishes that the DEVS formalism is closed under composition as defined in Chapter 3 and opens the way toward recursive decomposition and hierarchical construction of DEVS models.

We also note that the association mapping a DEVN to its result M_{DEVN} provides the standard of correctness for any simulation of a multicomponent DEVS (Section 8.7).

3. Language for Modular DEVS Specification

Figure 8.2 presents a language for specifying multicomponent DEVS models in modular form. Examples of models specified in the language are given in Section 6.2. These examples were constructed by converting DEVS models given previously in non modular form. A discussion of the language and its examples will be easier once the concept of modularization has been established.

4. Translating Non-modular Formalisms in Modular Form

Many formalisms are commonly employed which do not provide for modularity. From the perspective of the hierarchy of system specifications (Section 1 of Chapter 3), these formalisms specify models at the structured system level rather than at the modular coupled system level. Recall that at the coupled system level both model components and their coupling specification are given, while at the structured systems level, the coupling specification has been absorbed into the resultant system description. In other words, to go from a structured system representation to the higher level coupled specification conceptually requires the rediscovery of the wires which were used to connect up components. We shall call this introduction of input/output interfacing *modularization* (Section 1).

Recalling the language of Section 3, there are three non-modular situations that arise which are illustrated in Figure 8.3. In case (a), component 2 has *write access* to a variable v of component 1. This violates modularity since all actions must be effected explicitly via input/output. This situation is modularized by providing the component 1 with an external input, $v(x)$ which

3. Multicomponent models

The statements that may be used to specify a multicomponent DEVS in modular form:
$$DN = <D, \{M_i\}, \{I_i\}, \{Z_{i,j}\}, \text{SELECT}>.$$

To specify D:
        DECLARE COMPONENTS: {list of names}

To specify SELECT:
        TIE BREAKING ORDER IS: {list of component names}

[this is a commonly employed special form of SELECT; default order is that given in specification of D]

To specify I_i:
        DECLARE INFLUENCEES OF i: {list of component names}

To specify M_i:
  To specify X_i:
        DECLARE INPUT VARIABLES OF i: {list of name:range pairs}

[input variables include those in one-one correspondence with setable memory variables and phases; see translation map $Z_{i,j}$

To specify Y_i:
        DECLARE OUTPUT VARIABLES OF i: {list of name:range pairs}

[output variables are those employed in the translation maps $Z_{i,j}, j \in I_i$]

To specify S_i:
        DECLARE PHASES OF i: {list of names}
        DECLARE MEMORY VARIABLES OF i: {list of name:range pairs}

To specify $\delta_{\phi,i}$:

    {finite list of *statements* of the following type}

1) variable name: = {expression evaluating to range of variable}

2) if {predicate} then *statement* [else *statement*]

3) {phase name}. hold({non-negative real}) units
    [set your own σ to {non-negative real}]

4) go to {phase}

5) passivate [in {phase}]
    [set your own phase to {phase} and σ to ∞]

To specify $\delta_{ex,i}$:
in {phase} when receive {input variable(value)},*statements*

[when external event occurs at the {input variable} carrying {value} in its range then carry out action determined by sequential state {phase}, values of memory variables appearing in *statements*, and *elapsed time*].

in {phase} when receive {memory variable(value)}, memory variable = {value}
in {phase} when receive {activate__in(phase')},
  [if {predicate} then] go to {phase'}
in {phase} when receive {passivate__in(phase')},
  [if {predicate} then] passivate in {phase'}

Note: A triple (s,e,x) not specified explicitly above is interpreted to mean that the input x is ignored in the total state (s,e) (Section 4.3).

To specify ta_i:

[accomplished through the use of the hold statement]

To specify $Z_{i,j}$:
  send {output variable(value)} to {component j}
    [as {input variable}]

[{output variable} is an output variable of this component and {value} {component j} is an influencee and the option is employed if the input variable of j has a different name than the output variable of this component.]

  send {activate__in(phase)} to {component j}
    [as {j.phase}]

["suggest" to component j that it go to {phase} and set its σ to 0 where phase is computed from the current sequential state; j is an influencee and the option is used only if the PHASE set of this component is not a subset of that of j]

  send {passivate__in(phase)} to {component j}
    [as {j.phase}]

Fig. 8.2. The statements that may be used to specify a multicomponent DEVS in modular form: $DN = D, M_i, I_i, Z_i$ SELECT.

when received, causes the assignment of the value x to its variable v. When component 2 wishes to alter the value of v, it must do so by sending the appropriate signal $v(x)$, where v is the variable it wishes to write on and x is the value to be written. The output translation map from component 2 to 1 must be defined to enable this to occur (Section 2).

In case (b), component 1 has *read access* to the state of component 2 at the time that it computes its own transition (disallowed since transition is determined by input and *own* state). Modularity is achieved in this case, by augmenting the state of component 1 with a variable 2.u for retaining the prevailing value of variable u of component 2. In order to maintain the correctness of this memory, the component 2 must be modified so as to update the memory whenever it alters its variable u. This reduces the problem to case (a) which has already been solved. Note that this includes the case, where a component say 3 has write access to variable u of component 2. The modularization step of case (a) will cause an explicit input $u(x)$ from 3 to 2 to

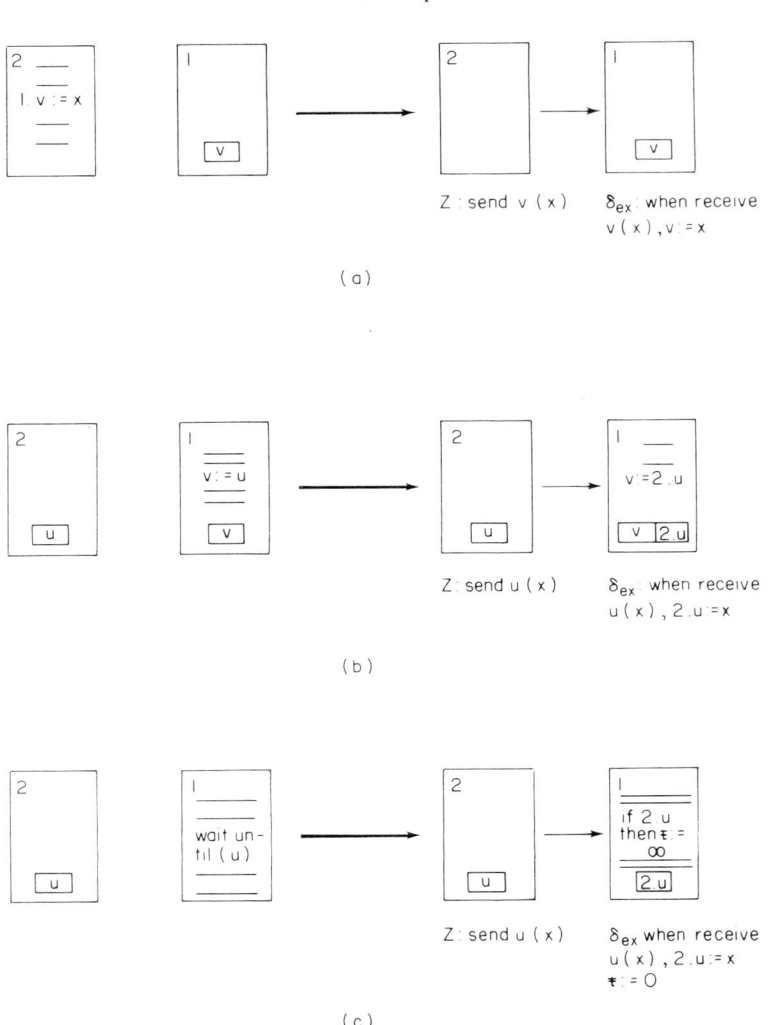

Fig. 8.3. Transformation from non-modular to modular formalism.

be defined. When 2 receives such an input, it must also pass on this signal to 1 besides updating its own state.

The most interesting situation is that of (c) where component 1 is blocked at a *waituntil* statement. Component 1 is modified so that it contains memory for the current state of each variable influencing the condition C of the waituntil statement.

The waituntil statement at phase P is replaced by the statement:

P. if $\sim C$ then passivate in phase P'

Thus if C fails to be true then component 1 enters a new phase P' in which it passivates itself. Whenever an influencing variable is altered, its component must notify component 1 of this fact (as in case (b) above). Also it must send a signal $activate\_in(P)$ which will cause component 1 to immediately transit from P' to P. After checking the truth of C, component 1 returns to P' if it is false or proceeds to the next statement if C has become true.

The above arguments may be formalized so that the following theorem may be proved:

Theorem. (Generality of Modular Form) Every DEVS model specified at the non modular structured system level may be simulated by a multicomponent DEVS in modular form.

This establishes that no generality is lost in restricting DEVS models to modular form. However, it is clear that it may be much more convenient and conceptually less demanding to work with the freer non modular formalisms. Since the proof of the theorem provides an algorithmic procedure to effect the conversion from non-modular to modular form, a resolution is possible. This is to provide a translator so that the user may specify the model in non-modular formalism and ask to have it converted to modular form when needed. For example, the process interaction formalism may be employed to specify a model which may later be modularized in order to disconnect the processes.

5. Examples: Multicomponent Models in Modular Form

To illustrate the specification of multicomponent DEVS models in modular form we shall present three examples. The first two are equivalent to the models specified earlier in non-modular formalisms. The procedure for converting models from non-modular to modular form was employed to convert these models into the latter form. A simplification step was then performed so as to render the resulting models as straightforward as possible. The third example, a neural network, is one that is naturally specified in modular form.

Predator prey cell space model

To specify the predator-prey model, uniformity similar to that embodied in the discrete event cell space of Section 3 of Chapter 6 is assumed. Indeed, a special formalism for modular specification of uniform cellular models can be defined (see Problem 2) that is geared to the specification of such models. The example illustrates the use of the input variable type activate_in(phase) which can cause the cell receiving this signal to be activated in the given {phase}. For example, when a cell in phase PREY' sends an activate_in(PREY) to a cell in the PREY neighborhood, and the latter is EMPTY, then it is activated in phase PREY. This corresponds to a prey migrating from a mature colony to an empty patch. Figure 8.4 illustrates the input/output ports ("wires") associated with such signaling.

Since uniformity prevails, we need give only the DEVS for the center cell at the origin (0,0) as shown in Figure 8.5.

Traffic congestion model

The traffic congestion example converted from the one in Section 6.2 illustrates a component's use of memory variables to keep track of the current states of its informational neighbors. For example in the model, a center cell must know the current states of the cells in N (cells to the right, above, and below) so that it can know when a transfer of the car it contains to one of these is possible. Accordingly there must be communication between cells in which a change in state is reported to all those who need to know it. Such communication is mediated by the input variable type my.state which causes the immediate updating of the receiving cell's representation of the sending cell's state. Again, due to uniformity, only the cell at the origin need be specified (Figure 8.6).

Also exemplified in the model is the handling of waituntil conditions. A

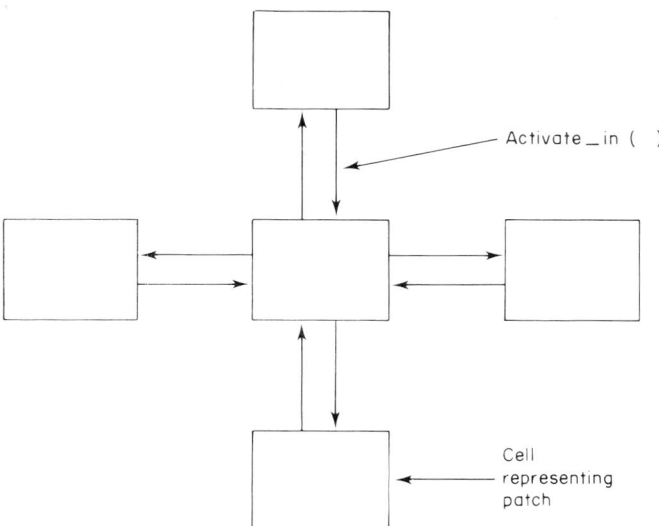

Fig. 8.4. Illustrating input/output parts for predator/prey model.

DECLARE COMPONENTS: {cell$(i,j)|i,j \in I$}

DECLARE INFLUENCEES OF CENTER CELL:N_{PREY} U N_{PRED}

DECLARE INPUT VARIABLE OF CENTER CELL:

DECLARE OUTPUT VARIABLE OF CENTER CELL:
activate__in:{PREY,PRED}

DECLARE PHASES OF CENTER CELL: {PREY, PREY',PRED, EMPTY}

Declare transition, time advance, and output translation functions:

::::::::::::::::internal transition function::::::::::

      PREY.   hold(PREY__GROWTH TIME)

      PREY'. hold(INTER__MIGRATION TIME)
              for each cell in N_{PREY}

                  send activate__in(PREY) to cell
            go to PREY'

      PRED. hold(PRED__GROWTH__TIME)
              for each cell in N_{PRED},

                  send activate__in(PRED) to cell
              hold(PRED__EXTINCTION__TIME)
    EMPTY. passivate

::::::::::::external transition function::::::::

in PREY or PREY': when receive activate__in(PRED), go to PRED

in EMPTY: when receive activate__in(PREY) go to PREY

:::::::::::::::::::::::::::::::::::::::

Fig. 8.5. Prey-predator model in modular form.

phase label is introduced (C) in which the waituntil condition is tested. If it is not satisfied then the center cell passivates in C', a second phase label introduced for this purpose. If the cell receives an activate__in(C) signal while in C', it is immediately activated and does a retest of the condition. Testing of the condition occurs only when it can change in truth value since only cells in N can send such activation signals, and only when they change state.

Neural network

Neurons are commonly modelled as discrete time logical devices [1, 6] but they can also be nicely expressed in discrete event formalism with potential

DECLARE COMPONENTS: {cell$(i,j)|i,j \in I$}
DECLARE INFLUENCEES OF CENTER CELL:$N \cup -N$
DECLARE INPUT VARIABLES OF CENTER CELL:
  activate__in:{C,OCCUPIED}
  cell.my.state:{0,1}, for each cell in N

DECLARE OUTPUT VARIABLES OF CENTER CELL:
  activate__in:{C,OCCUPIED}
  my.state:{0,1}

DECLARE PHASES OF CENTER CELL: {OCCUPIED,C,C',EMPTY}
DECLARE MEMORY VARIABLES OF CENTER CELL:
  cell.state:{0,1}, for each cell in N

Declare transition, time advance, and output translation functions:

::::::::::: internal transition function ::::::::

*)OCCUPIED. state := 1
  for each cell in $-N$, send my.state(1) to
    cell as center.state(1)

  C. if for each cell in N, cell.state = 1
    then passivate in phase C'
    else send activate__in(OCCUPIED) to first cell
      in preference order
    state := 0
    for each cell in $-N$
      send my.state(0) to cell as center.state(0)
      send activate__in(C) to cell
  EMPTY. passivate

:::::::::::::::: external transition function ::::::::

for each cell in N, when receive cell.my.state(x) then
  cell.state: = x

in C', when receive activate__in(C) go to C

in EMPTY, when receive activate__in(OCCUPIED) go to OCCUPIED

Fig. 8.6. Traffic congestion model in modular form.

benefit in the adequacy dimensions discussed in Section 8. When specified as DEVS components, these neurons can be interconnected so as to form neural network models of information processing in the brain.

Each neuron is modelled as a DEVS with three phases: (a) an absolute refractory phase in which the neuron finds itself immediately after firing and in which it is not capable of firing a second time; (b) a relative refractory phase in which the neuron is capable of firing if its input exceeds a threshold which attenuates with elapsed time in the phase; and (c) a resting phase which represents the asymptotic level to which the threshold tends. The DEVS has two memory variables; one for keeping track of the elapsed time in the refractory phase[20] and the second which does a temporal integration of the input. When a neuron fires it sends a pulse to its influencees. Such a pulse signal is an example of an interrupt (Section 5) since it causes rescheduling in the event that the receiving neuron is caused to fire.

This formulation is different from the usual formulation in the discrete time case where the sum of the input pulses at any time is considered to be the total input at that time. An integration over time is necessary in the discrete event case since pulses are assumed to have zero time duration so that a smearing of their influence is required if there is to be a non-zero input summation (Figures 8.7, 8.8).

Figure 8.7 specifies a *class* of neural net models since among others, the threshold, summation decay, pulse height, and neuron influencees are left as free parameters.

6. Equivalence of DEVS Formalisms

The theorem in Section 4 establishes that the DEVS multicomponent specification in modular form is at least as expressive as the non-modular form. The next theorem will show that in fact the two forms are equivalent (can specify the same class of DEVS systems) by providing a translation of the modular specification into the non-modular form. Actually the theorem shows more than that: it establishes that the next event non modular formalism is equivalent to the most general non modular formalism. This follows naturally since the theorem uses the next event formalism as the target non modular formalism into which the modular specification is mapped.

Theorem. (Translation of Modular to Non Modular Formalism). Given any multicomponent DEVS in modular form, there is a multicomponent DEVS in next event non-modular form equivalent to it in the sense of specifying isomorphic resultant DEVSs.

Proof. Let $DN = <D, \{M_i\}, \{I_i\}, \{Z_{i,j}\}, \text{SELECT}>$ be the given modular form coupled system specification. Construct a non-modular next event specification DE which has the same components and SELECT as DN: $DE =$

---

[20]Note that the elapsed time component of the total state of a DEVS does *not* keep track of the time elapsed in a phase, it only keeps track of the time since the last (internal, or external) event. However, employing this elapsed time component, we can readily keep track of elapsed times in phases, as is illustrated in this model.

DECLARE COMPONENTS: {neuron}
DECLARE INFLUENCEES OF NEURON: (open parameter)
DECLARE INPUT VARIABLES OF NEURON: {pulse:real}
DECLARE OUTPUT VARIABLES OF NEURON: {pulse:real}
DECLARE PHASES OF NEURON
    {FIRING, ABSOLUTE, RELATIVE, RESTING}
DECLARE MEMORY VARIABLES OF NEURON:
    elapsed__time__in__RELATIVE : non-negative reals
    sum : real

The transition, time advance and output functions of a neuron are given as follows:

:::::::::::internal transition function::::::::

FIRING.   hold(firing__time)
           (compute level of pulse,x)
           send pulse(x) to each influencee
ABSOLUTE. hold(absolute__refractory__period)
RELATIVE. elapsed__time__in__RELATIVE := 0
           sum := 0
           hold(relative__refractory__period)
RESTING. passivate

:::::::::::external transition function:::::::

in RELATIVE (with *elapsed time(e)*), when receive pulse(x),

    sum := Decay(sum,e) + x
    if sum > Threshold(elapsed__time__in__RELATIVE)
        then go to FIRING
        else add e to elapsed__time__in__RELATIVE

in RESTING, when receive pulse(x)

    sum := Decay(sum,e) + x
    if sum > Resting__level then go to FIRING

Fig. 8.7. Multicomponent DEVS version of neural network.

$< D, \{S_i\}, \{I'_i\}, \{T_i\}, \text{SELECT} >$. The sequential state set of component i, S_i is the same as that of the corresponding modular component. The influencees set of component i is obtained by adjoining i to the given I_i, $I'_i = I_i \cup \{i\}$. The transition function T_i is defined as follows:

$$T_i((s_1,\sigma_1),\ldots(s_n,\sigma_n)) = ((s'_1,\sigma'_1),\ldots,(s'_n,\sigma'_n))$$

where

$$(s'_1,\sigma'_1) = (\delta_{\phi,i}(s), ta_i(s'))$$

[the order of the influencees is such that the component i appears first in the

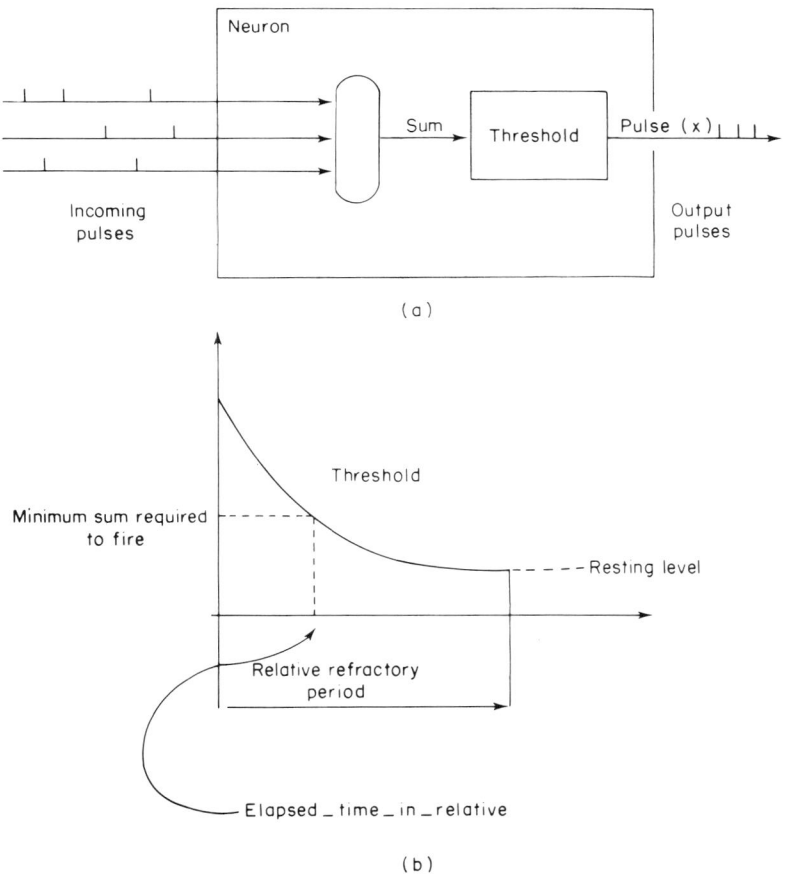

Fig. 8.8. Illustrating neuron DEVS component.

list, the internal transition of this component is accounted for here]

$$(s'_j, \sigma'_j) = (\delta_{ex,j}(s, ta_j(s_j) - \sigma_j, x_{i,j}), ta_j(s'_j))$$

where

$$x_{i,j} = Z_{i,j}(s_1)$$

[the effect of component i on its influencee component j is accounted for here]

Let M_{DN} and M_{DE} be the DEVSs associated with the multicomponent specifications DN and DE, respectively. Then it can be shown that M_{DN} and M_{DE} are isomorphic under a DEVS morphism h from the global sequential

states of M_{DE} onto those of M_{DN} defined by:
$$h(\ldots,(s_i,\sigma_i),\ldots) = (\ldots,(s_i,e_i),\ldots)$$
where $e_i = ta_i(s) - \sigma_i$.

Exercise. Verify that h is one-one and onto.

Note that h establishes a one-one correspondence between individual component states based upon the fundamental relation between time left σ and elapsed time e: $ta(s) = \sigma + e$. The proof of DEVS morphism shows that this correspondence is preserved under single global transitions of the two DEVSs.

Exercise. Verify that the correspondence is indeed preserved by tracing through the definition of global transition for each of the two DEVSs.

The theorem shows that the expressive power of the next event formalism is at least as great as that of the modular formalism. Recall however, that the process interaction formalism was presented as an extension of the activity scanning formalism which itself was presented as an extension of the next event formalism. Thus process interaction, the most inclusive formalism, has expressive power at least equal to the modular formalism. However, the Theorem of Section 4 established the converse. Therefore the modular and non-modular forms are equivalent in expressive power.

In particular, the next event formalism is equivalent to the modular formalism. It is therefore equivalent to the activity scanning and process interaction formalisms, and indeed the non-modular formalisms are all mutually equivalent.

The chain of inclusions is shown below:
  next event
    $\subset$ activity scanning
      $\subset$ process interaction
        $\subset$ modular formalism
          $\subset$ next event.

The results are summarized in the following Theorem.

Theorem. (Equivalence of the Formalisms: the three non-modular formalisms and the modular formalism). The Next Event, Activity Scanning and Process Interaction non-modular formalisms are all mutually equivalent in expressive power and indeed are equivalent to the modular DEVS multicomponent formalism (with respect to DEVS morphisms).

We hasten to note that as made clear in the discussion of formalism adequacy (Chapter 6), equivalence of expressive power does not necessarily imply equality along "softer" dimensions such as convenience, understandability, flexibility, etc. Indeed, the non-modular formalism is certainly more parsimonious than its modular counterpart. The latter requires that wires be cut into input and output ports and may require multiple redundant representation of variables (Section 4).

Perhaps the major implication of the foregoing equivalences is that the waituntil statement does not extend the range of DEVS systems which can be specified. Again this does not dispute the convenience and conceptual facilitation this primitive provides.

Note that the equivalence is established using the ability to immediately activate a component. Were this ability to be denied to the next event formalism then its expressive power might be severely impaired, depending on the level of morphism being considered. This question is taken up as Problem 2.

7. Hierarchical Specification of DEVS Models

The starting point for hierarchical model construction is the closure of the systems formalism under coupling. This means that one can couple component systems together, the resultants being systems that can in turn be coupled together. Any formalism for system specification which is closed under coupling has the nice property that the rules for coupling of its models are uniform in the sense that they apply in exactly the same way to any set of models no matter what level of nesting each represents. In particular, since the DEVS formalism is closed under coupling, the manner in which DEVS components are interfaced refers only to the DEVS specification at the I/O level and is the same whether the component is atomic or is itself a resultant of a DEVS multicomponent model.

Top down design resulting in hierarchically constructed models is the basic methodology of model construction compatible with the multifacetted systems approach. Simulation languages are being developed for supporting hierarchical model construction of large models or those representing real systems, such as communication networks, built to realize several layers of abstraction. Moreover, modularization of the model and experimental frame components of a simulation program requires such hierarchical facilities (Section 6.1).

The DEVS formalism provides a fundamental characterization of hierarchical construction with which to approach the design, and prove the correctness, of discrete event simulators intended to support top down modelling. Due to closure of the DEVS formalism under composition, the coupling of modular DEVS components is itself a modular DEVS. This fact implies that DEVS systems may be constructed recursively from DEVS components, each being atomic, or itself a coupling of DEVS components. Correspondingly, the simulation of a hierarchically constructed DEVS may be realized by a hierarchically constructed simulator. We take up this topic in Chapter 17.

Figure 8.9 presents a template for specifying hierarchical modular DEVS models while Figure 8.10 illustrates this formalism applied to the Subway Train model of Section 7.3.1 in modularized form. We note that at level 1 the components are listed and the input and output variables of the resultant system are given. Then for each of these components, its interface specification is given. The interface consists of specifying the influencees of the component, and its input and output variables. If the component is atomic then its internal and external transition functions (including the time advance and output translation map) are given following the conventions of Figure 8.2. If the component is the result of level 2 components, then the same template is applied to it, and so on.

In the case of the Subway model (Figure 8.11), the level 1 components are the lead, rear and passenger cars. At level 1 these components exchange information with each other concerning arrival and readiness to leave a station. Thus the interface of each component can be defined without knowing the details of its internal structure. As shown in Figure 8.11, *arrived*, *ready*, and *activate in*[21] signals are communicated between nearest neighbors. Each passenger car is composed of level 2 components viz., control, loading and unloading processes. Since each of these is atomic, the level 2 specification of a passenger car is given directly in the language for specifying multicomponent DEVS in modular form (Figure 8.2). Note that the processes inside a passenger car exchange information about the front and back doors which serves to co-ordinate them. The control process generates and receives the signals required at level 1 (arrival, readiness).

8. Summary: Adequacy of DEVS Formalisms

This chapter completes the introduction to discrete event formalisms on which the treatment of multifacetted modelling will be based. Thus it is appropriate to review the major features of the formalisms and to place them in some perspective from the point of view of adequacy.

Cell space models were introduced in Chapter 6 in both the traditional cellular automata and the new discrete event formalisms. The cell space structure manifests a uniformity in the constitution of components and their connectivity. This uniformity simplifies description since only one cellular component and its interaction with its neighbors need be specified. The

---

[21]Designing in a top down manner, the phase set of the component would not be known when writing the interface. However, the activate__in signal can be specified with the phase set left as a free parameter to be filled in later.

Multicomponent DEVS: Modularity and hierarchy 151

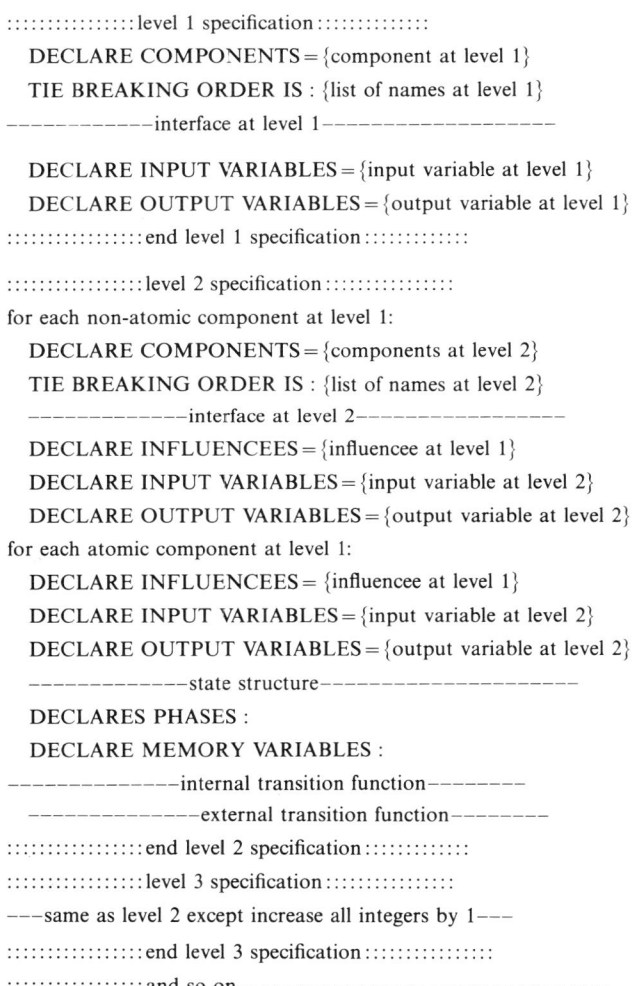

Fig. 8.9. Template for hierarchical modular DEVS specification.

DEVS cell space formalism was shown to be more expressive than that of the cellular automaton with respect to a strict correspondence in time and space and to have advantages in "soft" adequacy criteria as well. This relationship carries over to the structured system versions of the two formalisms where the uniformity constraint is dropped.

Generalization of the DEVS cell space to the multicomponent DEVS at the structured system level was discussed in Chapter 7. The three traditional

::::::::::::::::::::: level 1 specification ::::::::

    DECLARE COMPONENTS: {lead car, passenger cars, rear car}

:::::::::::::::: end level 1 specification ::::::::::::
:::::::::::::::: level 2 specification ::::::::::::::::

lead car

    DECLARE INFLUENCEE: first passenger car behind

        DECLARE INPUT VARIABLES: activate_in(p) : $p \in$ PHASES
                                                        ready(x) : $x \in$ {TRUE,FALSE}

        DECLARE OUTPUT VARIABLES: activate_in(p) : $p \in$ PHASES [22]
                                                          arrived(x) : $x \in$ {TRUE,FALSE}

passenger car

    DECLARE INFLUENCEES: {car ahead, car behind}

        DECLARE INPUT VARIABLES: activate_in(p) : $p \in$ PHASES
                                                          ready(x) : $x \in$ {TRUE,FALSE}
                                                          arrived(x) : $x \in$ {TRUE,FALSE}

        DECLARE OUTPUT VARIABLES: activate_in(p) : $p \in$ PHASES
                                                          ready(x) : $x \in$ {TRUE,FALSE}
                                                          arrived(x) : $x \in$ {TRUE,FALSE}

rear car

    DECLARE INFLUENCEES: {car ahead}

        DECLARE INPUT VARIABLES: activate_in(p) : $p \in$ PHASES
                                                          arrived(x) : $x \in$ {TRUE,FALSE}

        DECLARE OUTPUT VARIABLES: activate_in(p) : $p \in$ PHASES
                                                          ready(x) : $x \in$ {TRUE,FALSE}

:::::::::::::::: end level 2 specification ::::::::::::::::
:::::::::::::::: level 3 specification ::::::::::::::::

lead car (component of subway train)

    DECLARE COMPONENTS: {control process}

  control process (component of lead car)

    DECLARE INPUT VARIABLES: same as lead car

    DECLARE OUTPUT VARIABLES: same as lead car

    DECLARE INFLUENCEES: same as lead car

-------- state structure ----------------

    DECLARE PHASES: {STOPPED,MOVING,STOPPED'}

    DECLARE MEMORY VARIABLES: car_behind.ready

---------- internal transition function ----------

    STOPPED.   if ~ car_behind.ready
                  then passivate in STOPPED'
              send arrived(false) to car_behind

    MOVING.   hold(travel_time)
              send arrived(true) to car_behind
              send activate_in(MOVING) to car_behind
                  go to STOPPED

Multicomponent DEVS: Modularity and hierarchy

----------external transition function----------

in STOPPED', when receive activate__in(STOPPED)
  go to STOPPED

when receive ready(x), car__behind.ready = x

passenger car (component of subway train)

  DECLARE COMPONENTS: {control process, loading process
                          unloading process}

control process (component of passenger car)

  DECLARE INPUT VARIABLES: same as passenger car
    plus   front__door(x) : $x \in$ {OPEN,CLOSED}
           back__door (x) : $x \in$ {OPEN,CLOSED}

  DECLARE OUTPUT VARIABLES: same as passenger car
    plus   front__door (x) : $x \in$ {OPEN,CLOSED}
           back__door (x) : $x \in$ {OPEN,CLOSED}

  DECLARE INFLUENCEES: {loading process, unloading process}

-----------state structure----------------

  DECLARE PHASES: {STOPPED,MOVING,STOPPED',MOVING'}

  DECLARE MEMORY VARIABLES: car__ahead.arrived
                              car__behind.ready
                              front__door
                              back__door

-----------internal transition function-------

MOVING. if $\sim$ car__ahead.arrived
  then passivate in MOVING'
  send arrived(true) to car__behind
  send activated__in(MOVING) to car__behind
  send ready(false) to car__ahead

  front__door := back__door := OPEN
  send front__door(OPEN) to loading process
  send activate__in(MOVING) to loading process

STOPPED.if front__door or back__door = OPEN
  or $\sim$ car__behind.ready
  then passivate in STOPPED'
  send ready(true) to car__ahead
  send activate__in (STOPPED) to car__ahead

  go to MOVING

-------------external transition function-------

in STOPPED', when receive activate__in(STOPPED)
  go to STOPPED
in MOVING', when receive activate__in(MOVING)
  go to MOVING
when receive ready (x), car__behind.ready := x
when receive arrived (x), car__ahead.arrived := x
when receive front__door (x), front__door := x
when receive back__door (x), back__door := x

unloading process (component of passenger car)

DECLARE INFLUENCEE: control process
DECLARE INPUT VARIABLES: back__door (x)
DECLARE OUTPUT VARIABLES: my.back__door (x)
                           activate__in (p)
------------state structure------------
DECLARE PHASES: {MOVING,WORKING,MOVING'}
DECLARE MEMORY VARIABLES: back__door
----------internal transition function---------

MOVING.   if back__door = closed
          then passivate in MOVING'

WORKING. hold(unloading__time)
          back__door := closed
          send my.back__door (CLOSED) to control process
          send activate__in(STOPPED) to control process

          go to MOVING

------------external transition function----

when receive back.door (x), back.door := x
in MOVING', when receive activate__in (MOVING)
                           go to MOVING

loading process (component of passenger car)

DECLARE INFLUENCEE: control process
DECLARE INPUT VARIABLES: front__door (x)
DECLARE OUTPUT VARIABLES: my.front__door (x)
                           activate__in (p)
------------state structure------------
DECLARE PHASES: {MOVING,WORKING,MOVING'}
DECLARE MEMORY VARIABLES: front__door
----------internal transition function---------

MOVING. if front__door = CLOSED
                   then passivate in MOVING'

WORKING. hold(loading__time)

         front__door := CLOSED
         send my.front__door(CLOSED) to control process
         send activate__in(STOPPED) to control process

         go to MOVING

----------external transition function-------

when receive front.door (x), front.door := x
in MOVING', when receive activate__in(MOVING)
                          go to MOVING

rear car (component of subway train)

The rear car DEVS is the resultant of a coupling of control, loading and unloading process in the manner of a passenger car. The only difference is that the rear control process does not wait for a ready signal from the car behind (it is the rear car and knows it!).

::::::::::::::::::end level 3 specification::::::::::::::::

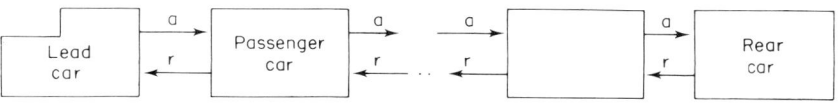

a = arrived : boolean
r = ready : boolean

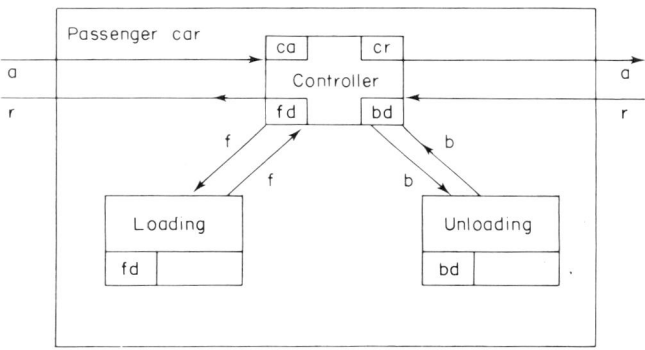

f = front door : { open, closed } ⎤
b = back door : { open, closed }   ⎦ inputs, outputs

fd = front door : { open, closed } ⎤
bd = back door : { open, closed } |
ca = car ahead arrived : boolean | state variables
cr = car behind ready : boolean   ⎦

Fig. 8.11. Modular version of subway train model.

world views: event, activity, and process, were formulated as distinct, though related versions of multicomponent DEVS formalism. The user-friendly language given there provides both for statements that are common to the three alternatives as well as those that are special to each one. This contrasts with the approach of Overstreet discussed in Chapter 18 in which a general discrete event representation language at the structured system level is given

---

Fig. 8.10. Hierarchical modular specification of subway train model. [22]We employ the convention that two variables with the same name are taken to be different variables if they appear as input and output variables. Thus, activate_in(p) as an input variable is distinct from activate_in(p) as an output variable.

whose descriptions are mapped into any one of the three world views. In our approach, the general DEVS formalism is at the I/O systems level (Chapter 4), and the world views are formulated as formalisms at the higher structured system level. The hierarchy of system specification levels thus provides the framework for our unification of the traditional world views. It also provides the basis for formulating the newer modular DEVS formalism.

The multicomponent DEVS in modular form is a specification at the coupled systems level presented in Chapter 3. Its closure under coupling is the basis for the hierarchical modular system specification of discrete event models with which the chapter closes.

It was shown that all formalisms, traditional and modular, introduced at the structured system level were equivalent in expressive power with respect to strict space and time correspondence. The formalisms may well differ with regard to "soft" adequacy criteria however. We shall list the advantages and disadvantages that appear to be associated with the formalisms:

event, activity, process formalisms:
  advantages:
    * supported by existing widely used simulation languages
    * may be easier in the conceptualizationstage of model construction since the system may be viewed as a "gestalt". "Wires" need not be identified and cut, and "locality" of description may be greater (see Section 18.3)
    * description may require less redundancy in usage of variables
  disadvantages:
    * "modelling in the large" not supported (see advantages of modular formalisms)

modular formalisms:
  advantages:
    * supports convenient assembly and disassembly of component models
    * supports top down design concepts
    * supports hierarchical coupled model construction
  disadvantages:
    * not supported by widely used simulation languages (but suitable languages have been recently introduced or are under development, Section 18.3)
    * may be more difficult in the conceptualization phase
    * may involve more redundancy of variables usage

To take advantage of the conceptualization advantages of the traditional world views yet obtain the "modelling in the large" advantages of the modular formalism, we have suggested the development of tools for the transformation of one formalism into the other. Such transformations can be based on the theorems in this chapter that establish equivalence between the formalisms.

PROBLEMS

1. Define a modular formalism for specifying cellular space models analogous to those in Chapter 6.
2. Provide modularized versions of the models constructed in Chapter 7 and its problem set.
3. This problem examines the effect of disallowing immediate activation in the next event formalism. For simplicity it restricts attention to the next event cellular space models. An NEVS (Chapter 6) $NE = <S, N, T, SELECT>$ is said to be *proper* if when $T((s_1, \sigma_1), \ldots, (s_n, \sigma_n)) = ((s'_1, \sigma'_1), \ldots, (s'_n, \sigma'_n))$ then $\sigma'_i > 0$ for $i = 1, \ldots, n$. Thus the local transition function cannot immediately activate (schedule with 0 time left) any neighbor cell.

Using arguments similar to those of Chapter 6, show that the plant model DEVS can not be simulated in real time with a structured system morphism by any proper NEVS.

References

[1] Arbib, M. A. (1964). *Brains, Mathematics, and Machines*, McGraw Hill, NY.
[2] Beauchamp, J. N. and R. C. Fields (1979). "Simulation Modelling by Stepwize Refinement". In *Proceedings of the Winter Simulation Conference*, San Diego, CA.
[3] Parnas, D. L. (1972). "On the Criteria to be used in Decomposing a System into Modules". *C.A.C.M.* **15**, No. 12.
[4] Unger, B. W. and D. S. Bidulock, (1981), "Modular Design of Multicomputer Systems". *Simulation*, July.
[5] Wulf, W. A. (1977). "Languages and Structured Programs". In *Current Trends in Programming Methodology*, Vol. 1. (ed, R. T. Yeh), Prentice Hall, NJ.
[6] Zeigler, B. P. (1976). *Theory of Modelling and Simulation*, Wiley, NY.

Chapter 9

AGGREGATION AND OTHER SIMPLIFICATION PROCEDURES

This chapter sets forth some procedures for model simplification that will serve as background for the upcoming discussion of multifacetted modelling methodology. As indicated in Chapter 1, model construction should be oriented to the attainment of model objectives which ultimately find their expression in the form of experimental frames. Now, model simplification is a form of model construction, differing from the latter in that one starts with an existing model rather than the data, and prior knowledge, of a real system. Thus a full accounting of model simplification should begin with the role of objectives and experimental frames in determining those features of the model that one wishes to preserve in doing a simplification. We shall however, delay such a discussion until Section 13.4, after the prerequisite concepts have been established.

We shall present four general forms of simplification procedures and the conditions under which they can be expected to result in valid simplifications. The presentation will be heuristic in nature (more rigorous theory can be found in the references cited in the discussion). The procedures operate on objects of a formalism and transform them into objects in the same or other formalisms. The first procedure applies to any specification at the I/O systems level and results in a reduction of the state space. The second employs aggregation and uniformity assumptions to a coupling of systems and produces a system specification of the same type. Both are therefore within-formalism transformations. The third approximates a multicomponent DEVS by a DESS (differential equation specified system) and the fourth maps a specification at the I/O systems level to a DEVS system at the same level. These procedures will be illustrated in the ecosystem example that follows this discussion.

1. State Space Reduction: The Congruence Relation

The most fundamental form of simplification operates directly on the state space of an I/O System $S = <T,X,\Omega,Y,\delta,\lambda>$. It is based on an equivalence relation π on the state set Q that is a *congruence* with respect to both the transition and output functions, i.e.

$$q \pi q' \text{ implies } \delta(q,\omega) \pi \delta(q',\omega) \text{ and } \lambda(q) = \lambda(q')$$

for all pairs of states q,q' and input segments $\omega \in \Omega$.

Given such a congruence one can construct an I/O system S/π that is behaviorally equivalent to S, i.e. $\text{IOFO}_S = \text{IOFO}_{S/\pi}$ by giving S/π a state set isomorphic with Q/π, the partition of Q induced by π and defining block-to-block transitions and block-to-output mappings ($\delta_{S/\pi}$ and $\lambda_{S/\pi}$, respectively) in a manner that mirrors the corresponding state transition and output functions in S([2], Chapter 10).

2. Aggregation: The Uniformity of Influence Principle

Consider a system specified as a coupling of systems which have been grouped into blocks. We have seen (Section 3.4) that the blocked system, in which each of the blocks is internally coupled and then the resultants are coupled, is isomorphic to the original one. Now if the coupled blocks are simplified and the simplified systems are coupled together, then the resultant is a simplified version of the blocked system and hence of the original system. The existence of a morphism at the coupled systems level (Section 3.4) constitutes the criterion for the validity of the simplification.

For convenience, we shall limit our consideration to discrete time systems specified in sequential machine formalism although the principle justifying valid aggregation applies generally. The principle can be stated as follows:

Principle of uniformity of influence

A partition of components leads to a valid aggregation if for all atomic components C and partition blocks B, the components in a block are all of the same type; moreover, the influence of components in a block B on a component C can be expressed as the *aggregate* influence of block B on component C and such influence *uniformly* affects all components in the same block as C.

The principle states sufficient conditions that validly replace the influence of components on each other by that of aggregated blocks on each other. It does not define such terms as "influence", "aggregate", and "uniform" so that its interpretation in particular formalisms must be given. Thus the principle serves as a heuristic for seeking valid aggregation rather than as an ironclad guarantor of validity.

To see the principle in operation in discrete time systems let each component be a sequential machine and consider a block B of components of the same type characteristic of the block $M_B = <X,Q,\delta>$. As shown in Figure 9.1, let the components form a parallel coupling[23] with each receiving the same input. Let the output of the composition be any map Z_B satisfying the condition:

(*) permutation invariance:

$$Z_B: \times x_i Q_i \to Y_B$$

is invariant under all permutations of its arguments if:

$$Z_B(q_1,\ldots,q_n) = Z_B(\pi(q_1,\ldots,q_n))$$

where $\pi(q_1,\ldots,q_n) = (q_{\pi(1)},\ldots,q_{\pi(n)})$ for any permutation π on $\{1,\ldots,n\}$.

Let $Q\_\_$DIST be the set of all distributions on Q, i.e. an element of $Q\_\_$ DIST is a mapping $f: Q \to [0, 1]$ with $\sum f(q) = 1$ (Section 3 of Chapter 11). Invariance under permutation is equivalent to the dependence of Z_B on the distribution of values of its arguments rather than on the list of values *per se*.

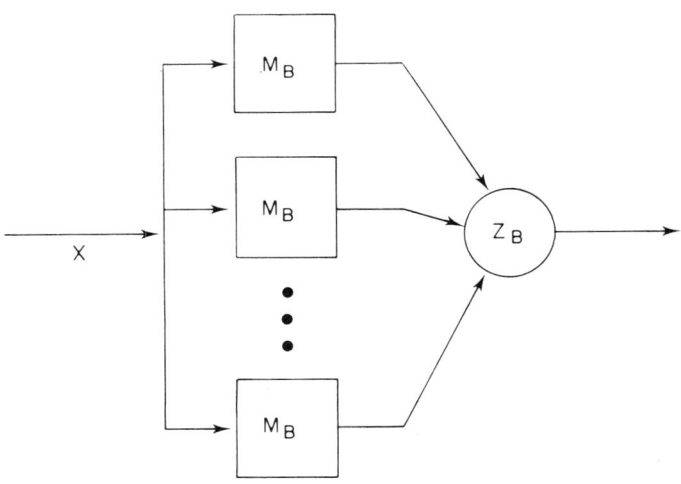

Fig. 9.1. Block of components of same type M_B.

---

[23]This does not prevent there being feedback within blocks in the ultimate application; it will however constrain the form of this feedback.

Lemma

Z_B is permutation invariant, if and only if, there is a map $Z'_B : Q\_\text{DIST} \to Y_B$ such that

$$Z_B(q_1,\ldots,q_n) = Z'_B(\text{dist}(q_1,\ldots,q_n))$$

where $\text{dist}(q_1,\ldots,q_n)$ is the distribution of the values $q_1,\ldots,q_n$.

Exercise. Prove the lemma noticing that a permutation of a list of values preserves the number of elements of each value. Show that for given permutation invariant Z_B the map Z'_B defined on Q_DIST by the above relation is unique.

Examples of permutation invariant mappings are the sum, the product, and statistics such as the average, maximum, minimum, and so on.

The resultant of the block composition can be validly simplified by a sequential machine based on the Markov formalism. The machine M'_B has the same input set X as any component, its state set is Q_DIST, and its output map is Z'_B. Its transition function is a set of Markov transition matrices indexed by elements of X:

$$M'_B = <X, Q\_\text{DIST}, \delta_B, Z'_B, Y_B>$$

where $\delta_B : Q\_\text{DIST} \times X \to Q\_\text{DIST}$

is given by

$$\delta_B(f,x) = fP(x) \quad \text{(matrix multiplication)}$$

where $P(x)$ is the transition matrix with typical element

$$P_{i,j}(x) = 1 \quad \text{if, and only if,} \quad \delta(q_i, x) = q_j$$

(the probability of transition from j to i is 1 if x takes j to i).[24]

Z'_B is the unique function defined on Q_DIST that exists according to the lemma.

That the simplification is valid is summarized in:

Lemma

There is a system morphism from the resultant of the block composition to M'_B constructed above.

The proof shows that the mapping

$$\text{dist}: \times x_i Q_i \to [0,1]$$

is a homomorphism that preserves both state transition and output functions.

---

[24] If each sequential machine is itself stochastic then $P_{i,j}$ may differ from the extremes 0 and 1.

Exercise. Prove the lemma.

Let us refer to the parallel composition of components in block B as the *base* block and its homomorphic image M'_B as the *lumped* block. Consider a series coupling of lumped block systems and a corresponding coupling of atomic components as shown in Figure 9.2 satisfying the uniformity of influence requirement:

Each component in Block 2 receives the same permutation invariant input from components in Block 1 viz, for each component $i \in B_2$, its influencers are all the components in Block 1, and the interface map Z_i is a permutation invariant map $Z_{1,2}: \times Q_{j,1} \to X_2$ (independent of i) where X_2 is the input set of any component of Block 2.

Reflecting this uniformity, lumped Block 2 has lumped Block 1 as its influencer and the interface map is

$$Z'_{1,2}: Q_1 \_\text{DIST} \to X_2$$

the version of $Z_{1,2}$ guaranteed to exist because of permutation invariance.

With the resultants of the given couplings of the base and lumped blocks being called the *base* and *lumped* systems respectively, the lumped system can be shown to be a valid simplification of the base system:

Aggregation Theorem. There is a system morphism from a base system satisfying the sufficient conditions to a lumped system constructed from it as above.

The proof follows the technique of Section 3.4 in establishing a morphism at the level of coupled systems for the base to the lumped coupled system

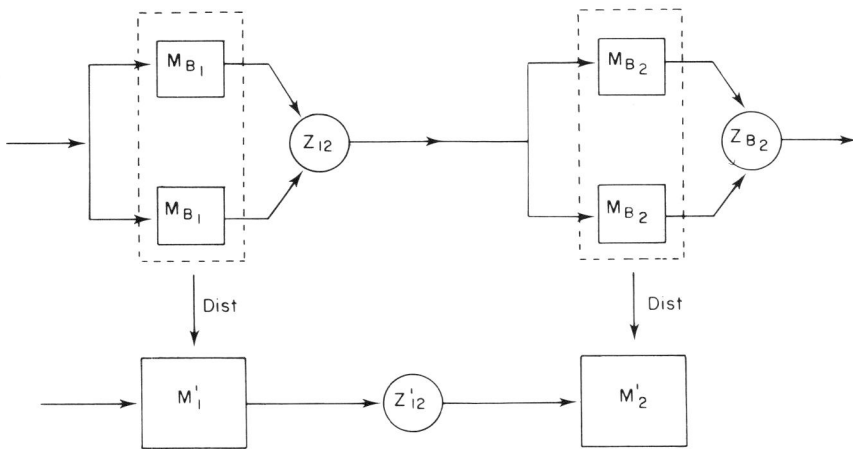

Fig. 9.2. Uniformity of influence and base to lumped model lumping.

specifications. The local aggregation mappings are the distribution maps applied to the respective blocks. The composite of the two distribution maps is the global mapping from the base system to the lumped system. It can be shown to be a system morphism on the basis that its components are system morphisms (from the previous theorem) and the interface requirements are met (by the uniformity of influence requirement).

The principles illustrated just now apply much more generally. A base system satisfying the following conditions can be validly simplified:

(*) *Uniformity of Blocks*: components can be partitioned into blocks each of which contains components of the same type.

(*) *Uniformity of Influence*: if *any* component in a block B receives input from a block B' (possibly $= B$) then *all* components of B receive input from B' and each of these inputs takes the special form: the influencers consist of all components in B' and the interface map is permutation invariant (restricted to B').

The Aggregation theorem can be proved for the base system and a lumped system constructed analogously to the two block case just illustrated.

The sufficient conditions can be loosened that the atomic components may have stochastic specifications and so that components can *sample* the activity of blocks, rather than getting a complete reading as the above requires. In this case, the simplification is justified by an approximate morphism, the error in which decreases as the block and influencer set sizes increase ([2], Chapter 12).

A second important illustration of the uniformity of influence principle is its use in simplifying linear and non-linear differential equation systems [5].

3. Random Phase-Random Space Approximation

The third simplification procedure is related to the second in that it requires uniformity in time as well as space. It applies to the next event cell space formation (Section 6.3) and maps into the differential equation formalism.

Models are assumed to be specified in the following form: Each cell has a finite set of sequential states, S. As usual the time advance function ta applied to a state $s \in S$, gives the maximum time $ta(s)$ that a cell will stay in s unless interrupted by other cells. At the occurrence of an internal event at the cell, it transitions to state $\delta_\phi(s)$ and influences the cells of its neighbors in the following restricted way:

A family of transition matrices $P_s(\ ,\ )$ is to be specified as a parameter of the model where $P_s(\ ,\ )$ has the following interpretation: Let an internal event be about to occur to cell c in state s. For each cell i in the neighborhood I_s (cells influenced in state s), let the sequential state of i be s_i. Then for every state $s_i' \in S$, there is a probability $P_s(s_i, s_i')$ that i will be caused to transition to state s_i' (of course, only one such s_i' will be selected). Note that the elapsed time that i has been in s_i does *not* play a role in this selection and that the selection for each neighbor is independent of the others.

Example. Consider a DEVCS in which each cells has two phases {ON,OFF} with respective neighborhood influencees specified by N_{ON} and N_{OFF}. The transition function is expressed as:

(*)ON. hold(ON__TIME)
    for each cell c in I_{ON},
        if c is in phase ON,
            then convert c to OFF with probability $P_{ON \to OFF}$
            else convert c to ON with probability $P_{OFF \to ON}$
OFF. passivate

Unless disturbed, a cell that is ON remains for a fixed period ON__TIME after which, it changes the phases of cells in its neighborhood according to two probability parameters, $P_{ON \to OFF}$ and $P_{OFF \to ON}$. Once set into the OFF phase, a cell remains there until turned on by some cell in its inverse neighborhood.

The internal transition function has only the transition ON to OFF. The external transition function satisfies the constraints required for the simplification procedure to work. There is only one transition matrix, that associated with phase ON:

$$P_{ON} = \begin{array}{cc} \text{ON} & \text{OFF} \\ 1 - P_{ON \to OFF} & P_{ON \to OFF} \\ P_{OFF \to ON} & 1 - P_{OFF \to ON} \end{array} \begin{array}{c} \text{ON} \\ \text{OFF} \end{array}$$

The time advance function is:

$$ta(s) = \begin{cases} \text{ON\_TIME} & \text{for } s = \text{ON} \\ \infty & \text{for } s = \text{OFF} \end{cases}$$

For such a family of models, the global state in which all cells are in the OFF phase is an absorbing equilibrium state. Since there is a finite probability that this state is reached the long run expectation of this system is to wind up in this state. However, how long the system may expect to have some cells ON depends on how much greater is $P_{OFF \to ON}$ than is $P_{ON \to OFF}$.

For a class of DEVCS models specified in the foregoing form, it is possible to write a system of differential equations that describe the rates of change in state occupancy under the following hypothesis:

Random phase-random space hypothesis
At any time the cell states are uniformly distributed in phase and in space.

More specifically, let $N_s(t)$ be the number of cells in state s at time t. The phase of a cell at time t is the elapsed time e that it has been in state s, where $0 \le e \le ta(s)$. The cells in state s satisfy the *random phase* hypothesis at time t if they are uniformly distributed in phase, i.e. the number of such cells with phase e in any subinterval of $[0, ta(s)]$ of size Δ is $\Delta \times N_s(t)/ta(s)$.

The random phase-random space hypothesis is an assumption made about the base model in constructing the differential equation lumped model. Whether such an assumption holds must either be proved analytically or be tested by simulation of the base model. A formal expression of this assumption can be obtained by employing the run control concept of experimental frames discussed in Sections 13.3, 13.4.

The run control variables selected for this purpose are the elapsed times e and the sequential states s of the cells. The acceptable run control segments are those for which the elapsed times and states are uniformly distributed as required at every instant in the domain of the segment.

The cells in state s satisfy the *random space* hypothesis at time t if they are uniformly distributed in space, i.e. the number of cells in state s in any subset of cells of size n is $N_s(t) \times n/N$ where N is the total number of cells.

Under the Random-phase hypothesis, at time t the number of cells that are in state s and will be in state $\delta_\phi(s)$ at $t+\Delta$ is $\Delta N_s(t)/ta(s)$. These cells subtract from the $N_s(t)$ and add to $N_{\delta_\phi(s)}$. Under Random-space hypothesis, these cells also cause the transitions of cells in their neighborhoods: the cells in state s' of which there are $N_{s'}(t) \times I_s/N$, are switched to state s'' with probability $P_s(s',s'')$ so that the number of cells in the neighborhood of s that are switched from state s' to state s'' is $[\Delta N_s(t)/ta(s)] \times N_{s'}(t) \times (I_s/N) \times P_{s,s''}$.

The above tallies can be substituted in the following system of equations, one for each state $s \in S$:

$N_s(t+\Delta) = N_s(t) -$ (number of cells leaving state s
due to internal transition)
$+ \sum_{s'}$ (number of cells leaving state
s' to enter state s
due to internal transition)
$+ \sum_{s',s''}$ (number of cells in state s'
causing neighboring cells to
convert from state s'' to state s)

Writing $dN_s(t)/dt = \lim_{\Delta \to 0} (N_s(t+\Delta) - N_s(t))$ yields a system of n differential equations, where n is the cardinality of S. Since $\sum_{s \in S} N_s = N$, only $n-1$ of the above equations are independent.

Example (continued). Under the random phase hypothesis the number of cells leaving the ON state between t and $t+\Delta$ is $\Delta N_{ON}(t)/\text{ON\_TIME}$. To get the number of cells that are turned from OFF to ON, multiply this number by $N_{OFF}(t)(I_{ON}/N) P_{OFF \to ON}$. Similarly for the number turned from ON to OFF multiply the same number by $N_{ON}(t)(I_{ON}/N) P_{ON \to OFF}$. Thus we have the equations:

$$dN_{ON}(t)/dt = N_{ON}(t)/\text{ON\_TIME}(-1 \\ + N_{OFF}(t)(I_{ON}/N) P_{OFF \to ON} \\ - N_{ON}(t)(I_{ON}/N) P_{ON \to OFF})$$

$$N_{OFF}(t) = N - N_{ON}(t)$$

where we have used the fact that $N_{OFF} + N_{ON} = N$, at all times.

Exercise. Set $dN_{ON}/dt = 0$ and solve the equations for N_{ON}. Solutions to the equations represent equilibrium solutions for the system under appropriate conditions. Check whether the equilibrium levels of N_{ON} and N_{OFF} make sense for special cases such as when $P_{ON \to OFF} = 0$ (see [1,3]).

Note that the above system can be formalized as a DESS with state set S__DIST. A formal proof of the validity of this simplification would show that the mapping of the global state set of the DEVCS onto S__DIST is a system morphism from the system specified by the latter to the system specified by the former. This validity is established within the experimental frame that expresses the random-phase random space assumption (see Section 13.4.1).

4. Representing Systems in DEVS Formalism

In contrast to the first three simplification procedures, the one to be considered, does not reduce the state space but actually expands it. Nevertheless, the procedure can be expected to result in a reduction in computational effort since advantage can be taken of the efficiency of discrete event simulation strategies (Chapter 17). The procedure takes an I/O system with an "event-like" input segment set and expresses it as a discrete event system. Let $S = <T, X, \Omega, Y, \delta, \lambda>$ be an I/O system. Its segment set Ω is *event-like* if there is a one-to-one correspondence between Ω and $\text{DEVS}(X)$ (the discrete event segments over X, see Section 12.4).

To be useful, the mapping of Ω into $\text{DEVS}(X)$ should have properties, such as being non-anticipatory, that make it conveniently computable. In the most straightforward case, the modeller has already formulated Ω as a discrete

Aggregation and other simplification procedures

event segment set. The set of piecewise constant segments STEP(X) is event like, in that a step change in level of input, is representable as the occurrence of an external event with value equal to the new level. Similarly, the set of piecewise polynomial segments is event like since a change in the polynomial description can be represented as an external event carrying a finite list of non-zero derivatives. Piecewise analytic functions would also qualify except that the list of derivatives is not finite so that mapping would not satisfy the computability requirements.

From here on we use DEVS(X) as the input segment set Ω. Now suppose that an external event x has occurred while the system is in state q. Then it will embark on a state trajectory STRAJ_{q,x_t} where x_t is the input segment of length t starting with x and otherwise identically ϕ. Let us assume that the output set Y is finite. Then along this trajectory the output will remain constant and then instantaneously change to a new value. We consider such changes as output events and construct a DEVS that can reproduce the input/output behavior of S viewed as transformations of DEVS input segments to DEVS output segments.

Let Q/λ denote the partition of the state set induced by the output function λ: blocks of this partition are characterized by having constant output values. As in Figure 9.3, let $t_{\text{next}}(q,x)$ be the time interval that the trajectory STRAJ_{q,x_t} remains in the block of Q/λ containing q. This is time to the next change in output value. Let $q_{\text{next}}(q,x)$ be the state in the block of Q/λ that is entered as the output changes value. The basic idea in constructing the DEVS to represent S is that we may schedule state changes using t_{next} for time advance and q_{next} for internal state transition.

The construction of the representing DEVS M_S proceeds as follows:
external event set
  The external event set of M_S is X.
sequential state set

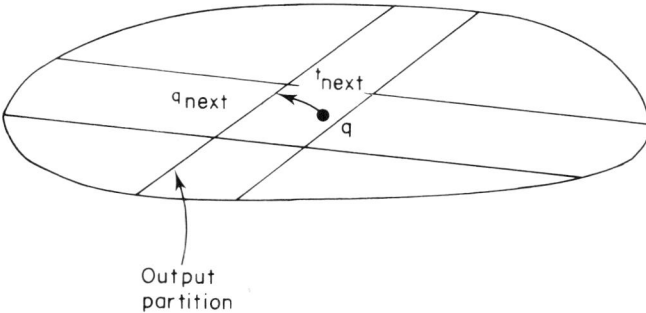

Fig. 9.3. Illustrating transformation into discrete event formalism.

The sequential state set of M_S is $Q \times X$; the pair (q,x) will record the fact that the last external input event was of type x and that q is the state of S at the last event (either internal or external).

output set

The output set of M_S is Y, the output set of S.

time advance function

The time advance function of M_S is t_{next} (note that its domain is the set of sequential states as required).

internal transition function

The internal transition function of M_S takes a sequential state pair (q,x) into the pair $(q_{next}(q,x),x)$ reflecting the update in state of S as its trajectory crosses the partition block containing q, and the continued operation under the same regime established by the last external input event x.

external transition function

The external transition function of M_S is defined as follows:

$$\delta_{ext}((q,x),e,x') = (\delta(q, x_e x'), x')$$

where $x_e x'$ is the input segment of length e beginning with x, ending with x' and otherwise identically ϕ. The explanation is that the arrival of an external event x' after the system has been operating under the regime of the last external event x, for an elapsed time e, causes it to update this state using the transition function of S and to record the new regime x'.

output function

The output function of M_S produces the output of S from state q, viz $\lambda(q)$ when an internal or external event occurs resulting in a transition to a state q.

That the DEVS is a valid representation of the original system is summarized in

Theorem: discrete event representation of systems

If the DEVS M_S is legitimate, then the system S_{M_S} that it specifies is such that there is a system morphism from it to the original system S. Moreover, since every total state of S_M represents a state of S, the two systems are behaviorally equivalent, i.e. have the same IOFO.

Exercise. Prove the theorem. Hint: show that the state mapping $h((q,x),e) = \delta(q, x_e)$ preserves transition and output functions.

The essence of DEVS representation then, lies in the construction of three functions t_{next}, q_{next}, and δ_{ext} for the time advance, internal and external transition function, respectively. In the usual case the system S is itself originally specified in some formalism, for example as a DESS. It is then necessary to work out its transition function δ sufficiently to express the three representing functions. In the DESS case this amounts to solving the system equations. If the system is tractable we may be able to provide complete solutions and express the functions in analytic form. Otherwise, we can use numerical simulation of the system to build tables for the functions. In this case, the results will be approximate depending on the discrete mesh employed to represent the continuous state space.

But it may be objected, once we have solved the equations we have its complete behavior so why should we bother with the DEVS representation at all? In the case of tractable models, the utility of the latter lies in the use of the model as a component in multicomponent DEVS — rather than having to numerically integrate the trajectories of this component, the more efficient discrete event strategies may be employed. This is especially true if all the components of the system are represented in this way, as in the example in Section 3.6 of Chapter 15.

In yet another approach, we may employ a more sophisticated scheme in which the tables are built up as the simulation proceeds rather than in the pre-simulation phase. In such a scheme local (step by step, or numerical integration) methods are employed as the basic simulation strategy. However, a supervisory system oversees the state trajectory, looking for boundary crossings and keeping track of elapsed time. The information so gathered is organized into the three tables required for DEVS representation. Also, when an event is detected, the tables are consulted to see if the current state (or its neighborhood) has been encountered before. If so local simulation is abandoned and the next event is scheduled using the information in the tables. Otherwise, it is continued and the information so gathered is placed in the tables. In this way, the simulation system can be said to learn how to make its simulation more efficient by building a model (the DEVS representation) of the original model [1].

5. Parameter Correspondences

We have seen that a simplification procedure maps from one class of models to a second. If each class is characterized as a model with parameters, then the procedure establishes a correspondence between the parameter spaces of the base and lumped models. This is fundamental to propagation of parameter information in "modelling in the large" concepts (Section 14.6).

For example, consider the random phase–random space procedure applied to the DEVCS class with cells having phase set {ON,OFF} (Section 9.3). The base model has parameters I_{ON} (influencees of center cell), ON__TIME (time in ON phase), $P_{ON \to OFF}$, and $P_{OFF \to OFF}$ (probabilities of switching neighbor cell states). The lumped model is a differential equation system that can be described as:

$$dN_{ON}(t)/dt = aN_{ON}(t) + bN_{OFF}(t) + cN_{ON}(t)N_{OFF}(t) \\ + dN_{ON}(t)^2 + eN_{OFF}(t)^2$$

$$N_{OFF}(t) = N - N_{ON}(t)$$

where a,b,c,d and e are parameters of the model.

The parameter correspondence maps the range of $(I_{ON}, ON\_TIME, P_{ON \to OFF}, P_{OFF \to ON})$ to the range of (a,b,c,d,e) thus:

$a = -1/ON\_TIME$
$b = 0$
$c = (I_{ON}/N)P_{OFF \to ON}/ON\_TIME$
$d = -(I_{ON}/N)P_{ON \to OFF}/ON\_TIME$
$e = 0$

In general, let the base and lumped models have parameter sets P and P', respectively. A simplification procedure sets up a *parameter correspondence* **P** between P.range and P'.range such that:

$(p,p') \in \mathbf{P}$ if, and only if, $M(p)$ is to be simplified by $M(p')$

where $M(p)$ and $M(p')$ denote the base and lumped models with parameter assignments $p \in P$ and $p' \in P'$, respectively.

Properties of the parameter correspondence are informative about the simplification procedure. For example, if the correspondence is many-to-one (maps more than one base parameter assignment to the same lumped parameter assignment) then the procedure has the properties of an information destroying abstraction (Section 2.2.2). The simplification procedures of Sections 1–3 (state reduction, aggregation, random phase) have this property. But the representation of systems in DEVS form (Section 4) does not since no information is lost in transforming a system into DEVS form.

The correspondence of the foregoing example is many-one. In particular, the pair of parameters $(I_{ON}, P_{ON \to OFF})$ in the base model are represented by the product $I_{ON}P_{ON \to OFF}$ in the lumped model. Thus, while the base model can distinguish the neighborhood size and transition probability parameters, the lumped model only represents their effect in product form. Thus a base model having a large neighborhood and a small transition probability and one having the reverse situation may be simplified to the same lumped model.

Also consider the question of whether the correspondence **P** is onto the range of P'. "Onto-ness" or surjectivity means that every point in P'. range participates in at least one pair in the correspondence. Thus if **P** is not onto, then some parameter assignments in the lumped model do not correspond to any assignments in the base model. A lumped model $M(p')$ that has such a missed assignment p' does not represent any base model $M(p)$ under the simplification procedure.

The parameter correspondence in our example above is clearly not surjective. The fact that we always have $b=e=0$ places very definite restrictions on the form of lumped model that can result from the simplification procedure: it cannot have linear and second order terms in $N_{OFF}(t)$. There are other constraints as well: parameter a must be strictly negative, and the ratio c/d must be negative or zero. Any assignment of parameter values to the lumped model that does not satisfy these constraints can not have been derived from a base model by means of the random-phase random-space simplification procedure.

A more complete illustration of parameter correspondence is given in [4].

Having pointed out some properties of the parameter correspondence set up by a simplification procedure, we shall return to consideration of their "modelling in the large" significance in Section 14.6.

6. Summary

A simplification procedure operates upon a system specification to produce another specification that is intended to preserve some behavior of interest and to be simpler in some respect. Four kinds of simplification procedures for simulation models have been discussed. The behavior of interest that one wishes to preserve in a simplification can be stated in terms of the experimental frames that are to be applicable to the simplified model. This discussion is picked up in Section 13.4 after the experimental frame concept has been more fully developed.

Parameter correspondences that arise from simplification procedures where introduced as a prelude to the discussion of parameter information propagation in Section 14.6.

References

[1] Hogeweg, P. and B. Hesper (1981). "Two Predators and One Prey in a Patchy Environment: An Application of MICMAC Modelling". *J. Theor. Biol.* **93**, 411–432.
[2] Zeigler, B. P. (1976). *Theory of Modelling and Simulation*, Wiley, NY.
[3] Zeigler, B. P. (1977). "Persistence and Patchiness of Predator-Prey Systems Induced by Discrete Event Population Exchange Mechanisms." *J. Theo. Biol.* **67**, 687–14.
[4] Zeigler, B. P. (1979). "Structuring Principles for Multifacetted System Modelling". In *Methodology in Systems Modelling and Simulation* (eds, B. P. Zeigler et al.), North Holland, Amsterdam.
[5] Zeigler, B. P. (1981). "Simplification of Biochemical Systems", In *Mathematical Models in Molecular and Cellular Biology* (Ed, L. Segel), Cambridge University Press, Cambridge.

PART 4

MULTIFACETTED SYSTEM MODELLING

Chapter 10

SPECIFICATION OF MODEL STATIC STRUCTURE

The preceding chapters have introduced the multifacetted modelling approach and a framework for the methodologies required to implement this approach. However the framework was stated in abstract conceptual terms and we must now begin the task to concretize these abstractions in the form of workable concepts and tools for supporting multifacetted simulation practice.

We shall start with basic concepts for structuring of models. Later we will be able to extend such concepts to organize families of models.

We have already described our basic formalism for modelling — the systems formalism of Section 3.1. Ultimately any formal model description should end up as such a system. However to write such a description directly is not a practical task for any but the simplest of models. Instead models are expressed in various special formalisms such as those of differential equations, automata or discrete event. Each such formalism can be viewed as selecting out a special class from the set of all systems. Once such a formalism is laid down so is the information common to the subclass of systems being referred to. Thus to express a model in such a formalism we need only give the information necessary to distinguish this model from the others in the class. In this way we can regard a special model formalism as providing a shorthand way of specifying a subclass of systems. And a model expressed in the formalism is a *system specification*, i.e. it indirectly selects a particular system from the set of all systems.

Any such system specification must eventually uniquely specify the sets and functions of some system description. We refer to the sets X (input), Q (states) and Y (outputs) and the output function λ as constituting the *static structure* of the system.

The remainder, namely T (time base), Ω (input segment set) and δ (state transition function) constitute the *dynamic structure*. The distinction is a formal one in that the first group does not contain any reference to the time base while the second group most certainly does. The static structure provides a framework for taking "snapshots" of the system while the dynamic structure provides the framework for the changes such "snapshots" would record over time.

Now each special formalism puts its own restrictions on the possible static and dynamic structures it wants to encompass. In this chapter, we concentrate on static structure specification.

1. Static Structure — Introduction

There are two fundamental concepts involved in specifying the static structure of models. These are the concepts of *variables* and of *decomposition tree*. Figure 10.1 gives an example of such a static structure to which the reader may refer.

A *variable* has a *name* and a *range set*. The name is a distinct identifier not possessed by any other variable. The range set is the set of values that the variable can assume. Recall Section 2.1 where we discussed a number of possibilities for such sets.

The name and range set play formal roles in specifying model structure. However to communicate the significance of the variable we should provide as well, *meaning* and *units*. The meaning documents what the variable is representing and what each of its range set elements signify. A variable which relates to real world measurement will also have units of measurement.

A *decomposition tree* represents a hierarchical decomposition of a model into components. Formally, a decomposition tree is a finite uniquely labelled tree structure. The root node is labelled by the name of the real system the model is intended to represent. The successor nodes of the root are labelled by names of first level components of the model. Each such first level component may itself have a decomposition and these second level components would be represented by successor nodes in the tree, and so on.

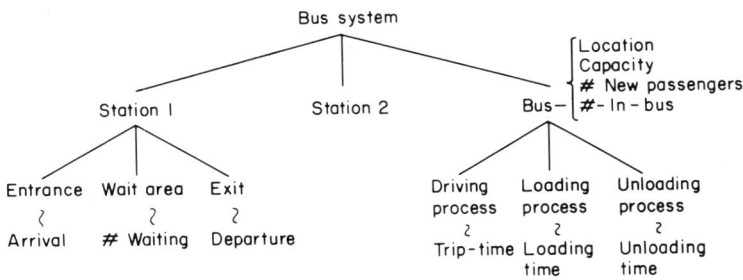

Fig. 10.1.

Clearly the decomposition tree is a "trace" of the system decomposition process which was mentioned in introducing Chapter 3. Since this process must stop, the decomposition tree is a finite one. Later we shall more fully discuss the relation between the decomposition tree and the model construction process (Chapter 15).

Now, the decomposition tree serves as a structure of pidgeon holes for the model's variables. That is each variable is assigned to one and only one node of the tree.

An attachment of a variable to a node signifies the fact that this variable represents an attribute of the component named by the node.

Before continuing with more formal exposition of static structure let us return to the bus example for illustration (Section 1.1.2). We shall discuss two models of the existing bus system. The first is relatively simple in that it does not keep track of passenger identities as does the second. Our discussion of this example will illustrate how static structures reflect *elaboration* of models as more detail is incorporated into them.

1.1. University Bus Service — Models of Existing System

The static structure of a model of the existing bus system is shown in Figure 10.1. The model does not keep track of passenger identities so that only aggregated counts of passengers in the various locations are kept. The variables representing these counts are attached to components to which these counts apply. For example, #_IN_BUS and #_NEW_ PASSENGERS describe the contents of the BUS. Unlike this one, subsequent models, oriented to the testing of passenger identification policies, will have to keep track of passenger identities.

Other important variables have to do with specifying various process times such as the times for loading and unloading of passengers. The inclusion of such variables in the static structure does not, however, contradict its definition! Indeed, such variables are either time-left or elapsed time variables necessary to properly formulate the state space of discrete event models such as this one (Chapter 8).

Thus, as is to be expected, the static and dynamic structures of any model are not independent of one another. On one hand, this means that the model construction process is more complicated than it would be if the specification of the two structures could be done separately or even one first and then the other (more formally if it could be given a parallel or sequential decomposition respectively). On the other hand, because of the interdependence, the static structural description can reveal much about the dynamic structure as well, a benefit from the communication/documentation point of view.

176   4. Multifacetted system modelling

To better understand the model, and the general situation, an informal description of the model's dynamic structure, is provided in Figure 10.2.

Exercise. Go through the dynamic structure description of Figure 10.2 and mark off any points that could not have been determined by inspection of the static structure, especially in view of the meanings given for the variables in the latter description.

Regarding the potential validity of the model one point should be noted. That is, that it cannot allow arrivals which enter a station during the time the BUS is there, to step on the bus.

Exercise, Imagine a real system in which this model assumption apparently holds and one in which it does not.

Now let us consider the static structure of a second model which keeps track of passenger identities, i.e. a passenger remains a distinctive entity throughout his sojourn in the model. Its static structure is displayed in Figure 10.3 and an informal description of the dynamic structure is given in

**   Persons arrive at the ENTRANCE of each STATION. Each arrival at a STATION is recorded by the variable, ARRIVAL, and causes an increment of #__WAITING in that station.

**   When the BUS arrives at the STATION (recorded by its LOCATION variable) it accepts the smaller of #__WAITING and CAPACITY (a parameter of the BUS). This *loading process* lasts LOADING__TIME, a random variable that depends on #__NEW__PASSENGERS (the number accepted by the BUS).

**   The *unloading process* goes on simultaneously and lasts UNLOADING__TIME, a random variable depending on #__IN__BUS. Departure of passengers is recorded by the DEPARTURE variable.

**   The BUS waits until both processes are complete before starting for the next station (*driving process*). Its time taken is TRIP__TIME, a random variable.

Fig. 10.2.   Informal description of dynamic structure for simple bus model.

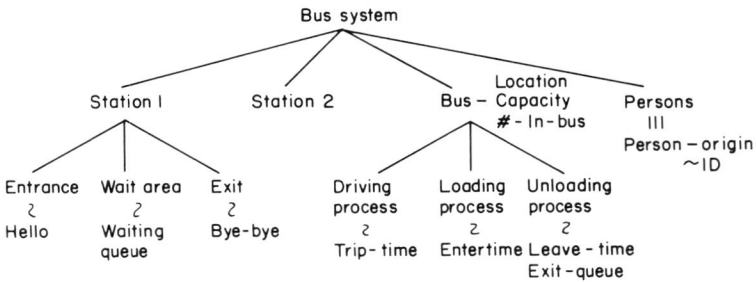

Fig. 10.3.

Figure 10.4. Note the addition of passenger identities requires refinement of many of the variables in the original model. For example, where before it was sufficient merely to keep a count of the passengers on the bus, now the range of this variable has been extended to take values representing lists of passengers.

Is this extra complication necessary to capture the existing bus system? The proper way to answer this question is first to phrase it correctly. The proper form is: are there objectives and hence experimental frames which are of interest to the modeller which are not applicable to the simpler model?

Some thought will reveal that without keeping track of passenger identities a model cannot compute a performance measure such as the amount of time taken by a passenger to complete his journey (from entrance to a station to departure at his destination). Since the problem statement (Section 1.1.2) implied that students were concerned about such matters, it would seem that an experimental frame of this nature is important for the study. This experimental frame is applicable to the more refined model but not to the simpler one. We shall return to consideration of this kind in Chapters 13 and 15.

Eventually, the static structure and informal description of dynamics must be transformed into a complete model specification in one of the formalisms presented in previous chapters. We take up this topic in its full generality in Chapter 15. However, having read Chapters 6–8, the reader is prepared to do the following exercise:

** PERSONS arrive at the ENTRANCE of each STATION. Each arrival is recorded by the variable HELLO, taking on the persons identification as a value.

** The PERSON immediately joins the queue of those waiting for the bus as indicated by inserting his ID to the end of the WAITING__QUEUE (which we shall refer to as LINE for short).

** When the BUS arrives at a STATION, the loading and un*loading processes* begin.

** In the *loading process*, the first PERSON in LINE enters the BUS, taking ENTER__TIME, a random variable. The second in LINE becomes the first and the process continues. The loading is complete when BUS is full (#__BUS equals CAPACITY). It passivates if LINE becomes empty (Λ) before that. (PERSONs arriving during the *unloading process* can still get on the BUS in this model.)

** In the *unloading process*, the disembarking passengers (all those who entered at the previous station) form an EXIT__QUEUE and each in turn disembarks with LEAVE__TIME, a random variable.[25]

** The BUS waits until both processes are completed before sampling TRIP__TIME to reach the next STATION.

Fig. 10.4. Informal description for more detailed bus model. [25]Random variables and parameters are discussed in Section 16.2.

Exercise. Write DEVS models in modular and non-modular form for the model described in Figures 10.1 and 10.2. Do likewise for its refined version in Figures 10.2 and 10.4.

2. Prestructures for Static Structuring

In the previous examples we have used some conventions about static structures which we should now explain. We will also introduce some new concepts which will make the development of such structures more convenient.

We indicated previously that a variable has a unique name, range, meaning and units; and it is attached to exactly one node in the decomposition tree. The decomposition tree is a finite tree whose nodes are labelled uniquely by component names.

Now it is not convenient that each time we add a new variable or component to the structure, we have to come up with a new name, never before used. Moreover often components like STATION 1 and STATION 2 have the same decomposition and variables except for their names. So that we would really like to have to define a *component type* like STATION just once and then have any component of that type such as STATION 1 or STATION 2 automatically have the decomposition and variables of that type. The same idea holds for variables where we would like to be able to define a *variable type* such as LINE and then have any variable of that type such as BUS.LINE or STATION.LINE have the same range, meaning and units of the type.

These considerations lead us to define what we call a *prestructure*. A prestructure will have the property that it *specifies* a static structure, i.e. it is a short hand formalism for static structures of models (recall Section 3.6).

A *prestructure* consists of a set of *variable types* attached to a *decomposition tree*.

A *variable type* has a unique name, range set, meaning and units.

The decomposition tree is finite labelled tree subject to the following restrictions:

(1) If a label appears more than once, then at most one of the like labelled nodes can have attached variable types and a successor decomposition. We call the one so honored the *component type* node and the other like labelled nodes are called *component nodes* of that type.

(2) No component node can appear in the subtree under its component type node.

To obtain a static structure from a prestructure one carries out the following *filling out* procedure which consists of two parts: copying and relabelling.

Specification of model static structure

(1) *Copying*: Copy the variable type names attached to each component type node to all the nodes of its component type. Likewise copy the subtree under the component type node to each of the nodes of the same type.

The copying operation is illustrated in Figure 10.5. One starts at the leaves of the prestructure tree and proceeds towards the root, accumulating new nodes to copy as one goes. Since no component node appears below the component type node there is no danger of having to reverse direction. Note that the result of the copying is a finite tree and the following property holds:

(a) *strict hierarchy*: No label appears more than once down any path of the tree. This means that no component can have a decomposition which eventually contains a component of the same type.

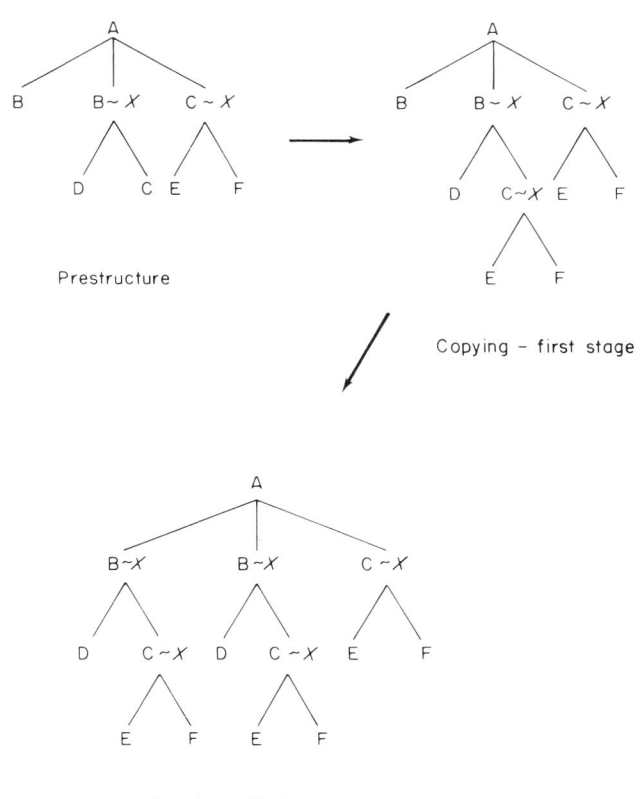

Fig. 10.5.

Exercise. Examine what would happen when copying as above in an example of a prestructure where restriction (2) did not apply.

Exercise. This property (a) sounds intuitively correct and, in view of the previous exercise, necessary to obtain finite decomposition trees. But it does put a restriction on the names we can give to components. Try for example decomposing a room into smaller rooms!

The tree also has the following property:
(b) *uniformity*: Any two nodes with the same label have the same attached variable types and isomorphic subtrees (isomorphic here means copies of one another). This just means that any two components of the same type have the same variable types and isomorphic decomposition structures.

The filled out tree however, is not uniquely labelled. Indeed all component nodes of the same type must somehow be distinguished from one another. This can be done in many ways. We give two such procedures — a simple and a more sophisticated one.

(2) *Relabelling*:
(a) *Numerical*: Go through the labels of the same type in some order and suffix the numbers 1,2,3,... in turn. This operation is illustrated in Figure 10.6a.
(b) *Relative.* Apply numerical relabelling only to like labelling brother nodes (successors of the same superior node). To distinguish any other pair of like labelled nodes there are two cases:
(b1) The superior nodes are distinctly labelled: In this case the nodes can be distinguished by suffixing the string 'in' followed by the label of the superior node to each. For example, ENTRANCE.in.STATION 1 is distinguished from ENTRANCE.in.STATION 2.
(b2) The superior nodes are not distinctly labelled: In this case, return to step (b) in order to label them distinctly and then return to the present pair to apply (b1).

This is a recursive procedure which must eventually terminate. This is so since any pair of upward paths in a tree must eventually converge so that the re-application of step b2 must halt at most when two brother nodes have been reached (and these have already been numerically distinguished). This operation is illustrated in Figure 10.6b.

Exercise. Apply the filling out procedure to the prestructures of Figure 10.1.

Prefixing

We now have ways of uniquely labelling the filled out prestructure tree. Such a tree has variable types which may be attached to more than one node.

Specification of model static structure 181

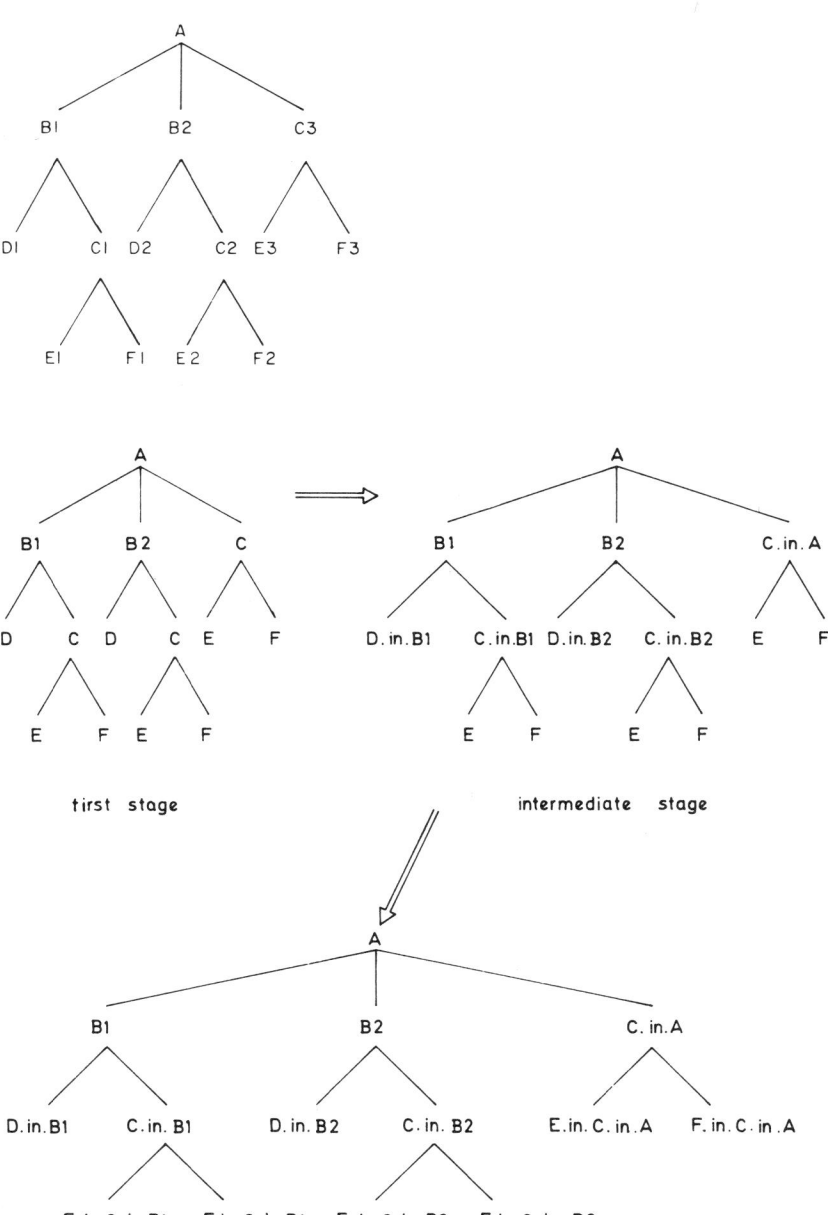

Fig. 10.6. (a) above; (b) below.

Such multiple appearances of the same variable type represent different variables of the same type. To distinguish these variables all we need do is prefix the node label to the variable type name. For example in Figures 10.5–6, the variable type X is attached to components B1, B2, C1, C2 and C3. This represents the attachment of variables B1.X, B2.X, C1.X, C2.X and C3.X to their respective components.

Applying this prefixing operation to a relatively labelled decomposition tree we obtain a variable name of the form for example, (F in C in B1).Y; this reveals that there is a variable of type Y which is attached to a component of type F which is a subcomponent of a component of type C which is a subcomponent of B1. We shall return soon to applications of this concept (Section 11.1.2).

Exercise. Derive the unique name for each variable in Figures 10.1 and 10.3. With practice you can do this by inspection, i.e. without first doing the filling out operation.

2.1. The University Bus System — Model for Identification Policies

We now refine the previous model of the bus system which kept track of passenger identities. This new model will add a variable to each person which identifies him as a student or non-student. The model represents the policy where non-students are allowed on the bus but must pay for the ride. The extra variables used for this purpose are shown in the static structural modification in Figure 10.7.

Exercise. Modify the informal description of dynamic structure of Figure 10.4 so as to represent the above policy.

2.2. Component Types — Multiplicities

We have seen that components of the same type may appear arbitrarily in a decomposition tree provided that strict hierarchy (Section 10.2) is maintained. So far we have discussed cases where a fixed number of components of the same type occur. For example, STATION 1 and STATION 2 represent two occurrences of the type STATION. There are cases however where we would not like to fix the number of occurrences once and for all. For example, Figure 10.3 represents a model structure in which an arbitrary number of components of type PERSON may appear. In the simulation of such a model, components of type PERSON are created (to represent arriving passengers) and destroyed (once these passengers have reached their destination). Thus, the actual number of such components varies with time.

Specification of model static structure   183

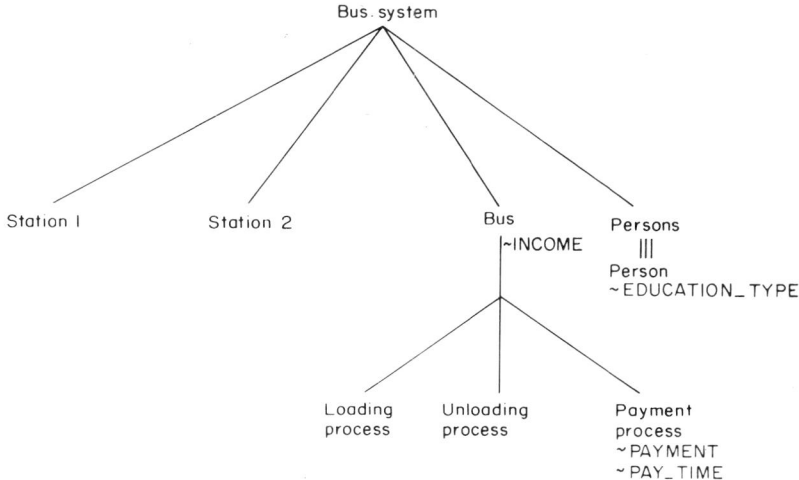

Fig. 10.7. Modification of static structure showing extra variables added to implement boarding control policy.

Now notice how we may represent this possibility while still retaining a decomposition tree with a finite number of nodes. This is done by adding a component of type PERSONS and having a special decomposition of the component, called a *multiple decomposition* into components of type PERSON. Notice that PERSONS has a variable type NUMBER, while PERSON is a distinct component type having other variable types.

Figure 10.8 depicts the general definition of multiple decomposition. Let *A* and *B* be component types. If *B* is a *multiple decomposition* of *A* (as indicated by the special marking of the arc joining their occurrences), then this is interpreted to mean that *A* consists of the number of components of type *B* indicated by the current value of *A*.NUMBER. Often a convenient name for *A* is the plural form *BS* (as in PERSONS). Not always, however, since collective noun forms may also be used, e.g. FLOCK has a multiple decomposition into BIRD. One may have hierarchies of multiple decompositions, e.g. FLOCKS → FLOCK → BIRD.

Exercise. Any army hierarchy consists of divisions, each division consists of regiments, each regiment consists of batalions, etc. Draw a decomposition tree which decomposes an army down to the individual soldier.

It is important to represent the distinction between a plural component (e.g. PERSONS) and its singular component type (e.g. PERSON) also from the point of view of the variables we may attach to them. We have seen that

184 4. *Multifacetted system modelling*

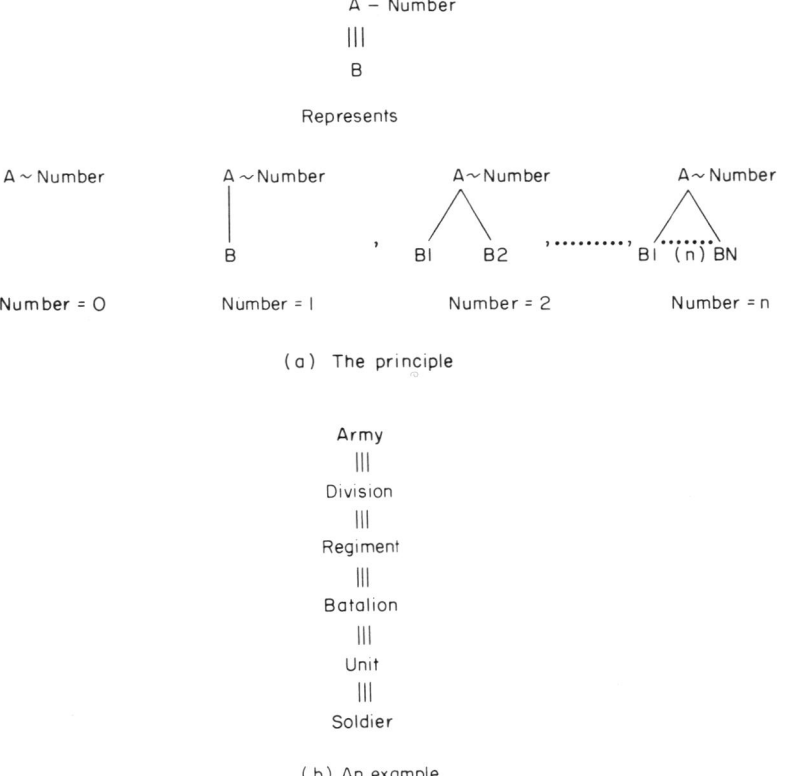

Fig. 10.8. (a) The principle. (b) An example.

NUMBER may be attached to the first but should not be attached to the second (unless it is itself plural, in which case NUMBER indicates the number of its singular components). On the other hand, attributes of the singular may or may not also apply to the plural. For example, PERSON has the variable EDUCATION-TYPE with range {STUDENT,NON-STUDENT} which does not apply to PERSONS. On the other hand, the variable CAPACITY applies both to a single BUS and to a FLEET of buses. We can see moreover, that the variable of a plural and its singular component may be related (e.g. FLEET CAPACITY is the sum of BUS.CAPACITY values). We shall return to this interdependence of definition and meaning soon (Section 11.3).

2.3. Specialization Hierarchies

The discrete event langauge SIMULA introduced two intriguing notions into programming languages: the class concept and class specialization hierarchy. The class concept enables the programmer to define a template for a class of objects which specifies the attributes and procedures individually owned by every instance of the class. For example, the class VEHICLE might specify that each vehicle has two variable types POSITION and SPEED. A particular instance of VEHICLE, v has the variable v.POSITION and v.SPEED. It should be clear that the class concept is modelled by the multiple decomposition terminology just introduced. Actually, the latter terminology extends this concept in that it allows one to associate variables such as NUMBER with a multiple component. If SIMULA had this ability, the user would not have to program a means of counting the number of objects of a class so far created (actually, this number is known to the simulation executive, it just is not available to the program).

However, beyond the ability to specify classes de novo, SIMULA provides the ability to construct classes by specializing other classes. A declaration such as:

LAND__VEHICLE class VEHICLE
begin
real TRACTION:
end;

tells the compiler than LAND__VEHICLE is to be a special kind of VEHICLE *inheriting* all the attributes of VEHICLE such as POSITION and SPEED, as well as unique ones, such as TRACTION which are defined for it in the above declaration. Since such specialization can be iterated indefinitely, a hierarchy of specialized class declarations results.

The specialization relation can be represented by the double arrow as illustrated in Figure 10.9. By iterating the relation, a *specialization* hierarchy is obtained in which successive levels of the tree represent more and more specialized types of the root class.[26] As in a decomposition tree, we may attach variable types to each node of the specialization tree. However, the interpretation is somewhat different: the variable types attached to a node represent the unique attributes associated with the specialized class labelled by the node. In addition, each node inherits all of the variable types attached to the nodes along the path leading from the node to the root.

---

[26] As in SIMULA we require that the relation form a tree. This prohibits the condition where a component is a specialized version of more than one superior component. Note that this disallows explicit representation of hybrids which inherit all the attributes of each parent. The present approach could however be extended to facilitate the handling of such convergences in the specialization relation.

186   4. *Multifacetted system modelling*

We emphasize that specialization is a distinct concept from that of decomposition introduced before: clearly to say that LAND__VEHICLE and AIR__VEHICLE are special types of VEHICLE is not at all to declare that VEHICLE can be decomposed into, and built up from, LAND__VEHICLE and AIR__VEHICLE. However, there is a way of mapping a specialization hierarchy into an equivalent decomposition tree which intimately involves the multiple decomposition mechanism, as we now show.

The basic idea, illustrated in Figure 10.9b is simple: if B and C are special types of A, then the multiple component As is decomposable in the multiple components Bs and Cs. Thus the multiple component VEHICLES is decomposable into LAND__VEHICLES and AIR__VEHICLES if LAND__VEHICLE and AIR__VEHICLE are declared as the special types of VEHICLE. Moreover, if the former are the only special types of the latter then we have:

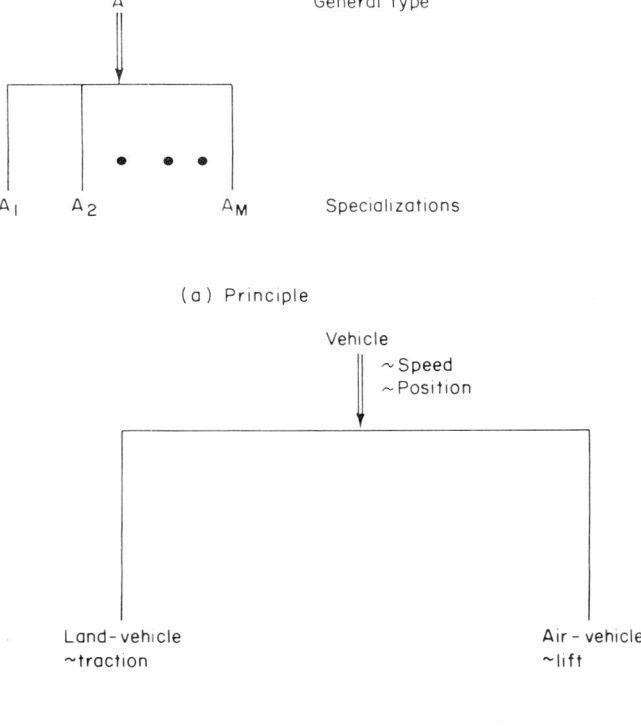

Fig. 10.9. The specialization relation.

NUMBER.VEHICLES = NUMBER.LAND__VEHICLES
+ NUMBER.AIR__VEHICLES

In other words, if we look at the set of vehicles existing in a model at any time, then we will be able to identify two distinct types of vehicles, the land and the air type. When we collect the elements of each type together in two sets LAND__VEHICLES and AIR__VEHICLES, we are providing a decomposition of VEHICLES, the set of vehicles at any time.

In general, the double arrow construction shown in Figure 10.9a is taken to mean that the special types $A_1,...,A_m$ of A specify a partition of As into A_1s,...,A_ms (a partition is a collection of mutually disjoint sets whose union covers the set being partitioned).

We have indicated that every attribute type owned by a SIMULA class is inherited by any of its specialized subclasses. This is represented in the decomposition tree by duplicating the variable types of A to each of its special types $A_1,...,A_m$ in carrying out the mapping illustrated in Figure 10.10. Any decomposition of the objects in general class A is also inherited by its special types. We shall return to this idea in Section 11.2.4 after developing the entity structure concept.

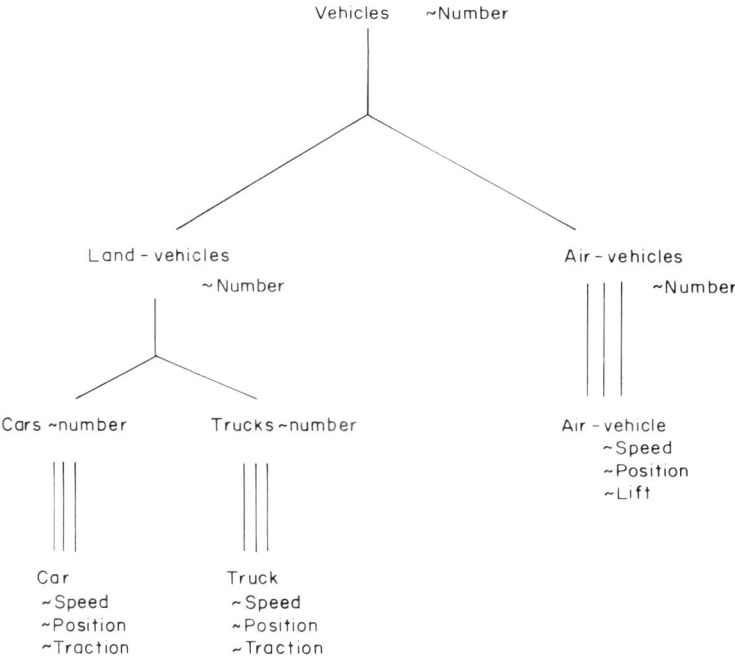

Fig. 10.10. Mapping specialization in the decomposition tree.

A specialization hierarchy can be mapped into a decomposition tree by iterating the above mapping at every node of the hierarchy as illustrated in Figure 10.10. Notice how the variable types propagate down in successive steps from general objects to successively more special objects.

Exercise. Specialize PERSON in the static structure of Figure 10.3 so that it has two types STUDENT and NON__STUDENT. Convert the new static structure into one without the specialization relation.

2.4. Basic Operations on Variables

In discrete event modelling (as well as in modelling, more generally) one often wants to treat a group of variables as a single unit. The reason for this is that such a "package" of variables is often the basis for constructing yet more complex variables. For example, in Figure 10.3 the variable STATION.LINE takes on values which are lists of pairs — each pair being a person's identity and his educational type. It then would be convenient to treat such a pair as a "package". Such a packaging mechanism should also clarify the interdependencies in definition of variables and provide the capability to consistently redefine the variables. This may be necessary, for example, when the range set of some variable is altered and this change ramifies through variables whose range set definitions are related to the altered one.

To do this let us recall that a variable has a name and a range set. Let V be its name and R_V its range set. We define an operator which returns the range set of a variable if it is defined. Thus,

$$V.\text{range} = R_V$$

For example, #-IN__BUS.range = [0,CAPACITY].

The advantage of this is that we can construct packages of variables without committing ourselves to give them definite range sets. These are "abstract" variables in the sense of Section 2.1 as we now shall see.

Let $V_1, V_2, ..., V_n$ be a list of variables. The *composite* of this list is defined as a variable with name $(V_1, V_2, ..., V_n)$ and range set

$$(V_1, V_2, ..., V_n).\text{range} = V_1.\text{range} \times V_2.\text{range} \times, ..., \times V_n.\text{range}$$

For example, (PERSON.ID,PERSON.TYPE) is the composite of the variables PERSON.ID and PERSON.TYPE (in that order). Moreover, (PERSON.ID, PERSON.TYPE).range = PERSON.ID.range × PERSON.TYPE.range. Thus a typical value of (PERSON.ID,PERSON.TYPE) is a pair (i,t) where $i \in$ PERSON.ID.range and $t \in$ PERSON.TYPE.range, e.g. (George,student).

Now we can use composite variables in combination with the other basic set theory operators of Section 2.1 to construct complex variables. For

example, we may define STATION.LINE as a variable having the range set
((PERSON.ID,PERSON.TYPE).range)*. (Recall that $A*$ is the set of all finite
sequences of elements of A.) And of course STATION.LINE.range (which
refers to this set) may be used to construct other composites.

Exercise. The variables in Figure 10.3 are all defined concretely. Which ones
would have to be redefined if the range of PERSON.ID was changed from
$a,b,c,\ldots$ to $1,2,3,\ldots$? Redefine the range set of each of these variables in terms of
PERSON.ID.range.

We introduce a relation on the set of variables which relates variables on
the basis of their range set definitions. A variable V precedes a variable W
if the range set of V is employed in the construction of the range set of W
using standard set theoretic operators. For example, PERSON.ID.*precedes*
(PERSON.ID,PERSON.TYPE). The *precedes* relation is transitive. Thus,
(PERSON.ID,PERSON.TYPE) *precedes* STATION.LINE, and by transitivity
PERSON.ID *precedes* STATION.LINE.

The *precedes* relation partially orders the variables in a digraph as
illustrated in Figure 10.11. The initial elements of this digraph, the variables
which are not preceded by any others, are called *basic*. All, and only, the
range sets of the basic variables must be specified in order to completely
determine the range sets of all the variables. For example, having assigned the
range sets of PERSON.ID and PERSON.TYPE, the range sets of
(PERSON.ID,PERSON.TYPE) and STATION.LINE are completely determined. Moreover, the digraph also displays the ramifications of modifying
the range set definitions of a variable. For example, if the range set of

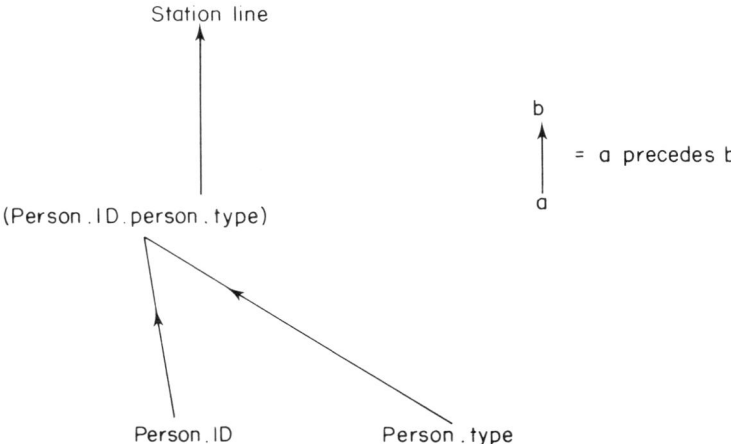

Fig. 10.11. Example of precedes diagraph.

PERSON.ID is altered then so is that of STATION.LINE since there is a path from the first to the second variable in the digraph.

Computer assistance could be developed in which the user would specify range sets in a language expressing the standard set theoretic operators. The *precedes* relation could then be derived algorithmically, and the set of basic variables could be reported to the user. By limiting specification of range sets to the latter variables a complete and consistent range set definition is guaranteed. If it becomes necessary to modify the construction of a range set, the computer would also be able to warn the user of the variables whose range definitions are affected by this change.

3. The Entity–Attribute–Set (EAS) Formalism

It has been suggested [1] that a good basis for standardizing the description of static structure is the approach of the widely used simulation language SIMSCRIPT. We shall present the SIMSCRIPT world view and formalize it in the terms we have introduced.

In every SIMSCRIPT program there is a PREAMBLE which sets up the static structure of a model. This is done by writing a sequence of statements of the following form: EVERY entity name HAS attribute names*, MAY BELONG TO set names *, AND OWNS set names *. Figure 10.12a shows how a bus system model with many stations and buses might appear in SIMSCRIPT.

A SIMSCRIPT *attribute* is a special case of our *variable*. It is a variable whose range set can be one of a very few types — real, integer, etc.

A SIMSCRIPT *entity type* is a special case of our *component type* in that it has attached variable types but there is no corresponding notion of decomposition in SIMSCRIPT. As in our concept, a particular entity has attributes of the types attached to its entity type.

A *set* in SIMSCRIPT is what we have called a list. The phrase "EVERY E OWNS S1, S2,...,Sn" declares in our terms that every entity of type E has attached to it the list valued variables of type $S1, S2,...,Sn$ in addition to its attribute type variables. Thus $E.S1, E.S2,...,ESn$ are variable types with range sets of the form $A_1^*, A_2^*,...,A_n^*$.

Now comes the interesting part of the formalization: specifying in our terms what are the sets $A1, A2,...,An$. The answer lies in the "MAY BELONG TO" phrase. Let S be a set owned by entity type E. Let $E1, E2,...,Em$ be the entities types which may belong to S (as declared in their respective "MAY BELONG TO" phrases).

To capture what SIMSCRIPT does, we now attach the variable type NAME to each entity type. The understanding must now be that each entity

EVERY PERSON HAS AN EDUCATION-TYPE, AN ORIGIN, A DESTINATION
AND MAY BELONG TO A LINE, AN ENTERING-LIST AND A LEAVING-LIST

EVERY STATION MAY BELONG TO A ROUTE
AND OWNS A LINE

EVERY BUS HAS A CAPACITY, A LOCATION, A SENSE
AND OWNS AN ENTERING-LIST, A LEAVING-LIST AND A ROUTE

(a) SIMSCRIPT preamble

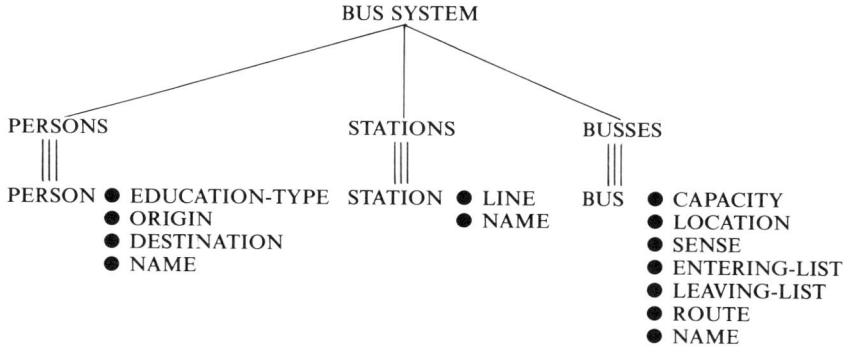

Range definition examples:
STATION.LINE.range = (PERSON.NAME.range)*
BUS.ROUTE.range = (STATION.NAME.range)*
PERSON.ORIGIN.range = STATION.NAME.range

(b) Corresponding static structure

Fig. 10.12.

is distinctly identified by the value of its name so that this name can be used as a pointer to the entity. Then the declarations that E OWNS S and that E_i MAY BELONG TO S, $i=1,...,m$, mean in our terms that attached to component type E there is a variable type with name S and with range A^* where A is the union of the name sets of the $E_i, i=1,...,m$.

Figure 10.12b shows the formal static structure corresponding to the SIMSCRIPT preamble in Figure 10.12a. We see that in our version a STATION.LINE is essentially a list of PERSON.NAME values. Each such name points uniquely to a PERSON and to the current values of his variables EDUCATION-TYPE, ORIGIN, and DESTINATION. Similarly, each BUS, ROUTE is a list of STATION.NAMES. Actually, because of implementation, SIMSCRIPT cannot represent this static structure as easily as we have indicated. The reason is that while an entity may belong to many sets at once, it cannot belong to more than one set of *the same type* at once. So in the SIMSCRIPT version no STATION can be on more than one BUS.ROUTE. For example, there cannot be a central station in the model. Of course, our portrayal of the SIMSCRIPT semantics does not represent, nor suffer from, this limitation. By adding new entities to pair up STATIONS and BUSES one can implement our freer version of the model in SIMSCRIPT. The merits of such a realization are arguable.

Exercise. Decompose a STATION into STOPS for each BUS.ROUTE passing through the STATION. Draw the static structure and write a SIMSCRIPT preamble that exactly represents this new formalization.

In Figure 10.12b we also use the abstraction mechanism to define DESTINATION and LOCATION ranges. SIMSCRIPT does not have this ability. As we have indicated only a limited number of range set types are available and one must be specifically chosen for each attribute.

4. Summary

Specifying the static structure of models was the theme of this chapter. Static structure is that part of a system description which provides a framework for taking snap shots of a system; the dynamic structure dictates how these snap shots change from one observation instant to the next. Static structure is specified by a hierarchical decomposition tree of components and their attached variables. A more convenient shorthand formalism called a prestructure, enables one to work with component and variable types. Likewise, a simple abstraction mechanism for defining range sets facilitates convenient, understandable and less error-prone model construction and modification. A wide spread form of static structuring, the Entity-Attribute-Set view of SIMSCRIPT, and the class specialization hierarchy of SIMULA, were represented in the formalism. From this perspective, certain strengths and limitations of these simulation languages could be readily pointed out.

PROBLEMS

1. Provide pre-structures for the models in Chapters 7 and 8. Convert these into full static structures. You may wish to use the Entity Structuring Program (Section 18.3) to do this in interaction with the computer.
2. Extend the static structure primitives to express the static structure of cellular space models in a convenient way.

References

[1] Markowitz, H. M. (1979). *Proposals For the Standardization of Status Description.* RC 7782(#33671), IBM Watson Research Center, Yorktown Heights.

Other Relevant Literature
The following offer related approaches to model representation and specification. The main difference with our approach is that a sharp distinction is not made between static and dynamic structural elements.

Nance, R. E. (1981). *Model Representation in Discrete Event Simulation: The Conical Methodology.* CS81003-R, VPISU, Blacksburg, VA.

Franta, W. R. (1980). "A Process Oriented Simulation Model Specification and Documentation Language." *Software Practice and Experience,* **10**.

Chapter 11

THE SYSTEM ENTITY STRUCTURE

This chapter further extends the framework for multifacetted system modelling introduced in Chapter 5. We indicated there that the computer should maintain model, data and experimental frame bases. Now we start considering how to organize such bases so that they effectively support the computer's assistance to the user. The central concept we shall introduce is the *system entity structure* (or just entity structure for short).

The basic idea is that a system entity represents the real system enclosed within a certain choice of system boundary. We have discussed in Section 1.1.1 how a statement of objectives orients modelling to such a system entity or conceptual part of reality. In multifacetted modelling we expect to deal with many such system entities and the experimental frames and models oriented to them. Thus it is natural to have the computer maintain a system entity structure and to organize the models and experimental frames around the structure.

How can we characterize this new structure? We already have a good start since we have available the concept of static structure for models (Chapter 10). Suppose that we had a collection of such structures. How would we put them together in a coherent way? In other words, how do we make the jump from "modelling in the small" to "modelling in the large"?

The basic idea is simple. Let us think of the labels on the nodes of the decomposition trees as labels for system entities rather then model components. Now we paste the trees together so that nodes with the same label become identical. This is illustrated in Figure 11.1. Each of the trees on the left represents a decomposition of a model and each label therefore represents a component. The amalgamated tree on the right is an entity structure in which each label represents an entity. In this structure, the entity X has a decomposition into entities A and Y, entity A has a decomposition into entities B and C, and so on.

The entity structure at any time thus represents the net result of system decomposition taking into account all the models created at that time. It is thus a template from which the decomposition trees of the existing models

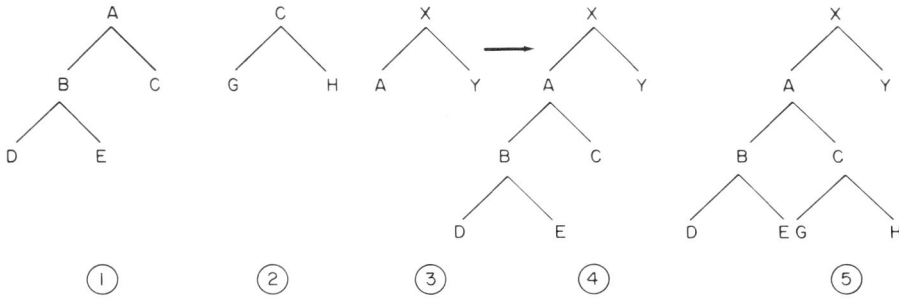

Fig. 11.1.

can be extracted. More than this it is a template for constructing models from those already existing. For example, the only extant model for X has components A and Y which are not further decomposed (tree #3). But if we composed the model of A represented by tree #1, with Y we would have a model represented by tree #4, which is also extractable from the entity structure.

The basic idea is simple, but its execution is not. We have glossed over many problems that arise in performing the amalgamation. The goal however should now be clear and we can turn the situation around by taking a *top down* approach. Suppose we lay down the axioms which an entity structure should obey. We then insist that models are constructed in conjunction with the development of the entity structure. Of course, we must also provide as much computer assistance as possible in maintaining consistency between the model structures and the entity structure.

Exercise. Point out some of the problems glossed over in amalgamating decomposition trees by the procedure given.

After laying down axioms for the entity structure we shall return to reconsider the amalgamation problem (Section 14.2.1).

1. The Entity Structure Axioms[27]

A *system entity structure* is a labelled tree with attached variable types which satisfies the following axioms:

---

[27] The axiomatic treatment of the entity structure is due to David Belogus.

uniformity: Any two nodes which have the same labels have identical attached variable types and isomorphic subtrees.

strict hierarchy: No label appears more than once down any path of the tree.

alternating mode: Each node has a *mode* which is either "entity" or "aspect"; the mode of a node and the modes of its successors are always opposites. The mode of the root is entity.

valid brothers: No two brothers have same label.

attached variables: No two variable types attached to the same item have the same name.

Except for the alternating mode axiom, an entity structure is formally identical with the model static structure obtained by filling out a prestructure. Indeed we shall use the prestructure shorthand to specify entity structures as well.

There is however a profound difference in the entity and model structures brought about by the addition of the alternating mode concept. Only *entity* nodes correspond to model components as we previously suggested. An *aspect* node on the other hand represents a decomposition. The alternating mode property states that an entity has zero or more decompositions (aspects) and each decomposition consists of zero or more component entities. We need to introduce the aspect concept since a system entity may correspond to a number of models, each with its own decomposition structure (amalgamation in the case of multiple decompositions was one of the problems glossed over earlier).

Figure 11.2 shows an entity prestructure for the university bus system problem after the system boundary has been extended to include the entire city. Note that the CITY is an entity which has currently three aspects: COMMUTER.ASPECT, DEMOGRAPHIC.ASPECT and SPATIAL.DECOMPOSITION. SPATIAL.DECOMPOSITION decomposes the CITY into DISTRICTS; the DEMOGRAPHIC.ASPECT decomposes it into PERSONS while the COMMUTER.ASPECT decomposes it into BUS__SYSTEM and PRIVATE__CAR__SYSTEM. Models of the CITY might exist based on each of these three decompositions. For example, a demographic model might recognize only PERSONS as components of CITY and proceed to provide dynamic structure to account for population growth or decline.

The entity structure is fully characterized by the axioms given for it. However, the interpretation of the axioms can not be specified by these axioms and must of necessity, be open to the choice of the modeller. Thus it may seem difficult initially to decide how to represent a concept of the real world: is it an entity? or perhaps an aspect? What should be kept in mind in making this decision are the following points:

The system entity structure

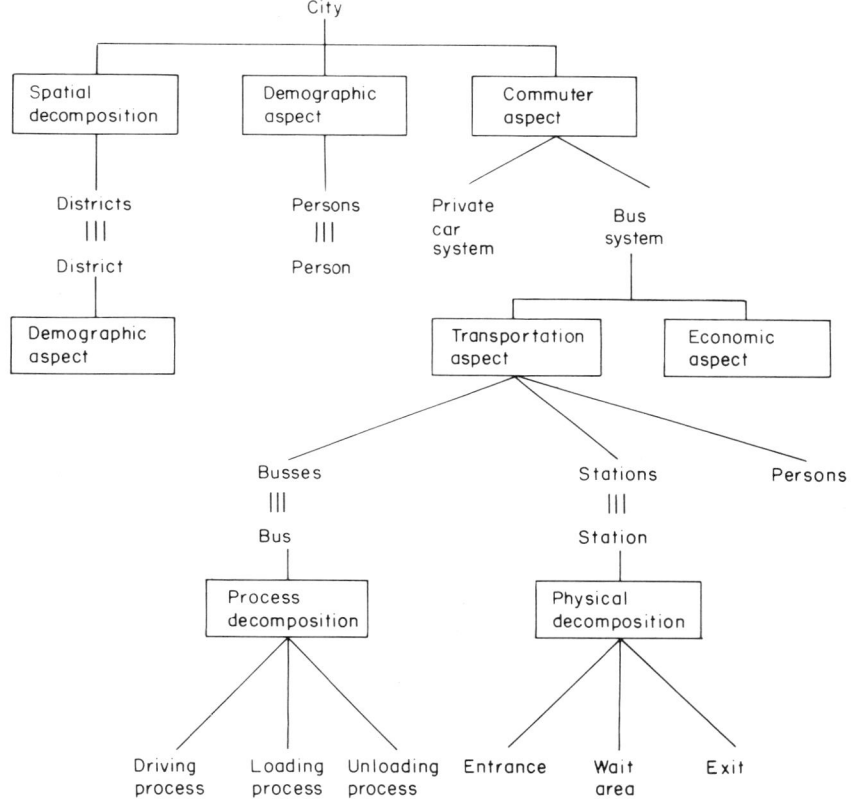

Fig. 11.2.

(a) An entity is intended to represent a real world object which either can be independently identified or is postulated as a component in some decomposition of a real world object.
(b) An aspect is intended to represent one decomposition out of many possible, of an entity. Thus the elements of an aspect are entities, the components of such a decomposition. The variables attached to an aspect are variables of that entity's superior entity that pertain to it only in the context of the aspect.
(c) The entities of an aspect represent distinct components of a decomposition. A model can be constructed by connecting together some or all of these components.
(d) The aspects of an entity do not necessarily represent disjoint decompositions. A new aspect can be constructed by selecting from these aspects as desired.

1.1. Items and Item Occurrences

Note than an entity or aspect may occur in more than one place in the entity structure (so long as the alternating mode and strict hierarchy axioms are obeyed). Using the filling out procedure with relabelling we can always distinguish the various occurrences if we so wish. We shall refer to an entity or an aspect by the more encompassing term: *item*. The mode of an item is thus either "entity" or "aspect". Each appearance of an item in the filled out structure is called an *item occurrence*. For example, the item DEMOGRAPHIC.ASPECT appears under CITY as well as under DISTRICT in the prestructure. Now DISTRICT itself is a singular component of DISTRICTS. If in a particular city there are 3 DISTRICTS, there will be a total of four occurrences of DEMOGRAPHIC.ASPECT in the filled out structure. This facility to represent multiple item occurrence has profound implications as we now shall see.

1.2. Variables: Rights of Access

Recall that variable types are attached to nodes in the entity structure. By the uniformity axiom, the same variable types are attached to every occurrence of a particular item. In other words, an item has its own variable types and carries these with it wherever it is located in the tree.

As in the case of model static structure, when a variable type V is attached to an item occurrence I, this signifies that a variable I.V can be used to describe the item occurrence I. We say that the variable I.V *belongs* to I. For example the variables BUS 1.LINE and STATION 2.LINE belong to the first and second occurrences of BUS and STATION respectively. Thus, while an *unqualified* variable type such as LINE may have multiple occurrences, a *qualified* variable such as STATION 2.LINE belongs to one and only one item occurrence.

However, in contrast to a model static structure, the entity structure represents many possible models. As we have seen, a model of an entity may or may not employ all of the substructure of the entity. In other words, there may be many models of the same entity which employ different decompositions and carry these decompositions to different levels of refinement. One thing we do know however is that these models can employ at most the variables belonging to the entity and to all of the item occurrences in its substructure.

In this way we arrive at the relation of a variable *pertaining to* an item occurrence. More formally, let an item occurrence I *subordinate* an item occurrence J if there is a path in the tree from I to J (in the direction of root to leaves). We also say that J is in the *substructure* of I in this case. Then a variable *J.V pertains* to *I* if *I* subordinates *J* in the tree. In other words, a variable pertains to an item occurrence if it is attached to any item occurrence in its substructure.

Due to the uniformity property, the above statement also holds when we replace "variable type" and "entity" for "variable" and "entity occurrence" throughout. Thus when an item is attached to a node it makes available all its pertinent variable types to the superiors of this node.

An example from Figure 7.2 should show the power of this concept. Variable types belonging to PERSONS all pertain to DEMOGRAPHIC. ASPECT and therefore also to CITY. But since DEMOGRAPHIC.ASPECT is also subordinate to DISTRICT and ultimately also to CITY, these PERSONS variable types also pertain to DISTRICT and ultimately to CITY. However, the different sources of PERSONS variable types give rise to distinct variables. To see this, recall that the above occurrences of PERSONS may be distinguished as PERSONS.IN.CITY and PERSONS.IN.DISTRICT (see Section 10.2). The variable type NUMBER attached to PERSONS thus gives rise to the variables PERSONS.IN.CITY.NUMBER (in other words the size of the population in the city) and for each i, PERSONS.IN.DISTRICT(i). NUMBER (in other words, the population of each DISTRICT). Note that *all* these variables pertain to CITY. Thus the uniformity axiom allows us to talk about PERSONS in the CITY overall and in each DISTRICT using the same basic characterization of persons.

Exercise. In Figure 10.3 (Chapter 10), enumerate all the variables pertaining to BUS__SYSTEM.

2. More on Specialization Hierarchies

The specialization concept introduced in Section 10.2.3 can be further expanded with the entity structure concepts in hand. Figure 11.3 illustrates a structure in which both specialization and decomposition relations appear. The interpretation of such a structure can be given by telling how to operate on the mixed structure so as to remove all instances of the specialization relation. The resulting pure entity structure should be equivalent to the original in the sense of providing the same variable types and substructure to each entity.

The elementary transformation required, shown in Figure 10.3. modifies the one previously given in Figure 10.10. Any variable types belonging to the general entity are removed and assigned to an aspect created specially for this purpose and added to the entity. The set of variable types pertaining to the entity is not changed in this procedure. Next all the aspects hanging from the

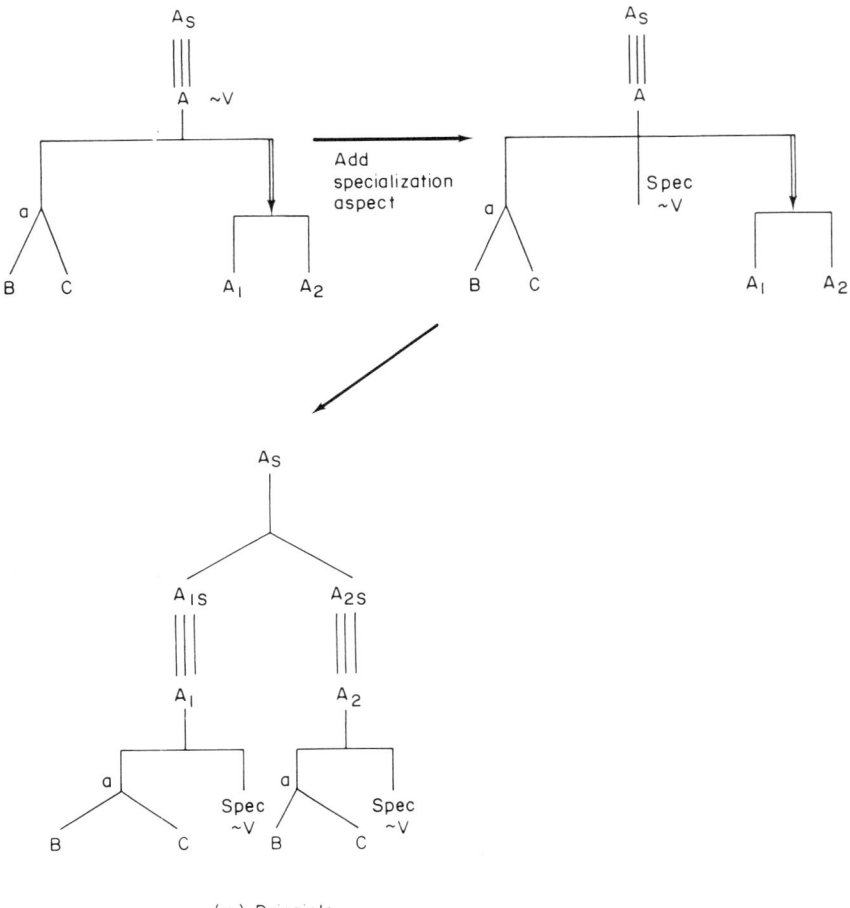

(a) Principle

general entity are copied to its specialization types. Recursive application of the two procedures continues until there are no specialization relations left. The result is that any special type contains all the substructure of its more general superiors. Also, the variable types pertaining to a special type include all those which have been accumulated in transferring the aspects created specially for this purpose.

The example in Figure 11.3b should help to illustrate the process. The variables of VEHICLE are attached to a new aspect SPEC created for this purpose. Next VEHICLES is expressed as a decomposition of LAND__VEHICLES and AIR__VEHICLES, which are multiple entities for the

The system entity structure

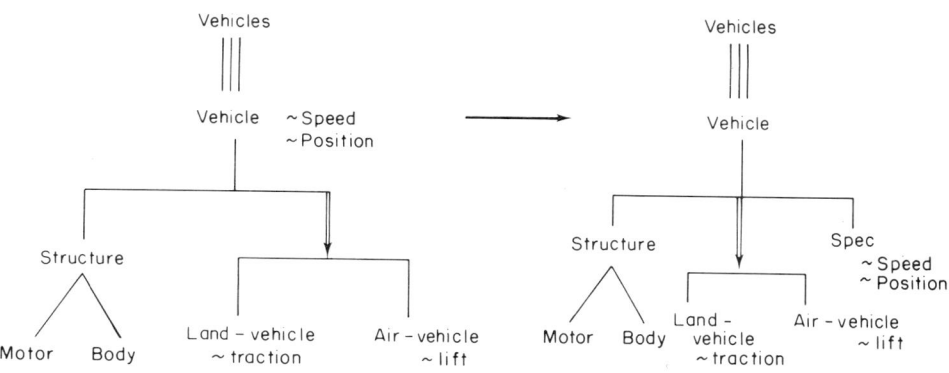

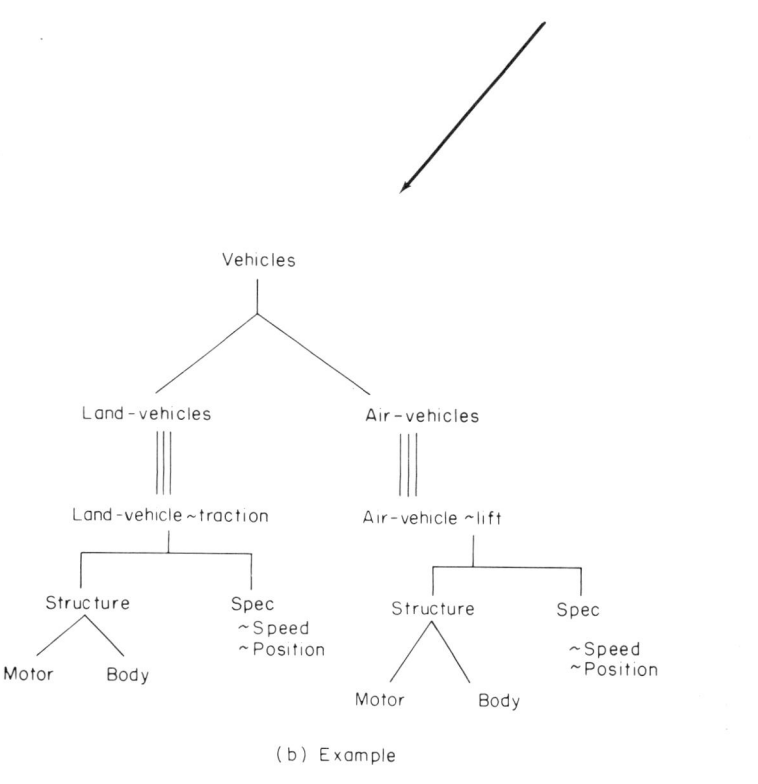

(b) Example

Fig. 11.3. Mapping specialization into the entity structure.

special types of VEHICLE. The SPEC aspect is attached to both of the special types of VEHICLEs as is all other substructure of VEHICLE. Thus for example, LAND__VEHICLE has the variables of VEHICLE as well as its STRUCTURE decomposition into MOTOR and BODY.

Now note that the uniformity axiom of the entity structure does just the right thing in regard to all such inherited structure. Whenever an item is modified, all its instances are identically affected. Thus if any item in the substructure of the general entity is altered then this alteration is automatically transferred to all its specializations in the hierarchy and conversely. The problem of maintaining consistency between all the entities related by specialization, which could otherwise prove rather nasty, is thus given a neat solution.

3. Semantic Structure on Variables

We have already seen that range sets of the variables organized by the entity structure may be related to one another through the "precedes" relation. There is a second relation on variables, called the *semantic structure* which places restrictions on values that variables can simultaneously assume. Such relations are required to hold because of the very meaning of the variables (hence the term "semantic") and so apply across all uses of the variables in the same or different models or experimental frames. If variables that are related by the semantic structure appear in different models, this constitutes a correspondence which must be preserved under the transition and output operations of the models (Section 3.3).

Consider for example, the variable USA.POPULATION and the variables STATE(i).POPULATION, for $i=1,...,50$. They are linked by the semantic relation:

$$\text{USA.POPULATION} = \sum_i \text{STATE}(i).\text{POPULATION}$$

if a demographic model for predicting population growth in the USA keeps track of the populations of each individual state then it must use the above equation to compute the population of the whole country. Suppose that an aggregated model keeps track only of total population, i.e. USA.POPULATION is its only state variable concerning population. If the aggregated model is to be judged a valid simplification of the first then the trajectories of the two models over the same observation interval must be related by the above relation. So semantic relations form the basis for judging the consistency of model behavior such as the validity of an aggregated model relative to its disaggregated counterpart.

Let us formally define what is meant by a semantic relation. Let A and B be variables where we especially note that each could be composites of more elementary variables (Section 10.2.4). A *semantic relation* between A and B is a relation

$$R \subset A.\text{range} \times B.\text{range}$$

which constrains the values that should be simultaneously assumed by A and B in the sense that if at any time, A takes on value a, and B takes on value b, then (a,b) should be in R.

In the above example, let USA.POPULATION play the role of A and the composite (STATE(1).POPULATION,...,STATE(50).POPULATION) — {STATE( ).POPULATION} for short — play the role of B. Then the relation R is given by the function

$$f: \{\text{STATE}(\ ).\text{POPULATION}\}.\text{range} \to \text{USA.POPULATION}$$

where

$$f(p_1,\ldots,p_{50}) = \sum_i p(i)$$

This is an example of many kinds of semantic relations of an aggregation nature that hold between a multiple component and its singular components.

A generic form of semantic relation between a multiple component and its individual components arises as follows. Let V be a variable pertaining to the individual component. Then a variable type $V\_\text{DISTRIBUTION}$ belongs to the multiple component which has as range set, the set of mappings from V.range to the non-negative integers. The value of such a mapping at a value in V.range should represent the number of components whose variable V has that value at the present time.

For example, if STATE has the variable POPULATION with range set {low,medium,high}, then STATES has the variable POPULATION_DISTRIBUTION with range $= \{f | f: \{\text{low,medium,high}\} \to I_0^+\}$. A typical value of POPULATION_DISTRIBUTION in this case would be represented as a triple (l,m,h) which represent the number of individual states having the low, medium, and high populations respectively.

Table 11.1 displays a host of such generic semantic relations involving a multiple component and its individual instances.

Often such aggregate relations are obtained by including weighting factors in the relation. For example, suppose that each STATE has variables POPULATION_DENSITY and AREA. Then USA.POPULATION should be equal to the sum of STATE(i).POPULATION_DENSITY $\times$ STATE(i).AREA. Note that if POPULATION_DENSITY is measured in units of persons/square meter and the units of AREA are square meter, then the units of both USA.POPULATION and the weighted sum agree. Such dimensional analysis can be employed more generally to generate semantic relations.

Table 11.1. Some generic semantic relations arising from multiple decomposition. I is an entity with real variable V. Is is the multiple entity for I with the composite variable $\{I(\ ).V\}$. Variable types belonging to Is are shown in the first column, their semantic relation to $\{I(\ ).V\}$ in the second, and their interpretation in the third.

Variable type of Is	Semantic relation to $\{l(\ ).V\}$	Meaning
NUMBER	$\|\{l(\ ).V\}\|$	Number of components
V_SUM	$\sum_i l(i).V$	Sum (in case range $.V$ is numerical)
V_AVG	$V\_SUM/NUMBER$	Average
V_MAX	$\max_i \{l(i).V\}$	Maximum
V_MIN	$\min_i \{l(i).V\}$	Minimum
$\#\_v$	$\|\{i/l(i).V=v\}\|$	Number having value v
$V\_DISTRIBUTION$	$f{:}V.\text{range} \to l_0^+$ $f(V) = \#\_v$	Distribution over V

Similar semantic relations arise from variables of the multiple components of a generic entity and those of its specializations. An example was given earlier in Section 10.2.3. Recall that VEHICLE is specialized to LAND_VEHICLE and AIR_VEHICLE so that VEHICLES is decomposed into

VEHICLES.NUMBER = LAND_VEHICLES.NUMBER + AIR_VEHICLES.NUMBER

Table 11.2 presents some generic examples of this kind.

Table 11.2. Some generic semantic relations arising from specialization. A is an entity with real variable V. As is the multiple entity for A. A_i is the ith specialization type of A, and A_is is its multiple entity. The first column displays a variable of As and the second gives its semantic relation to the corresponding variables of the A_is.

Variable type of As	Semantic relation
NUMBER	$\sum_i A_i s.\text{NUMBER}$
V_SUM	$\sum_i A_i s.V\_SUM$
V_AVG	$\sum_i (A_i s.V\_AVG \times A_i s.\text{NUMBER}) / \sum_i A_i s.\text{NUMBER}$
V_MAX	$\max_i \{A_i s.V\}$
V_MIN	$\min_i \{A_i s.V\}$
$\#\_v$	$\sum_i A_i s.\#\_v$
$V\_DISTRIBUTION$	$f(v) = \sum_i f_{A_i s}(v)$

4. Summary

This chapter introduced the entity structure concept. An entity structure is similar to a model static structure except that it must represent the amalgamated structures of a family of models not just a single one. Accordingly, it has the possibility to represent a multiplicity of aspects (i.e. decompositions) for a given entity (i.e. subsystem). The variables which pertain to an entity (i.e. can be used to construct models and experimental frames for it) are those which belong to it (directly) or to any of its subentities. Methodologies for such constructions are the topic of ensuing chapters.

Also introduced in this chapter, for later use, was the concept of semantic relation. Such relations hold between variables independently of their appearance in models or experimental frames. Aggregation relations of this nature arise naturally from the multiple entity and specialization properties of the system entity structure. Semantic relations are employed as the basic means for establishing the derivability relation on frames and the applicability relation between frames and models.

Chapter 12

OBJECTIVES-DRIVEN METHODOLOGY: EXPERIMENTAL FRAMES

We have by now acquired some appreciation for how a model is constructed and represented in a discrete event formalism. Recall however, from the introductory discussions in Chapter 5 that objectives and experimentation play a role in the modelling enterprise at least equal in significance to model construction. In current modelling and simulation practice however, the statement of objectives is not formalized and cannot play its proper role in a computer supported methodology. This chapter demonstrates how the statement of objectives can be operationalized in a process whose product is the formulation of experimental frames. In this process, called *objectives-driven methodology*, initial objectives lead to asking specific questions about the real system which in turn require that suitable variables be defined. Ultimately such a choice of variables is represented in experimental frames which also express constraints on the trajectories of these variables.

The discussion opens with an example situation for which modelling objectives arise in a decision making situation. This leads to a presentation of the objectives driven methodology and to a complete definition of the experimental frame concept. The chapter closes with a general examination of the role of objectives driven methodology within systems design methodology.

1. Example: Experimental Frames for the University Bus System

The process of going from objectives to questions and to variables and experimental frames can best be illustrated by example. Let us therefore return to the university bus system example (Section 1.1.2) to see how this process works.

Recall that the primary objective of the committee appointed to study the current situation and make recommendations was to bring about satisfactory bus service for the students. Let us call this the *passenger satisfaction* objective and consider some of the questions and frames that arise.

Passenger satisfaction objective

Recall that the university's problem arose because of complaints by students that they were waiting too long to get on a bus. In fact, often several buses might pass before a student might finally succeed in getting on one because non students were increasingly using the bus service. Thus a student is interested in such questions as: How long will it take me to make a trip between stations? How many round trips will the bus make before I can expect to get on? How long must I wait in line until I get on the bus? Of course every passenger, non student as well as student, is interested in such questions. This explains why we used the title "Passenger Satisfaction" for the objective of primary concern.

Now consider a systematic way in which a passenger can answer such questions for himself. Every time he arrives at a station he marks down the station and time of his arrival in a special pad. Call this time $t_{arrival}$. When the bus arrives he makes an X. If he manages to board the bus, he notes the time at which this happens, say t_{board}. If not, he awaits subsequent arrivals of the bus, making an X each time until he finally succeeds in boarding.

When he arrives at his destination he notes the time, say $t_{departure}$.

Each such episode thus provides the following information:

*transit time: (time to make the trip)

$$= t_{departure} - t_{arrival}$$

*number of missed buses: (number of buses passing before passenger boarded)

$$= \text{number of Xs} - 1$$

*waiting time in line $= t_{board} - t_{arrival}$

Carried out often enough, say over a month's duration, this procedure will enable a passenger to build up a picture of the service that he can expect to get. For example, he can take the list of transit times that he has acquired and find its minimum, maximum and average. Such statistical operations are well known and other than the discussions of Section 11.3, we shall not go into any detail about them here. Our present concern is more fundamental — we are interested in how the elementary data are acquired in a real system or model from which such statistics are ultimately computed.

Let us summarize what has been done up to this point. We started with an objective: Passenger Satisfaction. We developed a set of questions that give more substance to the objective: How long does it take for a passenger to reach his destination? Of this, how much is time spent waiting for the bus?, etc. We then defined a set of variables (transit time, number of missed buses, waiting time) and suggested a procedure for measuring these variables.

It is this kind of mental process:

objectives

→ questions
　→ variables
　　→ measurement

that a modeller must carry out, *before* he begins actual model development, to achieve meaningful results. We call such an approach *objectives-driven* methodology since it starts from the top with a statement of objectives and proceeds in stages to a solution intended to meet these objectives.

Let us refine the presentation of the objectives-driven methodology:

objectives
　→ questions
　　→ interest variables
　　　(performance indexes,
　　　resource utilization measures,
　　　comparison variables)
　　→ mediating variables
　　　→ experimental frame
　　　　→ realization
　　　　　(measurement)

Consider the several changes that have been made. We have distinguished two types of variables: interest variables and mediating variables. An *interest variable* is one that is of direct interest to us in meeting our modelling objectives. Such a variable may be a *performance index* that measures how well the real system of interest performs a given task; or it may be a *resource utilization measure* that measures the efficiency with which certain resources of interest are employed in the operation of the system. Or such a variable may be a *comparison* variable, i.e. a variable that can be employed to compare the behavior of a model with that of the real system or another model.

Usually an interest variable does not correspond directly to measurements than can be made directly on a real system. Therefore, we seek more fundamental variables that do correspond to observable quantities and from which the interest variables can be computed. These new variables are called *mediating variables*. Such a variable may not be of direct interest to the objectives but is employed as an intermediate variable for computing an interest variable. To define an interest variable we specify a set of mediating variables, called its *influencers*, and a processing algorithm which maps time segments of the influencers into time segments of the interest variable. Such an algorithm is a *transducer* in the terminology of Section 4.5. Thus mediating variables are those that we would expect to be able to directly measure in a real system or observe in a model while interest variables are those that we could compute from such observations and are of interest with respect to performance, resource utilization, or comparison. (We do not exclude the possibility that an interest variable is also a mediating variable, i.e. it is itself directly measurable in the real system.)

In our example, TRANSIT__TIME is a performance index since it is a measure of quality of bus service undoubtedly of interest to the passenger. It might also be employed as a comparison variable when trying to calibrate or validate a model with respect to a real system. In the measurement scheme we discussed above, it is not a mediating variable however. This is because we have defined it in terms of arrival and departure time measurements. What about WAITING__TIME? As a component of TRANSIT__TIME it is certainly relevant to the passenger satisfaction objective. But should it be a performance index in its own right? The answer would be yes if waiting causes a definite inconvenience not fully accounted for in increased transit time. For example, if students could not do any useful work while waiting, then minimizing waiting time would be certainly of interest to them. Indeed, they might be more interested in this variable than in TRANSIT__TIME since they might be able to do some work once on the bus.

Finally, consider that on objective criteria, it might seem that #__MISSED__BUSES is not a performance index of interest but if we think of the psychological anxiety involved in missing a bus, such a usage might well be appropriate.

We see that the choice of relevant variables to represent objectives is not necessarily an obvious one. Indeed, it requires us to put some thought into what it is that we wish our models to do for us. The same idea holds for the formulation of each stage in the objectives-driven methodology. We are forced to consider how well the choices we have made for the current stage implement those we have made for the predecessor. Often, we may be forced to consider possibilities we had not thought about at that stage or to reconsider choices that we had already made.

All this mental effort pays off if it leads us to construct models that are well oriented to the objectives that motivated the model construction in the first place (recall the discussion of Section 1.1).

So far we have not discussed examples of resource utilization measures. Consideration of the operation of the bus provides a natural context in which to do so.

Bus operation objective

While satisfying student service is the primary objective of the committee, the bus is, literally, the vehicle by which this objective is, or is not, attained. Thus characteristics of the bus operation should be of interest to the committee. Consider, for example, the size of the bus. A larger bus having greater capacity would tend to reduce passenger waiting time but might be slower and more expensive to operate. Conversely, a smaller bus might pick up fewer passengers each trip but make a faster, less costly round trip. Thus the following are some questions that arise: How much of the capacity of the bus is being used? What is the time taken to make a round trip? How long does the loading of passengers take?

Such questions relate to the utilization of resources at the university's disposal — how efficiently and at what cost can buses be employed to carry out the task of transporting passengers from station to station. They lead us to consider such resource related measures as BUS.UTILIZATION, ROUND__TRIP__TIME, and STANDING__TIME.

Consider suitable mediating variables for such questions. By observing #__IN__BUS, the number of passengers in the bus, we can obtain the fraction of utilized capacity (#__IN__BUS/CAPACITY) at any time. UTILIZATION is then the average of the foregoing fraction over some observation interval of interest.

ROUND__TRIP__TIME can be obtained by observing time segments of the BUS.LOCATION variable. STANDING__TIME at a station however requires a second variable since LOCATION[28] does not mark the departure of the bus from a station. Let us then define a new boolean variable, STANDING which will be TRUE just when the bus is standing at a station. From segments of this variable, the standing time can be observed.

2. Definition of Experimental Frames

Having seen how questions lead to interest and mediating variables, we are ready to fully consider the experimental frame stage of the objectives-driven methodology introduced above. Recall from Section 5.5 that an experimental frame should characterize the circumstances under which a model or its real system counterpart are subjected to experimentation. These constraints on observation and control should be in accord with the objectives and questions that the modeller or experimenter has in mind. The choice of variables is a major concern in formulating such a frame. But more than this, the variables must be categorized into input, output and run control categories and constraints placed on the time segments of these variables. We shall proceed to the formal definition of the frame concept before discussing the interpretation of its elements.

An *experimental frame* is a structure

$$< T, I, C, O, \Omega_I, \Omega_C, SU >$$

---

[28] As defined LOCATION takes on only two values {STATION 1, STATION 2} and so cannot distinguish between the BUS located at a station or between stations.

where
> T is a time base
> I is a set of variables, the *input variables*
> C is a set of variables, the *run control variables*
> O is a set of variables, the *output variables*
> Ω_I is a set of segments, the *admissible input segments*
> Ω_C is a set of segments, the *admissible control segments*
> SU is a set of summary mappings

subject to the constraints:
> Ω_I is a subset of all time segments over the crossproduct of the input variable ranges
> Ω_C is a subset of all time segments over the crossproduct of the run control ranges
>   The summary mappings have as domain the I/O data space defined by the frame (defined below).

Let X be the *input value set* defined by the frame, i.e. X is the crossproduct of the ranges of the input variables. Then the first constraint states that

$$\Omega_I \subset (X, T)$$

Let Y be the *output values set* defined by the frame, i.e Y is the crossproduct of the ranges of the output variables. Then (Y, T) is the set of all segments over the output space.

The *I/O data* (also called the behavior) space defined by the frame is the set of all pairs of I/O segments:

$$\mathbf{D}_E = \{(\omega, \rho) | \omega \in \Omega_I, \rho \in (Y, T), dom(\omega) = dom(\rho)\}$$

Any input/output data that is acquired in an experiment on a model or a real system within the frame lies in $\mathbf{D}_E$.

The final constraint determines the roles of the elements of SU in the frame. Such an element is a mapping whose domain is $\mathbf{D}_E$ into a much smaller dimensional space since it is intended to be a summarization of the input/output pairs observed within the frame. For example, in the case where such a mapping computes a histogram, a finite dimensional vector would result; if a statistic such as an average is computed, the range would further shrink to a scaler (one dimensional vector space).

In sum, we see that an experimental frame selects out a subset of input/output pairs for consideration from all those conceivable. This subset, called its I/O data space, constitutes a focus of attention determined by the objectives which motivate the modelling effort. The frame also specifies data reduction procedures intended to reduce the data (input/output segment pairs) acquired in the frame to manageable and understandable dimensions.

3. Development of Experimental Frames

Let us return to elaborate the objectives-driven methodology having the formal definition of experimental frame in hand. The process now appears as follows:

    objectives
        → questions
            → interest variables
                (performance indexes,
                resource utilization measures,
                comparison variables)
            → mediating variables
                → experimental frame
                    (input variables,
                    output variables,
                    control variables,
                    input segments,
                    control segments,
                    summary mappings)
                → realization
                    (measurement)

To this point, modelling objectives have led to the posing of specific questions about the behavior of a system and these have led to the selection of variables that are of interest to the modeller/decision maker. So as to be able to measure these so-called interest variables, new variables called mediating variables, that correspond to more fundamental measurements, are added. We emphasize that the choice of such variables is a heuristic one that depends on the experimenter's understanding of the behavior he is trying to measure.

Since the mediating variables are the ones considered to be directly observable in the real system it is they that must now be categorized in terms of their relation to the interface of the experimenter with the system. This interface is formalized essentially by the input and output specifications in the experimental frame. Recall from Section 5.5 that *frame input variables* are variables that the experimenter wishes to treat as variables that influence the system and may be under his control or that of the environment. Also, *frame output variables* are variables that are to be treated as under the influence of the system and measurable by the observer. Thus we must decide for each mediating variable whether it is to be an input variable or an output variable. This choice is essentially a guess as to the actual role of the variable in the system. (Hopefully, this guess is a well-educated one when prior knowledge of the real system is available.)

In making the assignment, we should be guided by its consequences. If a mediating variable is designated as a frame input variable, its values will be determined by the input segment specification of the frame. If it is designated as a frame output variable, then we expect that a system (or model) to which the frame applies will determine the values of this variable.

From the definition of summary mappings of a frame, we see that such mappings operate on the I/O data space of the frame. Thus it is natural to formulate the interest variables arising from the objectives as summary mappings of the experimental frame that represents these objectives.

In the next section we return to the bus system example to discuss the choice of input, output, and summary mappings for experimental frames based on modelling objectives. In subsequent sections, we shall continue the discussion of experimental frame concepts.

3.1. Bus System Experimental Frames Example Continued

Choices for the input, output and summary mappings for $E_{\text{Passenger Satisfaction}}$ and $E_{\text{Bus Operation}}$, experimental frames formulated for the corresponding objectives, are shown in Figure 12.1.

As an example of how these choices were made, consider the transit time question. An experimental frame picks out a subset of input/output segment pairs from all those conceivable. What kind of I/O data is sufficient to determine the TRANSIT_TIME of a passenger? As said above, we need to be able to identify the passenger uniquely so as to record his t_{arrival} and

Passenger Satisfaction
input variables:
    HELLO (attached to both STATIONs)
output variables:
    BYE (attached to both STATIONs)
summary mappings
    TRANSIT_TIME
    WAITING_TIME
    #_MISSED_BUSSES

Bus Operation
input variables:
    HELLO (attached to both STATIONs)
output variables:
    BYE (attached to both STATIONs)
    #_IN_BUS
    STANDING
summary mappings
    UTILIZATION
    ROUND_TRIP_TIME
    STANDING_TIME

Fig. 12.1. Experimental frames for passenger satisfaction and bus operation objectives.

$t_{\text{departure}}$. It is natural then to choose as input and output variable types HELLO and BYE respectively as shown in Figure 12.2. HELLO is defined as a variable which takes on the value (x,student) when a student named x arrives at a station. BYE (short for BYE.BYE, Figure 10.3) takes on the value (x,student) when a student named x gets off the bus. Thus by observing an I/O segment pair we can obtain xs transit time if x appears in both input and output segments, i.e. arrived at a station and departed from the next one during the observation interval.

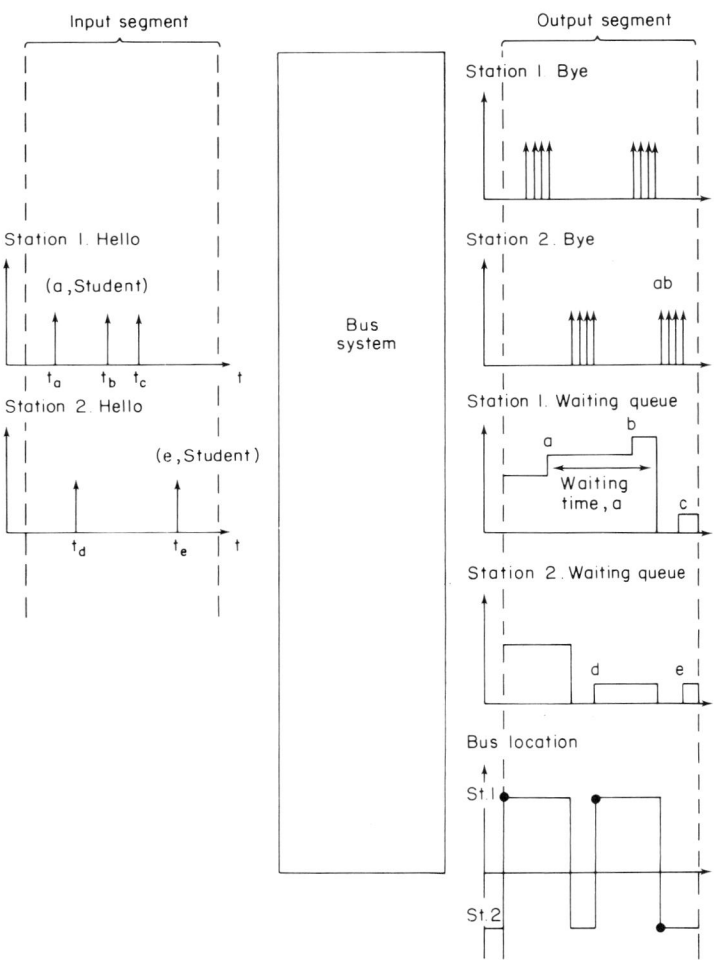

Fig. 12.2.

Now consider the WAITING_TIME variable. As shown in Figure 12.2, a suitable output variable type here is LINE since we can determine how long x had to wait if during the observation interval x appears and then disappears from the list of elements in LINE.

Finally in the case of #_MISSED_BUSES we need both LINE and a variable such as BUS.LOCATION as output variables. The number of buses missed by x is one less than the number of times BUS.LOCATION has the value of xs STATION during the period that x is included in LINE.

3.2. Separation of Model and Experimental Frame

Note that the mechanisms for measuring the performance variables represented by $E_{\text{Passenger Satisfaction}}$ are slightly different than those of our earlier discussion. In fact, earlier we had each passenger gather the data for calculating his transit and waiting times, and number of missed buses. The frame as formulated relieves the passenger of this responsibility assigning it in effect to some "demon" capable of recording and processing appropriate input/output segment pairs. (Later we shall show how this demon can be realized in a simulation program (Chapter 16).)

Exercise. Formulate an experimental frame for the passenger satisfaction objective that corresponds to the passenger based measurement scheme. Hint: associate mediating variables such as ARRIVAL_TIME and DEPARTURE_TIME with each passenger.

In general, we prefer to minimize the responsibility of entities in model to gather data about themselves. There are two reasons for this. One is that a frame should be equally applicable to a real system as to a model. Except in special circumstances we would not expect a real system component to gather the data about itself that is of interest to us. Secondly, even though it is possible to arrange for model components to do this (indeed this is commonly done in simulation programming), this usually requires associating attributes and computations with the components that have nothing to do with model dynamics but are there only for measurement purpose. As discussed in Chapter 5, it is desirable that such interleaving between model and experimentation specifications be eschewed in favor of a separation of the two concepts.

The general principle, of which the above is a special case can be stated thus:

Principle of model experimental frame separation

> Any data gathering/reduction (statistics, performance measurement, etc.) or behavioral control (initialization, termination, etc.) that is conceptu-

ally not carried out in the real system should not be placed within its model but rather formulated as part of the experimental frame.

Conversely, any dynamic structure that is supposed to correspond to mechanisms existing in the real system should be placed within the model.

This principle is fundamental to the approach to simulation program design to be developed later (Chapter 16).

4. Input Segment Specification

We now return to consider the part of experimental frames concerned with specifying the set of admissible input segments.

When modelling with the discrete event formalism, we employ discrete event segments (Section 4.2) as the basic mechanism to specify input segment restrictions. Recall that to specify such segments one must distinguish an element in the input value set as denoting the *non event*; the rest of the elements are considered to signal *external* events. When such a non event element does not already exist we require that it be added. For example, PERSON.NAME.range may be defined as A^* where $A = \{a,b,c,...,z\}$, so each PERSON would have a textual name such as "John". In this case it would be natural to designate the empty string symbol Λ as the non event. On the other hand, if PERSON.NAME.range is A (each PERSON has a single letter name) then we would have to add a non-event symbol ϕ.

There is a second consideration in the case of multiple input variables. If we wish to have events occurring independently in each component variable, we require that *each* has a non event symbol. For example, consider STATION 1.HELLO and STATION 2.HELLO, each having range A^*, the input variables for the bus system frames discussed above. The range of the composite variable, (STATION 1.HELLO,STATION 2.HELLO) is the input value set X, in this case. The pair (Λ,John) means that "John" is arriving at STATION 2 while no one is arriving in STATION 1. Similarly, (Fred,Λ) signals Fred's arrival at STATION 1 with no simultaneous event at STATION 2. The non-event symbol in this case is (Λ,Λ).

Because it occurs so often, it is worthwhile defining an operation on a variable which adds to its range set, a special symbol (or identifies one already existing) as signifying the non occurrence of the other elements. Called ϕ-closure, it can formally be defined as follows:

$$\phi\text{-closure}(V,V.\text{range}, V.\text{meaning}) = (V,V.\text{range}',V.\text{meaning}')$$

where $V.\text{range}' = V.\text{range}^\phi$
$V.\text{meaning}' = V.\text{meaning} \sim [\phi$ is the non-occurrence symbol$]$
($\sim$ here denotes string concatenation).

Objectives-driven methodology 217

Recall that X^ϕ identifies a symbol [such as (Λ,Λ)] already existing in X as the non-event symbol ϕ, if such a symbol exists, or adds the symbol ϕ to X if not.

Now we can define general classes of segments for use in discrete event, and other, model/frame specification.

Piecewise constant or step segments
  $\text{STEP}(X) = \{\omega | \omega \in (X,T),$
    there is a subset $\{\tau_1,...,\tau_n\} \subset dom(\omega)$
    such that ω is constant in the successive
    intervals defined by the subset.

Discrete event segments
  $\text{DEVS}(X) = \{\omega | \omega \in (X^\phi, T)$
    there is a subset $\{\tau_1,...,\tau_n\} \subset dom(\omega)$
    such that $\omega(t) = \phi$
    except for $t \in \{\tau_1,...,\tau_n\}$

The subset $\{\tau_1,...,\tau_n\}$ in either case is the set of *event times* (note that it may be empty). At these times a change in value of ω occurs.

Exercise. Design an I/O system that upon receiving an input segment in DEVS(X) will produce an output segment in STEP(X) that corresponds to it in the sense of having the same event times and the same values of X at these times. Also, design an I/O system that will do the reverse transformation.

As suggested by the foregoing exercise, it is always possible to convert between the STEP and DEVS forms of segment specification and therefore either form is suitable by itself for theoretical purposes. Thus we shall adhere to the DEVS form in the sequel. In practice, the extra processing required to convert from one form to the other, may make it more convenient to work with both forms.

We also extend both definitions to apply to lists of variables. We write STEP (V_i) to mean STEP $((\{V_i\}).\text{range})$. Similarly, DEVS (V_i) means DEVS $((\{V_i\}).\text{range})$, where each of the V_i has been ϕ-closed, as has the composite $(\{V_i\})$. A typical element of DEVS (STATION( ).HELLO) is shown in Figure 12.2.

Given a choice of input variables I, we can specify that the input segments be restricted to discrete event segments by setting $\Omega_I = \text{DEVS}(I)$. This is illustrated in the specification of $E_{\text{passenger satisfaction}}$ where an input segment represents a series of arrivals of persons at the two stations.

Of course further restrictions may be desired. For example one may be interested in discrete event segments in which the event times are close together, or alternately are sparsely distributed. In the case of $E_{\text{passenger satisfaction}}$, this would mean a high (respectively, low) rate of passenger arrivals, for example, representing rush hour versus early morning conditions, respectively. Moreover, one may want to specify the process by which the arrivals are generated.

There are a number of ways to go about placing these constraints on the input segment set. One could define separate frames for each type of condition of interest. For example, one could specify $E_{\text{passenger satisfaction, high}}$ and $E_{\text{passenger satisfaction,low}}$ restricting arrivals to high (low) rates, respectively. Each of these frames is called a *sub-frame* of $E_{\text{passenger satisfaction}}$ and is related to it by derivability (Section 13.1).

Alternatively, we may parameterize experimental frames in the same manner that models are parameterized (Section 16.2.3). For example, let us specify that a Poisson process is to generate the passenger arrivals. Then such a frame is parameterized by λ, the interarrival rate constant which determine the mean interarrival time of the generated segments. Specifying a high (low) value for λ will generate segments with closely (sparsely) spaced interarrivals.

Just as we have models specified in a particular formalism we can have experimental frames understood to be specified in various formalisms. We say that $E < T, I, O, C, \Omega_I, \Omega_C, SU >$ is a *DEVS frame* if T is the reals and Ω_I and Ω_C are subsets of DEVS(I) and DEVS(C), respectively.

4.1. Input Segment Generation and Acceptance

So far we have been tacitly assuming that input segments are generated in a simulation. However, the experimental frame concept has to be formulated in such a way that it is equally interpretable as governing experimentation on a real system or on a model. Let us now look at some implications for this in the case of input segment specifications.

An *experiment* in a frame E is a pair $(\omega, \rho) \in \mathbf{D}_E$. It can represent "applying" an input segment ω and observing an output segment ρ in response. The term "applying" is certainly appropriate in the case of experimentation with a model, i.e. via simulation, but it may not be a good description of experimentation on a real system. Only if all of the input variables are under experimental control can we fully determine the input segment that a real system will receive. More generally the input segment is determined (at least in part) by the system's environment. For example, in $E_{\text{passenger satisfaction}}$, normally passenger arrival at the stations is not under experimenter control. Thus we shall develop a more flexible view of experimentation in which input segments may be either generated or accepted, concepts that were introduced in Section 4.5.

In the case of generation, we realize the set of admissible input segments by employing a special form of I/O system called a *generator* (Section 4.5). Its task is to generate as its output segment, a segment belonging to Ω_I, when started from a suitable state, and run for a desired observation interval. An experiment in this case consists of feeding the output of the segment generator to the system or model under study and observing the latter's output over the same interval.

Employing the DEVS formalism to specify generators provides a means of specifying input segments classes in high level parametric form. For example, the generator in Figure 4.2 of Chapter 4 has two random variables as parameters: the interval time between events, and the identity of the event. To generate a particular class of segments, particular random variables are chosen. Thus, such a generator can generate a wide variety of input segment classes.

Exercise. Give examples of choices for the generator parameters in Figure 4.2. Hint: since the range of the interarrival time random variable is the positive reals, all that must be specified is its distribution, e.g. negative exponential; for the identity random variable, both the range of the variable (e.g. HELLO.range) and the probability distribution (e.g. uniform over the range) must be given.

Now consider the case of experimentation by *acceptance*. Here we are in a position where data are available from which we must select those of interest to us.

Recall from Section 5.1, that we regard the real system as a source of data. Working within an experimental frame, the data potentially available are the pairs of input and output segments considered by the frame. At any time we may have collected a subset of these data. In $E_{\text{passenger satisfaction}}$, for example, we may have placed observers at STATION 1 and STATION 2 to record the identity, time of arrival, and time of departure of each passenger. However all of this collectable data are not necessarily of interest to us and we may define a subframe to express our limited objective. For example, if we specify high rate of arrival input segments, e.g. $E_{\text{passenger satisfaction, high}}$ then only some parts of actually collected I/O pairs may fit this description.

Let us rephrase what we have just said more formally. Suppose we have specified a frame E with input variables I, output variables O, and input segments Ω_I. The data space for such a frame is $\mathbf{D}_E$ as we have defined it in Section 2. However, in the case of experimentation by acceptance we cannot acquire such data directly. Let F be a frame having the same variables as E but having no input segment restriction. Then $\mathbf{D}_F$ represents the space of acquirable real system data and $\mathbf{D}_E$, the subset of interest, must be extracted from it. The acceptance process is simple conceptually — we *accept* an observed I/O segment pair (ω,ρ) from $\mathbf{D}_F$ if, and only if, $\omega \in \Omega_I$ of E. A special form of I/O system that can perform the latter test is called an *acceptor* (Section 4.5).

Exercise. Design an acceptor that will accept all, and only, segments whose passenger arrival rate is sufficiently high, i.e. the accepted set is:

$$\Omega_I(h) = \{\omega | \omega \in \text{DEVS(HELLO)}, \text{max.interarrival}(\omega) < h\}$$

where max.interarrival(ω) is the largest of the interarrival times $(\tau_{i+1} - \tau_i)$, between successive events in ω. Hint: modify the transducer DEVS of Figure 4.4 so that it accumulates the elapsed time between arrival events, and saves the current maximum.

To summarize, when realizing an experimental frame we can engage in experimentation by generation or acceptance. In the generation mode we may couple an input segment generator to the real system or model to generate one or more input segments. In the acceptance mode the frame selects those I/O segment pairs from those observed whose input segment belongs to the admissible input segment set of the frame.

4.2. Run Control Variables: Initialization, Termination

We come to the last part of the experimental frame concept in need of explanation. Recall from Section 3.1 that, in the case of experimentation with a model, specifying the input segment is not sufficient to uniquely determine the output segment — an initial state must be given as well. However, in the case of experimentation on the real system there is no concept of initial state. Yet we would still like to be able to narrow down the range of possible responses. Since experimental frames must be interpretable both for models and for real systems we must come up with a concept that has the role of restricting the initial state in the case of a model but is meaningful for real systems too. This concept is that of *run control variables* (or just "control variables" for short in the proper context).

Run control variables serve not only to initialize experiments but to terminate them as well. Suppose for example we are interested in experiments starting with an empty BUS at STATION 1 with no one waiting to enter it; and terminating when the BUS arrives at STATION 2. Call such a frame $E_{\text{travel time}}$ since in it we can ask questions about the bus travel time from STATION 1 to STATION 2 uncorrupted by its standing time at a station. Appropriate run control variables for $E_{\text{travel time}}$ are: BUS.LOCATION, #__IN__BUS and STATION 1.#__WAITING.

The conditions under which an experiment is initiated are:
  BUS.LOCATION = STATION 1
  #__IN__BUS = 0
  STATION 1.#__WAITING = 0
These are called *initialization conditions*.

The condition under which an experiment is terminated is:

BUS.LOCATION = STATION 2.

This is called a *termination condition*.

For neither of the bus system models of Section 10.1.1 do the above run control variables constitute a set of state variables (defined in Section 16.2.1). In general initialization conditions will not uniquely specify an initial state of a model. Thus a family of possible experiments, rather than a single one, is what is determined by an experimental frame.

For many simulation practitioners it may seem questionable to specify initialization conditions that do not uniquely fix the initial state of the model. The alternative approach of uniquely setting the initial state in the experimental frame is fine in the context of "modelling in the small" (c.f. Section 1.2). However, in "modelling in the large" we wish to apply experimental frames to more than one model, e.g. to compare the behavior of two models within the same frame (Section 13.4). For a given frame, there is a virtual infinity of models to which it may apply. For most of these models, the initialization conditions of the frame do not uniquely determine the initial state. Thus were we to insist that initialization conditions uniquely fix the initial state of any model to which a frame applies, this would greatly limit the applicability of the concept. (There is a cost for the extra flexibility so afforded: we need a third component, besides model and frame, called execution control, to select initial states from those satisfying the initialization conditions (Section 16.2).)

4.3. Run Control Segments

It turns out that initialization and termination conditions are not the most general, or useful way of controlling experiments. To see this, first let us develop those concepts more formally.

Let C be a set of control variables specified by a frame E. For example, C = {BUS.LOCATION, #_IN_BUS, STATION 1.#_WAITING} in $E_{travel\ time}$. The *control space* thus set up is defined as Z, the crossproduct of the range sets of the run control variables.

A set of initialization conditions has the effect of specifying a subset, call it INITIAL, of the control space. For example, in $E_{travel\ time}$

$$INITIAL = \{(STATION\ 1,0,0)\}$$

a singleton subset of Z.

All experiments must start in INITIAL.

Similarly the termination conditions specify a subset, TERMINAL, of the control space. An experiment terminates as soon as this TERMINAL subset is entered. Another way of saying the same thing is that an experiment is continued so long as the TERMINAL set is not entered. If we define a set called CONTINUATION as the complement of TERMINAL (all points not in TERMINAL) then an experiment continues while the model or real system remains in the CONTINUATION set. For example in $E_{\text{travel time}}$

$$\text{TERMINAL} = \{(\text{STATION } 2, c, n_1) | c, n_1 \in I_0^+\}$$
$$\text{CONTINUATION} = \{(\text{STATION } 1, c, n_1) | c, n_1 \in I_0^+\}$$

Clearly if any experiments are to be possible we must have INITIAL ⊂ CONTINUATION. This is indeed the case in our example. Now consider any experiment allowed by a frame with such INITIAL and CONTINUATION sets. As traced out in the control space, it must originate in INITIAL and remain in CONTINUATION. Such a trace is called a *control segment*. The set of all such segments constitutes Ω_C, the set of admissible control segments. Formally,

$$\Omega_C = \{v | v \in (Z, T), v(t_{in}) \in \text{INITIAL}, v(t) \in \text{CONTINUATION}, t \in \text{dom}(v) = <t_{in}, t_f>\}$$

Now we come to realize that specification of a set $\Omega_C \subset (Z, T)$ is what a frame wants to do. Specification by means of INITIAL and TERMINAL sets is one, convenient, but limited way of doing this.

For example, suppose we wish to specify a new frame E_{empty} in which we can ask questions about the time it takes the BUS to go from any STATION to any other (including itself) provided no passengers ever board on route. Such a frame is simply specified by selecting #__IN__BUS as the run control variable and as admissible segments those in which #__IN__BUS remains equal to 0. That is,

$$\Omega_C = \{v | v(t) = 0 \text{ for all } t \in dom(v)\}$$

Suppose we want to restrict this frame further so that experiments start and end at the same STATION. Then we choose $C = \{\text{BUS.LOCATION}, \#\_\_\text{IN}\_\_\text{BUS}\}$. A typical segment in Ω_C,

$$v: <t_{in}, t_f> \to (\text{BUS.LOCATION}, \#\_\_\text{IN}\_\_\text{BUS}).\text{range}$$

would be specified as follows:
(i) $v(t) = (v_1(t), v_2(t))$
(i) $v_1(t_{in}) = v_1(t_f)$ (start and return at same station)
(ii) $v_2(t) = 0$, for all $t \in <t_{in}, t_f>$

Note that since the *conditions at initialization and termination are the same, we cannot use the INITIAL and CONTINUATION subsets to specify* Ω_C.

The appropriate I/O system for checking run control segments is the acceptor (in contrast to the case of input segment specification, it is not meaningful to generate these segments externally to the model or real system). Such an acceptor would allow an experiment (ω, ρ) into $\mathbf{D}_E$ only if it acccepted the run control trajectory generated during this experiment, v say, i.e. $v \in \Omega_C$.

In the case of Ω_C specified by INITIAL and CONTINUATION subsets of the control space Z, the acceptor can take the following special form. Consider such a device receiving $v \in (T,Z)$ as its input segment. In its initial state q_0, it checks whether the initial input value $v(0)$ is in INITIAL. If so, it immediately transits to state q_1 and stays there so long as the current input value $v(t)$ is in CONTINUATION. Both q_0 and q_1 are acceptance states. If the initial input is not in INITIAL the system transits immediately from q_0 to a dead state q_3, as it does likewise from q_1, if an input value in TERMINAL is received.

Exercise. Give a complete DEVS specification of the above acceptor. Hint: see Figure 4.3 and Table 4.1.

To implement the run control concept, an acceptor for a set of run control segments can be linked to a system or model so as to monitor its trajectory in run control space. For the case where the control segments are determined by initialization/termination conditions, the operation is as follows: so long as the trajectory continues to be an admissible one, the acceptor accepts it. As soon as the trajectory becomes unacceptable (i.e. moves into the TERMINAL subspace) the acceptor puts out a signal to terminate the experiment. In the general case, the acceptor may alternate between acceptance and non-acceptance decisions during an experiment. The I/O results of the experiment are accepted if at its end, the acceptor is in an acceptance state. A formal specification of this operation is given in Section 13.1.

Table 4.1 in Chapter 4 lists several acceptors that represent commonly employed run control segment specifications. For example, detection of equilibrium is commonly employed in stochastic simulations. Such a detector can take the form of a *reached_constant* type of acceptor (in practice, more sophisticated algorithms for equilibrium detection have been developed). In Chapter 13, we shall discuss some "modelling in the large" applications of the run control concept.

Exercise. Consider a variable V which is to be employed as a run control variable such that constancy of V's value is taken as equilibrium. Describe the segment sets $\Omega_{C,\text{until}}$ and $\Omega_{C,\text{while}}$ for the respective experimental frame conditions: *run until equilibrium reached* and *run while in equilibrium* and define corresponding acceptors. Hint: take the test of equilibrium to be the existence of n consecutive equal values, where n is an integer.

5. Realization in Experimental Frames

We can now summarize what has been said concerning the realization of experimental frames employing system specifications. Figure 12.3 depicts the situation. The experimental frame E is realized by a system S_E which is coupled to the model or system under study. S_E consists of generators, acceptors and transducers that realize the input segment, control segment, and summary mapping parts of the frame specification.

More formally, we say that a system S_E *realizes* an experimental frame E if it is a parallel composition of systems S_I, S_C, and S_O such that S_I is either a generator, or an acceptor, of Ω_I, S_C is an acceptor that accepts Ω_C and S_O is a parallel composition of transducers, each of which realizes a summary mapping in SU. Since S_E is a coupling of systems and each of the components systems may be realized itself as a coupling of systems, the result is a hierarchically specified system (Section 15.3).

The more common frame specifications may be realized by choosing finite memory systems (Section 4.5.1) as atomic components and restricting the coupling of these components to series/parallel schemes (feedback is rarely required beyond that realized in the finite memory components). Also, just as models are compactly specified employing modelling formalisms, so may

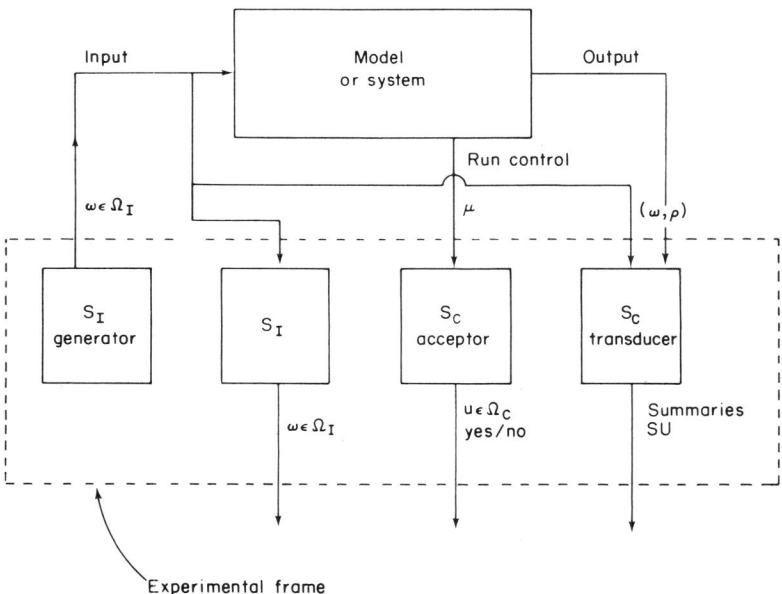

Fig. 12.3. Realization of experimental frames.

these formalisms be employed to specify the frame components. Indeed, the frame and model formalisms should agree for easy coupling.

Coupling of the frame realization S_E to the model requires that certain interface requirements be satisfied. Satisfaction of these requirements is embodied in the "applicability" relation between frames and models (Section 13.1). The most direct form of applicability requires the presence of the input, output and control variables specified by the frame as variables in the model. This form is sufficient for the present discussion.

There remains the translation of the frame system S_E into program coded form. We shall return to fully consider the construction of simulation software according to model/experimental frame separation principles in Chapters 16 and 18.

5.1. Summary Mappings and Transducers

We have by now discussed all the elements of experimental frame specification. We have seen that input, output and run control variables may be designated. In experimentation, input variables are those variables that will be treated as influencing the system under study; output variables are those capable of direct measurement and mediating the computation of variables of interest for the modelling objectives; and run control variables are those capable of direct measurement that serve to monitor the course of an experiment. Two sets of segments operationalize these roles: the admissible input segments determine the kind of stimulation of interest as input to the system; control segments determine the system trajectories that are of interest for the objectives of the frame. Note that there are no output segment specifications since these are determined by the system during experimentation. In the process of explaining these ideas, we also discussed the more concrete forms of input segment and control segment specification, viz, generators and acceptors. We also gave examples of the computations involved in computing interest variables from mediating variables (Section 12.3). When we move to the experimental frame stage, these computations are expressed as the summary mappings of the frame. The appropriate concrete form of specifying such mappings is the *transducer*. Recall that such an I/O system maps input segments into output segments (or final values in the case of a final value transducer). Several examples of useful transducers are presented in Table 4.1 of Chapter 4. We shall now give an example of the specification of summary maps by means of the transducer concept.

We shall develop finite memory transducers to realize the summary maps for experimental frames of the coin tossing model (Section 7.2.1). This is a simple illustration of the kinds of counting and summary processes that are typically involved in statistical computation.

The objective of the coin tossing model (Section 7.2.1) is to estimate the relative frequency of heads (and tails) outcomes as a function of the model parameter τ. Indeed, we expect that the relative frequency of heads should be τ, the coin bias.

We note that the termination of a toss occurs when the phase of the translation process (which we denote by the variable TRANSLATION.PHASE with range: {START,STOP}) enters STOP. Observing this variable provides the information necessary to count the number of tosses, N.TOSSES. We are interested in the phase variable of the spin process (SPIN.PHASE with range: {HEADS,TAILS}) only when the TRANSLATION.PHASE is STOP. Thus the composite variable (TRANSLATION.PHASE,SPIN.PHASE) provides the information necessary to count the number of tosses which turned out heads, N.HEADS.

Accordingly the experimental frame specifies as output variables: {TRANSLATION.PHASE, SPIN.PHASE}. It specifies as summary mappings the computation of N.TOSSES from TRANSLATION.PHASE and the computation of N.HEADS from (TRANSLATION.PHASE, SPIN.PHASE). Figure 12.4a shows the transducers which realize the latter computations.

A more inquisitive frame expresses an interest in estimating the transition probabilities: head to head, head to tail, tail to head, and tail to tail. The first and third provide all the needed information and are expected to equal τ. To obtain the corresponding relative frequencies requires that the result of the previous toss be saved for comparison with the current outcome. Accordingly, a transducer component is required which stores the last outcome: this is a finite memory DEVS with input order $n=1$ and output order $m=0$. The saved toss outcome is compared with the current one and the transitions: HEAD to HEAD, TAIL to HEAD are detected and counted (Figure 12.4b).

Exercise. Design transducers for the summary maps of the Passenger Satisfaction and Bus Operation frames of Sections 12.1 and 12.3.1 (see also Section 4.5.1).

6. Wymore's Tricotyledon Theory of System Design

We have emphasized the place of modelling and simulation within a broader context of management, control and design. In the case of the latter, a well formalized theory exists that helps us to make explicit links between modelling and design.

While the process of formulating objectives and expressing them as experimental frames is not well understood in general at the present time, it is possible to get a better insight into this process in the case of systems design.

Objectives-driven methodology 227

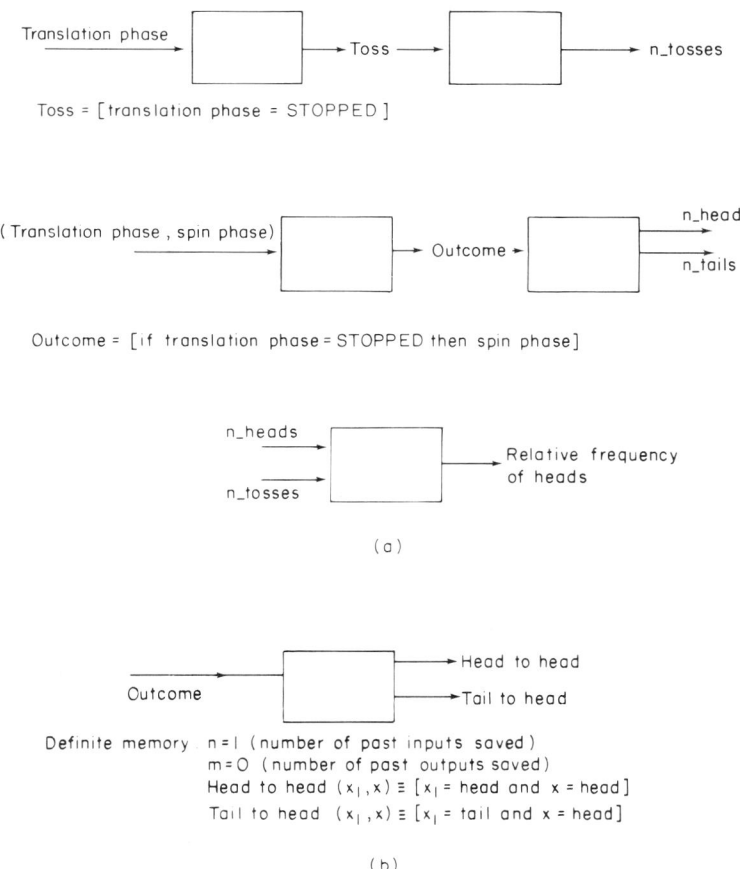

Fig. 12.4. Transducers for coin tossing frame.

In this case, there exists a formal theory that lays out the elements involved and that can be employed to understand modelling and simulation methodology as a component of systems design methodology. Links between the two methodologies are the subject of this chapter.

Figure 12.5 summarizes, in outline form, the inputs and outputs of Wymore's [1] theory of system design (also called the tricotyledon theory). Wymore conceives of a client with a problem consulting a systems design team. The end result of this consultation is comprised of the following elements: The general case is slightly more involved and is considered in Section 13.1.

228   *4. Multifacetted system modelling*

*An Input/Output specification of the desired end system. This is a specification, similar to the IORO (Section 3.1), of the input/output relationships that the system is expected to realize.

*A merit order based on the I/O Specification. This is a means of evaluating the performance of any system that satisfies the I/O specification. The general idea is that the I/O Specification presents the minimal constraints that any end system must satisfy but there is still great flexibility in how "good" or "bad" its performance is within these constraints. Practical means of evaluating performance involve the definition of performance indexes, sensitive to such aspects as how fast, how reliable, how error-free, etc., does the system carry out the desired tasks. Such separate performance indexes are "rolled up" into a single index that orders the systems satisfying the I/O Specification (the so-called I/O Cotyledon).

*A characterization of the class of buildable system designs. A buildable system design is a system that can be built by coupling of components that belong to the "technology" (to be discussed soon) together with the coupling recipe whose resultant it is (Section 3.2).

*A merit order on the class of buildable system designs (also called the Technology Cotyledon). This ranks buildable system designs according to utilization of resource criteria such as capital, operating and maintenance costs, complexity of structure, efficiency, impact on the environment, etc. The same "roll-up" concepts apply to compact the various resource utilization indexes into an overall figure of merit.

*A trade-off merit order on the set of feasible system designs (the Feasibility Cotyledon) which resolves any conflicts between the I/O and U/R (utilization of resource) merit orders. A feasible system is one that is buildable in the technology and implements a system that satisfies the I/O Specification. A feasible system design is the triple consisting of a feasible system, the system it realizes, and the implementation mapping (system morphism) linking the two. The trade-off concept acknowledges the existence of constraints which often force an improvement in one merit order to be attained only at the expense of a worsening of another.

*A test plan specifying a finite set of experiments to estimate the actual performance of a built system, the conformation of the built system with the feasible system design (model), and the acceptability of the built system as a solution to the problem posed by the client.

As shown in Figure 12.5, these outputs of the system design methodology can be interpreted ("interfaced") as inputs to the modelling and simulation methodology. Specifically, the I/O Specification, performance and resource utilization indexes are starting points for the construction of experimental frames that reflect the system design objectives. The models to which the frames are intended to apply are those expressed as the feasible system

Objectives-driven methodology

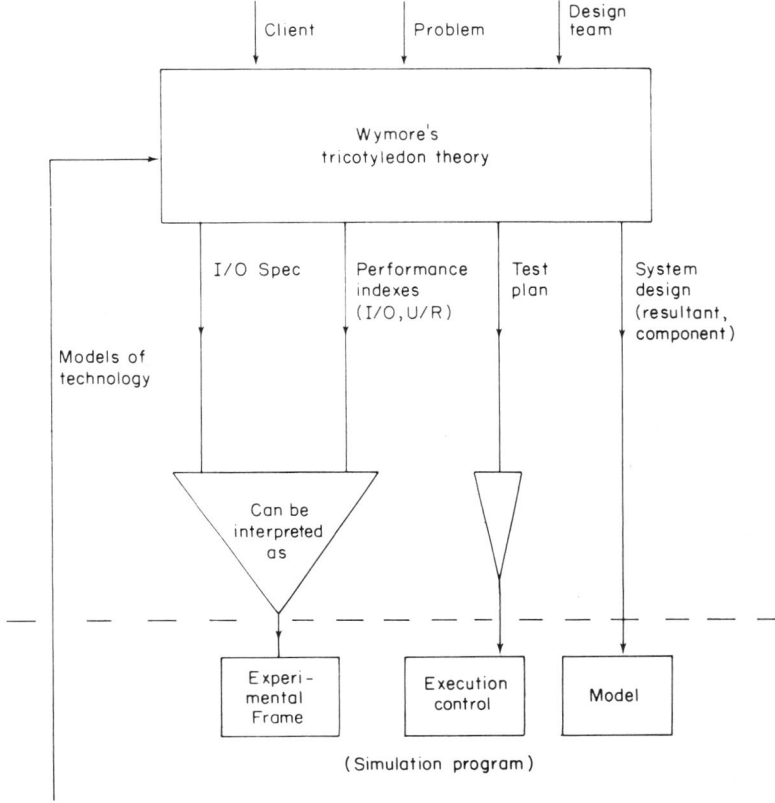

Fig. 12.5. Wymore's system design methodology interfaced with modelling methodology.

designs. Note however, that modelling methodology provides a feedback to the design methodology in the form of the class of models that represent the technology available to the system design team. Feasible designs represent potential solutions to the design problem that can be synthesized from this technology. The test plan dictates the form of finitary testing applicable to such feasible designs or their real system counterparts. It can be interpreted as specifying the third component of a simulation scheme — execution control. This component is responsible for selecting the finite set of experiments of interest to the user from the set of all those made possible by the application of a frame to a system (whether real or model).

6.1. Example: University Bus System Revisited

As an example, let us interpret the university bus system problem within Wymore's framework. The client is the university administration and the system design team is the committee that has been convened. The system design problem (viewed within its initial limited scope) is that of building a transportation system that will provide satisfactory service to the university students. Wymore's philosophy is to avoid narrowing the possible solutions to the given problem by prematurely narrowing the choice of available technology. However, the committee, feeling its latitude quite constrained by its mandate, has already narrowed its technology to the class of bus systems similar to that already in place. It has for example eliminated the possibility of using helicopters or conveyor belts to provide the transportation service. Of course, there remains extensive freedom of choice even within the chosen technology.

The modelling subtask of this committee is to provide valid models of the bus system within the boundaries defined in Section 1.1.2. This includes representing the input demand for service (the arrival rates of passengers at the stations, their types (student, non-student, etc.) and the operation of the buses *per se*. As we have seen, in objectives-driven methodology, the input demand (also called workload) gets formulated as the input variables and segments of experimental frames. The bus technology is then formulated as the class of models to which such inputs can be applied, including a model of the system in place, as well as possible replacements for it.

Recall that we have defined two kinds of objectives. The Passenger Satisfaction objective is formulated as Wymore's I/O Specification and performance merit order, i.e. to the "technology free" statement of the desired I/O behavior of the system to be designed. The I/O Specification requires that passengers arriving at one station eventually appear at the other (presumably having been transported there). The performance indexes include measures such as transit time and waiting time that we have defined. The experimental frames defined for the Passenger Satisfaction provide the framework for evaluating the performance of the system design as measured by the performance indexes.

The Bus Operation objective discussed earlier reflects considerations involved in Wymore's Technology Cotyledon. Specifically, such measures as utilization of bus capacity, round trip time, etc. are examples of the U/R indexes that are defined on systems buildable in the chosen technology. Experimental frames defined to realize the Bus Operation objective provide the framework for evaluating the existing system, feasible models, and the system finally built in terms of the given utilization of resource measures.

6.2. Cotyledons and their Experimental Frames

In general, each of the three cotyledons can be viewed as motivating the construction of corresponding experimental frames. The I/O Cotyledon[29] and Technology Cotyledons give rise to experimental frames that are the basis for performance evaluation and resource utilization assessment, respectively. The Feasibility Cotyledon gives rise to frames that, in addition to elements of the other frames, also contain trade-off algorithms for "rolling-up" performance and utilization of resource figures of merit.

In the bus system example, trade-off considerations will arise when the committee wish to evaluate system alternatives taking into account measures of both passenger satisfaction and bus system performance. They might define a *Trade-Off Experimental Frame* that combines the Passenger Satisfaction and Bus Operation frames and contains as well summary maps for computing an overall figure of merit. (See [1] or [2] for a complete discussion of how this can be done in a systematic fashion.)

In addition to frames for system evaluation, there are frames for validation. One such class arises from conformance testing — testing that the system as built conforms to the feasible system design that gave rise to it. This falls within the concept of validation as we have defined it (Section 5.4) since it involves comparison of model and real system data within the same experimental frame. However, conformance testing may be conducted from the standpoint of high confidence in the model so that the emphasis is on determining whether the fabrication of the real system was correctly executed.

Another class of validation frames arises from the requirement to have adequate models of available technological components. Validation of such models would be carried out in order to increase the confidence that the coupled system models of the Technology Cotyledon do in fact represent buildable system designs.

Problem 3 explores the use of experimental frames in system design.

---

[29]Interfacing of the I/O Cotyledon and experimental frames is achieved as follows: Wymore's I/O Specification can be regarded as specifying an observation frame (Section 3.1), a set of input segments, and an I/O relation. The observation frame and the input segments are directly accommodated into the experimental frame. The I/O relation has no corresponding frame element as the latter has no mechanism for restricting input/output pairs. However, the intention of the frame concept is to set up the framework in which pairs of systems can be compared. In this case, the performance indexes (incorporated in the SU components of the frame) and the test plan (incorporated in the execution control) specify the modelling and simulation elements for comparing feasible system designs and real world implementations.

7. Summary

An objectives-driven methodology was presented in which modelling objectives are translated in step-by-step fashion into experimental frames. This leads to an indepth consideration of the experimental frame concept and its realizability in simulation schemes. The principles of model/experimental frame separation were stated to guide the modular construction of frames and models.

An experimental frame represents a set of questions about a system in the sense that it sets up the data space in which such questions may be answered. Different objectives in dealing with a system may thus require the formulation of different experimental frames. In addition to providing input/output specifications, a frame may also specify run control constraints which restrict the conditions under which I/O segments may be obtained.

Formulation of the experimental frame concept was made "neutral", between real systems and models so as to be interpretable to both real world and model data acquisition. Such acquisition may take the form of direct experimentation or may be made via reduction of data acquired in other more inclusive frames.

Wymore's theory of systems design was employed to show how the objectives-driven modelling methodology constitutes an important component within system design methodology.

PROBLEMS

1. Specify experimental frames for the Elzas Negotiation Methodology based on Figure 1.6 and the discussion in Section 1.3.

2. In general, control segment specification by means of initialization and termination conditions does not suffice to express all cases of interest. However, there is a special case where additional control variables can be added so that the control segment specification in the expanded space is equivalent to that of the original but is given in terms of initialization and termination conditions. We say that $\Omega_C \subset (Z, T)$ is *terminating* if once a segment has been rejected, no continuation of it can be accepted. More formally, for all $\omega \in \Omega_C$, $t \in dom(\omega)$,

$\quad$ if $\omega_{t>}$ is not in Ω_C then ω is not in Ω_C

2.2. Show that if Ω_C is specified in the form of initialization/termination conditions, then it is terminating.

Suppose that Ω_C is terminating. Show how to augment the control variables so that the admissible segments in the new space are specified by INITIAL and TERMINAL conditions and an equivalent effect in terms of controlling experimentation is obtained. Hint: Let A be an acceptor the segment set Ω_C. Show that $v \in \text{DEVS}(Z)$ is accepted if, and only if, the associated state trajectory consists only of accepting states. Add to the original control variables C a variable whose range is the state set of A and incorporate the acceptor A as part of the model. Show that in this control space the admissible control segments are specified in terms of initialization and termination conditions, and the same effect on model experimentation will be achieved as in the original control space.

Note however that the incorporation of the acceptor computation strongly violates the separation of model and experimental frame principle if there is no corresponding mechanism in the real system to do this computation.

3. The designers of the RNC (Revolutionary New Computer) have engaged your services to construct a modelling and simulation system which will enable them to investigate new architectures for parallel processing. They want to design an arithmetic unit for the RNC which will consist of a number of adders, subtracters, and routers. Instructions in the form "OP,x,y" where OP could be "add","subtract","multiply", and "divide" arrive to the unit. Addition and subtraction are handled directly, multiplication and division require repeated addition and subtraction. For the latter, after each add or subtract operation, the instruction is sent to a router which decides whether it is complete and should leave the system or not yet complete in which case it is sent to an adder or subtracter for further processing. The designers will be playing with different numbers of components and interconnection patterns. However, the components will always have no capability for queueing of instructions. Because of this finite capacity, instructions may be rejected (in some designs provision will be made to recycle rejected instructions to the RNC control unit, in others they will just be lost).

The RNC design team want you to provide a simulation system which will enable them to specify experimental frames with the following output portions:
(a) Processing Time: time taken for an instruction to be completed (transit time) — average, maximum, median per instruction type and overall.
(b) Utilization of components: number of components which are busy — average, maximum per component type and overall.

For the input part of a frame, the team want to be able to choose between two possibilities.
(a) *Workload*: rate of arrival of instructions, operation mix (% adds, % subtracts, % multiply, % divides), and interarrival distribution.

(b) *Specific Instructions*: specify an observation interval, a finite sequence of instants $t_1,...,t_n$ and corresponding instructions $I_1,...,I_n$ which arrive at these instants.

For the control part of a frame the team wish to examine the behavior of a proposed architecture under the following conditions:
(a) *start up*: start from an empty system, run until a given number (e.g. 1000) of instructions have been completed (not counting rejections).
(b) *controlled rejection*: run so long as the rate of rejection is kept below a given level (e.g. run so long as no instructions are rejected).
(c) *correctness*: run so long as every instruction which is not rejected is correctly computed (e.g. the answer given to ADD,x,y is $x+y$).

A particular frame will be specified by a user by selecting one of the input possibilities, one or both of the output possibilities, and any combination of the control possibilities. Then he will prescribe the parameter values required to completely determine the frame.

3.1. Define the input variables and input segments for each of the input possibilities.

3.2. Define output variables and the summary indexes required in order to realize each of the output possibilities. Adhering to the principles of model/experimental frame separation (Section 12.3.2) construct a set of mediating variables that is as small as possible yet sufficient to provide the time segments from which the index variables and statistics can be computed.

3.3. Define the control variables and control segments (in the form of the conditions they must satisfy) for each of the control possibilities.

3.4. For each of the above components, identify whether it is associated with performance evaluation or utilization of resources assessment.

4. Assume that you will write a SIMSCRIPT program which will represent the instructions as temporary entities and the components as permanent entities with associated processes, i.e. adders, subtracters and routers are each a permanent entity class. Do *not* write such a program but *do* show quite explicitly how the above set of experimental frames can be implemented as a segment of the program distinct (except for the preamble) from that of the model of the arithmetic unit. Use monitored variables and note that attributes of entities may be monitored (e.g. LEFT ROUTINE $A(I)$ monitors the attribute A and accepts the index I of the entity instance whose A is currently changing). You should assume that the only variables available for monitoring are those required to implement the model. Since the model

definition is not available, you should therefore minimize the number of entity attributes and other variables that are monitored (if a computation can be done in the experimental frame section rather than in the model that is where it should go) (see Section 16.1.1).

References

[1] Wymore, W. (1980). *A Mathematical Theory of Systems Design*, Tech. Rept, College of Engineering, University of Arizona, Tucson.
[2] Wymore, A. W. (1967). *A Mathematical Theory of Systems Engineering: The Elements*, Wiley, NY.

Other Relevant Literature
The notion of experimental frame has evolved since its initial conception. The following is a trace of this evolution:
Zeigler, B. P. (1976). *Theory of Modelling and Simulation*, Wiley, NY. Experimental frames defined in terms of a fixed base model representing a real system.
Oren, T. I. and B. P. Zeigler (1979). "Concepts for Advanced Simulation Methodologies". *Simulation* 32, 69–82. Frame concept is unlinked from base model so as to be applicable in principle to arbitrary model and real system experimentation.
Zeigler, B. P. (1979). "Structuring Principles for Multifacetted System Modelling". in *Methodology in Systems Modelling and Simulation* (eds, B. P. Zeigler *et al.*), North Holland. First formal definition of experimental frame and relationships it is involved in (see Chapter 13). Present definition elaborates, but basically does not deviate, from this definition. Elaboration is in treatment of run control specification.

Chapter 13

EXPERIMENTAL FRAMES: OPERATIONAL CONCEPTS

The previous chapter defined the experimental frame concept and gave examples of its use in the context of objectives-driven methodology for model development. This chapter further extends the theory of experimental frames in preparation for their use in a full "modelling in the large" methodology.

The discussion begins with some technical developments necessary for rigorous operation with experimental frames: the derivability relation for variables and frames, the applicability relation between experimental frames and models, and the input-output data generated by a model within a frame applicable to it.

Employing these concepts, we proceed to deepen our understanding of the role that experimental frames play in stating relations we seek to establish between two systems. Two cases in point are discussed. In *component isolation* one attempts to identify a component of a real system by experimental means. In this case the experimental frame concerns the relation between the system and its component. In *model simplification*, a relatively complex model is to be replaced by a relatively simple one. Here, the frame concerns the relation between the two models. In either case, we see how the frame concept emerges as a means of relating pairs of models. This is a step toward "modelling in the large" which concerns the manipulation of a collection of models.

1. Derivability Concepts: Motivation and Approach

As we have seen, the experimental frame concept is a mechanism for formulating modelling objectives in manner that constrains experimentation within a relative small scope. The objectives-driven methodology results in experimental frame specifications that are conceptually independent from models which generate input-output data. However, this independence creates the problem of how to tell when a model is suitable to generate the data requested in an experimental frame. Such a problem is non-trivial if we consider that model and experimental frame may have been separately retrieved from model and frame bases.

Thus one essential task required to operationalize the multifacetted methodology is characterization of the applicability relation between frames and models. But there is yet a more basic relation called *derivability* on experimental frames that must be dealt with first. Derivability is a relation that partially orders frames according to the extent of experimentation they embody. Although such a measure of scope is interesting in itself, it turns out that the applicability relation is readily defined in terms of it.

Figure 13.1 depicts the situation: $\mathscr{E}$ is a set of experimental frames and $\mathscr{M}$, a set of models; *derivability* is a binary relation on $\mathscr{E}$ and *applicability*, a relation on $\mathscr{E} \times \mathscr{M}$. We shall show how to associate a *scope frame* E_M with each model $M \in \mathscr{M}$. The scope frame represents the full range of experimentation that can be done on the model, more informally, the widest set of questions that it is intended to address. Then we shall define a frame E to be applicable to a model M if E is derivable from its scope frame E_M. Intuitively, E is applicable to M if the experimentation that it embodies is somehow included within the full range of experimentation possible with M.

More formally, we shall define E' is *derivable from* E (written $E' \leq E$) in such a way that the I/O data space of E' is derivable from that of E employing reduction procedures that are specified by the pair (E,E'). The fact that E is applicable to M will then mean that the data that M generates can be reduced to data lying within the I/O space of E; in other words, that the experimentation embodied in E can be performed on M.

Thus our first order of business is to formalize the derivability relation. This turns out to be surprisingly difficult. The reader may wish to skip this development on first reading. Our characterization of the reduction procedures underlying the derivability concept will be based on a more fundamental concept of derivability, that applying to variables.

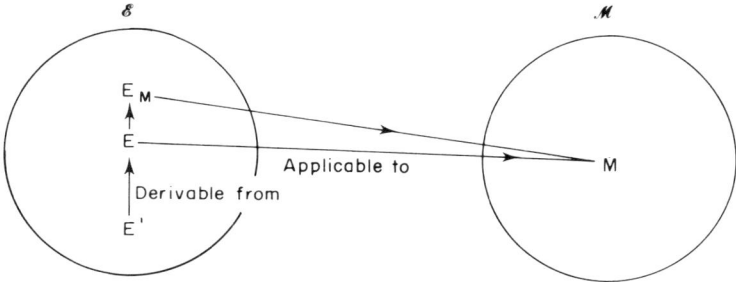

Fig. 13.1. Applicability and derivability relations.

1.1. The Semantics Structure of Variables and Derivability

Imagine the set of all variables attached to the system entity structure at some time. As we have indicated in Section 11.3 these variables are related through "semantic" relations that restrict the values that can be simultaneously taken on by pairs of variables. We refer to the set of all these variables together with the semantic relations as the *semantic structure* associated with the entity structure.

Let $\mathscr{V}$ be the set of variables and $\mathscr{R}$ be the set of semantic relations in the semantic structure. Then with each relation $R \in \mathscr{R}$ are associated sets of variables $V_1,...,V_n$ and $W_1,...,W_m$ such that $R \subset (V_1,...,V_n)$.range $\times (W_1,...,W_m)$.range. Now we close $\mathscr{V}$ and $\mathscr{R}$ simultaneously under all of the following:

composite variables:
  If $V_1,...,V_n$ all belong to $\mathscr{V}$ then so does $(V_1,...,V_n)$

identity relations:
  If $V \in \mathscr{V}$ then the identity relation $i_V \in \mathscr{R}$, where $i_V = \{(v,v) | v \in V.\text{range}\}$

extension to composites
  Let $V_1,...,V_n$ and $W_1,...,W_m$ be subsets of $\mathscr{V}$ and let $R_1,...,R_p$ be the set of all relations $R_i \in \mathscr{R}$, such that R_i holds between subsets $V_{i,1},...,V_{i,v}$ and $W_{i,1},...,W_{i,w}$ of the variables $V_1,...,V_n$ and $W_1,...,W_m$, respectively. Then the extension $(R_1,...,R_p) \in \mathscr{R}$ where $(R_1,...,R_p)$ holds between $(V_1,...,V_n)$ and $(W_1,...,W_m)$ such that $((v_1,...,v_n),(w_1,...,w_m)) \in (R_1,...,R_p)$ just in case $((v_{i,1},...,v_{1,v}),(w_{i,1},...,w_{i,w})) \in R_i$, for each i, $1 \leq i \leq p$. In other words, the extension is a relation which holds between the composites of two sets of variables in such a way that it takes account of all of the relations holding between subsets of these variables.

composition of relations
  If R holds between U and V, and S holds between V and W, where $R, S \in \mathscr{R}$ and $U, V, W \in \mathscr{V}$ are all distinct, then $RS \in \mathscr{R}$ (where RS is the composition of R and S, i.e. $(u,w) \in RS$, if and only if, $(u,v) \in R$ and $(v,w) \in S$).

We are particularly interested in the case where the constraints expressed in the semantic structure result in functional relations between variables. We say that V is *derivable* from W, where V and W are variables in $\mathscr{V}$, if a functional relation $f \in \mathscr{R}$ holds between W and V, such that it maps W.range onto V.range. (All mentions of $\mathscr{V}$ and $\mathscr{R}$, here and in the sequel refer to their closures under the above operations.) In particular, every variable is derivable from itself. Also a finite composition sequence of onto functional semantic relations results in an onto functional semantic relation, hence gives rise to a derivability relation. Derivability is obviously a transitive relation.

Extending the derivability relation to sets of variables, we say that $V_1,...,V_n$ is *derivable from* $W_1,...,W_m$ if each V_i, $1 \le i \le n$, is derivable from $W_1,...,W_m$.

For example, in Table 11.1, V_AVG is derivable from NUMBER and V_SUM, each of which is derivable from the composite $\{I(\ ).V\}$, so V_AVG is derivable from $\{I(\ ).V\}$.

Exercise. Let V and W be real variables belonging to a entity I such that W is derivable from V. Taking the variables as those in Table 11.1 for both V and W, display the closure of the semantic structure under the above operations. For example, show that $\{I(\ ).W\}$ is derivable from $\{I(\ ).V\}$, and W_DISTRIBUTION is derivable from V_DISTRIBUTION but W_SUM is not derivable from V_SUM.

1.2. The Experimental Frame Derivability Relation

As indicated the concept of experimental frame derivability plays a central role in advanced "modelling the large" activities of multifacetted modelling methodology. As a preamble to the formal characterization of this concept, let us note the informal criteria that must be accounted for. We want our definition of the relation E' is *derivable from* E for frames E and E' to satisfy:
(1) There are procedures defined by the pair (E,E') for mapping the data space $\mathbf{D}_E$ into the data space $\mathbf{D}_{E'}$.
(2) Any restrictions on data acquisition specified by E' are over and above those specified by E.
(3) There are no restrictions on data acquisition in E that are not implied by those in E'.

Requirement (2) means roughly that any experiment that can be performed in E' should have some corresponding experiment that can be performed in E. Requirement (3) says moreover, that there should not be any bias placed by E on the results of such a corresponding experiment other than that implied by E'. Thus (2) requires that E contain all the potential data for E', while (3) requires that this data not be biased in a manner not predictable from the knowledge of E'.

These requirements will become clearer as we proceed to express them in more precise terms.

1.3. Induced Segment-to-Segment Mapping

We begin consideration of derivability with the first requirement above, viz, that the pair (E,E') specify a unique mapping from the I/O data space of E into that of E'. Such a map takes a pair $(\omega,\rho) \in \mathscr{D}_E$ into a pair $(\omega',\rho') \in \mathscr{D}_{E'}$. To be completely general we could allow an arbitrary assignment of a transducer to each pair (E,E') to perform the mapping. Instead, we shall define a special form of mapping than can be determined from the semantic structure.

For derivability of E' from E we shall require, among other conditions, that the input and output variables of E' are derivable (in the semantic structure) respectively from the corresponding variables in E. Let the input and output sets of the frames be X, Y and X', Y' respectively. The derivability condition just stated implies the existence of onto mappings $f: X \to X$ and $g: Y \to Y'$. Let $\bar{f}$ and $\bar{g}$ be the *pointwise* extensions of these mappings to segments. For example, $\bar{f}: (X,T) \to (X',T)$ according to the definition:

$$\bar{f}(\omega)(t) = f(\omega(t))$$

for each $t \in dom(\omega)$. We call $\bar{f}$ the *induced segment-to-segment mapping* arising from the derivability of input variables of E' from those of E. Similarly, the derivability of output variables induces the segment-to-segment mapping of output segments, $\bar{g}$.

We associate the pair of induced mapping $(\bar{f},\bar{g})$ with the pair (E,E'). This pair is the basis for the mapping of the I/O data space of E into that of E' that we desire. To fully formulate this mapping however, first requires us to characterize the selection processes set up by the input and run control segment specification portions of experimental frames.

In the sequel, we will not distinguish between a map f and the segment-to-segment mapping, $\bar{f}$ that it induces. We shall also restrict our discussion to frames having the same time base T for convenience. Our treatment can be readily extended to derivability of frames over different time bases.

1.4. Principle of Derivability

Let A and B be sets related by a function $f: A \to B$. Let $\mathscr{A}$ and $\mathscr{B}$ be subsets of A and B, respectively representing admissible subsets. We wish to characterize the two conditions:

(*) The restrictions on B are over and above those on A.
(**) There are no restrictions on A that are not implied by those on B.

The characterizations are illustrated in Figure 13.2 and formalized as follows:

Characterization of (*)

The admissible set $\mathscr{A}$ as mapped into B by f is $f(\mathscr{A})$. Since every element in $\mathscr{B}$ must be derivable from some element in $\mathscr{A}$, we must have

$$\mathscr{B} \subset f(\mathscr{A})$$

or in other words, f maps $\mathscr{A}$ onto $\mathscr{B}$. Thus the restrictions on B namely, $\mathscr{B}$ are over and above those on A namely, $f(\mathscr{A})$.

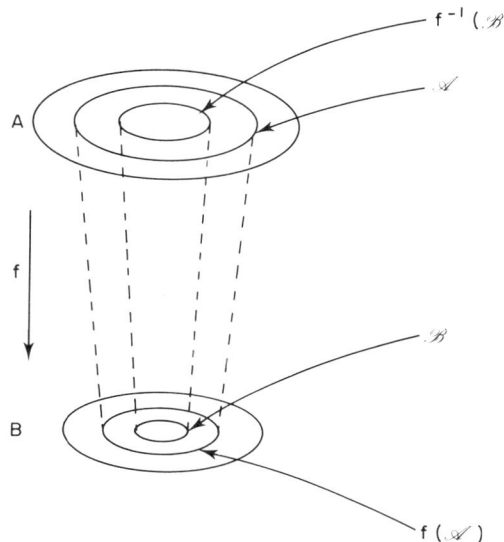

Fig. 13.2. The principle of derivability.

Characterization of (**)

Let $b \in \mathscr{B}$. Then by the characterization of (*), there is an element $a \in \mathscr{A}$ such that $f(a) = b$. This element cannot be a special element in the equivalence class of elements of A mapping to b since there can be no restrictions on A that are not implied by those in B. Thus we arrive at the condition:

$$\text{for each } b \in \mathscr{B}, f^{-1}(b) \subset \mathscr{A}$$

i.e. $\mathscr{A}$ includes all the inverse images of the elements of $\mathscr{B}$.

Condition (**) can be expressed in short as $f^{-1}(\mathscr{B}) \subset \mathscr{A}$. Since it implies condition (*), the two conditions are satisfied if (**) is satisfied.

Example. Let $A = \{0,1\}^2$, $B = \{0,1,2\}$, and $f: A \to B$ be the aggregation $f(a_1, a_2) = a_1 + a_2$. Consider the restriction on B that its elements be less than 2, i.e. $\mathscr{B} = \{0,1\}$. Then $f^{-1}(0) = \{0,0\}$ and $f^{-1}(1) = \{(1,0), (0,1)\}$. So if $\mathscr{A} = \{(0,0), (1,0), (0,1)\}$ then both starred conditions are satisfied. However, if we remove $(1,0)$ from $\mathscr{A}$, then condition (**) is not satisfied because both $(1,0)$ and $(0,1)$ must be allowed if there are to be no further restrictions on A than implied by those on B. Moreover, if we remove $(0,1)$ as well, then $\mathscr{A}$ does not satisfy condition (*) either, since then $f(\mathscr{A}) = \{0\}$ and the restrictions on B ($\mathscr{B} = \{0,1\}$) are more relaxed than those induced by $\mathscr{A}$.

1.5. Application to Input Segment Derivability

We now apply the principle of derivability to the problem of characterizing when the set of admissible input segments of one frame is derivable from that of another.

Let I and I' be sets of input variables such that I' is derivable from I in the semantic structure. Let $f: (X,T) \to (X',T)$ be the induced segment-to-segment mapping between the corresponding segment spaces. Then the input segment set $\Omega_{I'}$ is *derivable* from Ω_I under the conditions:

(*) $\Omega'_I \subset f(\Omega_I)$

and

(**) for all $\omega \in \Omega'_I, f^{-1}(\omega) \subset \Omega_I$

Example. Let X be a set of customer identifiers, let $X' = \{1\}$, and let arrival: $X^\phi \to X'^\phi$ be such that arrival$(x) = 1$, i.e. the processing performed is to extract only the fact of arrival or non-arrival of customers. Let f be the point-wise extension of arrival to the mapping of DEVS(X) to DEVS(X'), so that f maps a segment of arrivals onto a similar segment in which all customer identifiers have been replaced by ones (1s).

Let $\Omega_{I'}$ be the set of all segments with equi-spaced arrivals. Then condition (*) requires that Ω_I include the set of equi-spaced arrivals, possibly among others. The equivalence class of segments that map to a given segment $\omega \in \Omega_{I'}$ is the set of segments in DEVS(X) that have the same timing of arrivals. The conditions (**) requires that any assignment of customer identifiers to such an arrival sequence results in a segment that is in Ω_I. For example, if Ω_I were to specify that only customers having small processing times[30] are allowed to arrive, then this constraint would violate (*), i.e. would represent a restriction in E that is not implied by any restriction in E'. Note that such a bias on the type of customers that can arrive might lead to misleading results if the data gathered in E were employed instead of data gathered in E'.

Note that if I' is a subset of I, then I' is derivable from I in the semantic structure (see Problem 2). In particular, I' may be empty in which case Ω'_I is vacuously derivable from Ω_I. Also if $I' = I$, then Ω'_I is derivable from Ω_I, just in case, Ω'_I is a subset of Ω_I.

1.6. Application to Run Control Segments Derivability

The case of run control segments is similar to that of input segments with respect to derivability. The main difference arises from the fact that control variable trajectories do not appear in the I/O data available in a frame. We

---

[30] This presumes that the set of identifiers X, has a component for processing time specification.

shall see that this means that a notion of equivalence, stronger than derivability, must be applied in this case.

We say that sets of segments Ω and Ω' are *equivalent* under the processing map $f: (X,T) \to (X',T)$ if the condition (**) holds in stronger form:
$$(***)\ f^{-1}(\Omega') = \Omega$$
This means the restrictions placed on Ω and Ω' are exactly the same when translated through the mapping f. To see this, note that (***) directly implies the weaker condition (**) which states that there are no restrictions embodied in Ω that are not implied by those in Ω'. But (***) also implies the strong form of (*) which states that there are no restrictions embodied in Ω' that are over and above those of Ω. This is shown in the following exercise:

Exercise. Show that (***) implies the strong form of (*):
$$f(\Omega) = \Omega'$$

1.7. Derivability of Experimental Frames

With the principle of derivability in hand we are able to formally specify the derivability relation for experimental frames. Let $E = \,<T,I,O,C,\Omega_I,\Omega_C>$ and $E' = \,<T,I',O',C',\Omega'_I,\Omega'_C>$ be experimental frames. We omit specification of the summary mappings since these operate on the data space defined by the other components and therefore redundant if the latter raw data is retained as we are presuming. However, the formalization to be given can be readily extended to include the case where only processed data are retained after experimentation. (The consequent derivability relation is much weaker in this case.)

We define $E' \leq E$ (E' is derivable from E) if:
(a) The input variables I' are derivable from the input variables I.
(b) The input segments of E' are derivable from those of E, i.e. conditions (*) and (**) are satisfied for Ω_I and Ω'_I with respect to the induced segment mapping set up by the input variable derivability.
(c) The output variables O' are derivable from the output variables O.
(d) The run control variables C' consist of two classes C'_1 and C'_2. The control variables in C'_1 are derivable from the control variables C. The control variables in C'_2 are derivable from the input and output variables I and O.
(e) The projection of the set Ω'_C on the segment space over C'_1, call it $\Omega_{C'_1}$, is equivalent to Ω_C with respect to the induced segment mapping set up by the derivability of C'_1 from C. Also, the projection of the set Ω'_C on the segment space over C'_2, call it $\Omega_{C'_2}$, is derivable from $\mathbf{D}_E$ with respect to the induced segment mapping set up by the derivability of C'_2 from I and O.

The list of conditions, while long, is necessary and sufficient, to capture the criteria given in Section 13.1 for derivability. The conditions (a), (c) and (d) requiring derivability of variables set up the procedures for mapping the data space of E to that of E' (criterion 1). Conditions (b) and (e) formalize the requirement that any restrictions on data acquisition in E' be over and above those in E (criterion 2) and that there be no restrictions in E that are not implied by those in E' (criterion 3).

To explain the conditions relating to the run control variables in more detail: Condition (d) requires that for E' to be derivable from E the control variables of E' are derivable from the variables of E. Moreover, a distinction is made between those derivable from the control variables of E and those derivable from the input or output variables of E.

Consider first a control variable in C'_1. It is derivable from the control variables C but not from the I/O variables of E. Since it is not derivable from the latter variables there is no way to check whether a restriction placed on trajectories in C'_1 is satisfied by examining the I/O data available in E. Such a restriction must therefore already have been governing the data acquisition in E. In other words, there are no additional restrictions on variables in C'_1 over and above those in C. Thus the sets of admissible segments in C'_1 and C must be equivalent, as specified in condition(e).

Now consider a control variable in C'_2. It is derivable from the I/O variables of E. Whether the restrictions imposed on such a variable are satisfied in an experiment can be ascertained by examining the I/O data in $\mathbf{D}_E$. Such restrictions can therefore be over and above those imposed by E. Thus, condition (*) should hold between $\Omega_{C'_2}$ and $\mathbf{D}_E$. By definition, $\mathbf{D}_E$ is constrained only by Ω_I, recall Section 12.2. condition (**) should also hold so as to obviate bias that might be placed on input segments in E that is not implied by the control segment specification of E'. We conclude that $\Omega_{C'_2}$ must be derivable from $\mathbf{D}_E$ as required by condition(e).

Example. Consider an experimental frame E' for a multi-server queueing network in which ARRIVAL and DEPARTURE of customers are the input and output variables, respectively. The total number of customers in the network, represented as the sum of all queue lengths, TOTAL_QUEUE is the control variable. The control condition is that this sum remain less than some upper bound, U. Consider a finer frame E that has the same input and output variables and has individual QUEUE.LENGTHs as control variables.

E' is derivable from E if the control condition in E is precisely that the sum of QUEUE.LENGTHs is less than U. E' is not derivable from E if any bound other than U is placed on the sum of QUEUE.LENGTHs since then condition (***) requiring the two sets of control segments to be exactly the same in the primed space is violated. Also E' is not derivable from E if a condition is placed on the individual QUEUE.LENGTHs such as that each must be bounded by an upper bound U/N where N is the number of queues. Such a constraint violates condition (**) since it represents a restriction in E not implied by that in E'.

However, control variables derivable from ARRIVAL and/or DEPARTURE may be added to E' to further restrict its data space. E' will be derivable from E provided that any constraints placed on these additional control variables are consistent with constraints on the ARRIVAL segments.

Exercise. Suppose that E' has a control variables ARRIVAL' and DEPARTURE' that are derivable from ARRIVAL and DEPARTURE using the identity erasing mapping of the previous example. Moreover, E' places the restriction on segments in the (ARRIVAL',DEPARTURE') space that the arrival rate must be equal to the departure rate. This is a requirement for input/output equilibrium. Are there any requirements placed on the input segments of E so that condition(e) for derivability will hold? What about the case where both rates are required to equal some specific value?

1.8. Frame-to-Frame Processing

As indicated earlier, the derivability relation on frames sets up an associated mapping on the corresponding I/O data spaces. For frames E' and E with E' derivable from E, let (f,g) be the pair of induced segment-to-segment mappings on the input and output segment spaces respectively. Then any I/O segment pair, $(\omega,\rho) \in \mathbf{D}_E$ is mapped to the pair $(f(\omega),g(\rho))$ in $(X',T) \times (Y',T)$. In the case that E' has no control variables, then $(f(\omega),g(\rho))$ belongs to $\mathbf{D}'_E$ just in case $f(\omega) \in \Omega_{I'}$ (Figure 13.3).

If E' has control variables, let h be the segment-to-segment mapping induced by the derivability of the run control variables C'_2 from the input and output variables of E. Then h maps the pair $(\omega,\rho) \in \mathbf{D}_E$ to a trajectory v in the run control space C'_2. The pair $(f(\omega),g(\rho))$ is mapped into $\mathbf{D}_{E'}$ if, and only if, $f(\omega) \in \Omega'_I$ and $h(\omega,\rho) \in \Omega_{C_{2'}}$. In other words, the I/O segment pair that results from the segment-to-segment mapping is allowed into the I/O data space of E' when, and only when, the mapped input segment is an admissible input segment of E', and the associated run control segment is admissible in the control space of E' derivable from the I/O variables of E.

The mapping just described and illustrated in Figure 13.3 is called the E-to-E' *processing*, and fulfils the criteria laid down originally in Section 1.

Exercise. Show that derivability is transitive, i.e. if $E'' \leq E'$, and $E' \leq E$, then $E'' \leq E$. Also show that the composition of the E-to-E' processing and the E'-to-E'' processing give the same results as the direct E-to-E'' processing.

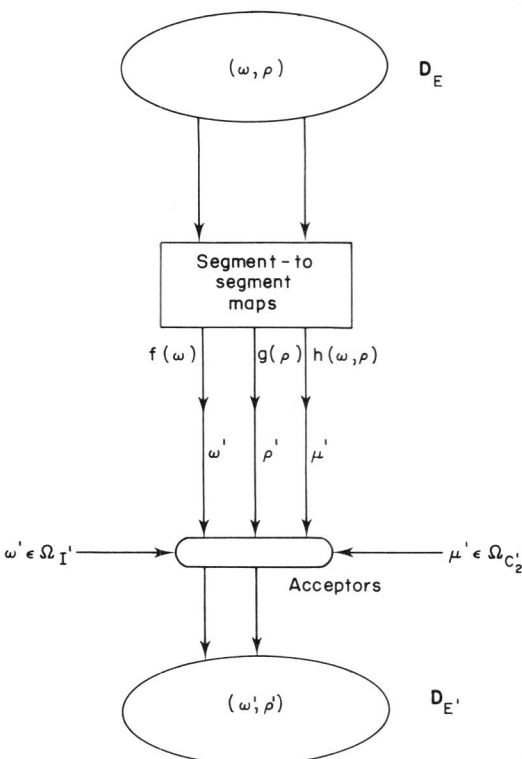

Fig. 13.3. Frame-to-frame processing.

1.9. The Scope Frame of a Model

Let M be a discrete event model whose input and output sets are structured by input and output variables I_M and O_M, respectively. We associate an experimental frame with M as follows:

$$E_M = <T,I,O,C,\Omega_I,\Omega_C>$$

where

T is the reals (the time base of a DEVS)
$I = I_M$
$O = O_M$
$C =$ the empty set
$\Omega_I = \text{DEVS}(I_M)$
$\Omega_C =$ empty set

We call E_M, the *scope frame* of M. Essentially, the scope designates the input and output variables of M as its input and output variables. Otherwise, it places only the standard constraints on its admissible input segments, and no run control restrictions. The scope frame embodies the full freedom to experiment with the model M consistent with its being a discrete event model. We shall discuss augmentation of the control specification C and Ω_C in the context of model simplification in Section 3.4.

As a bonafide experimental frame, E_M has an associated I/O data space $\mathbf{D}_{E_M}$. The input/output pairs that M generates within this space are just those in the I/O relation associated with the system that M specifies, which we call R_M. Thus R_M is a particular subset of the data space associated with its scope frame. Another model might have the same input and output variables as M, hence the same scope frame, but might generate a different subset of the scope data space as its I/O relation.

1.10. Applicability of Experimental Frames to Models

We can now define the applicability relation between experimental frames and models. We say that E is *applicable* to M if E is derivable from the scope frame of M. In symbols,

$$E \text{ is applicable to } M \equiv E \leq E_M$$

It is convenient also to employ the term "accommodate" for the converse of the applicability relation. Thus M *accommodates* E means E applies to M.

Exercise. Show that:

$$E' \leq E \text{ and } E \text{ is applicable to } M \text{ implies } E' \text{ is applicable to } M$$

The I/O data generated by M in a frame E applicable to it also falls out of the applicability definition:

$$R_M/E = E_M\text{-to-}E \text{ Processing}(R_M)$$

i.e. the input/output pairs observed within frame E are those that result from applying the E_M-to-E Processing to the I/O pairs that the model generates in its scope frame. Note that R_M/E is defined only for a frame E that is applicable to M since we require the existence of the E_M-to-E Processing that is guaranteed by the applicability relation.

Note that the definition of R_M/E provides an operational means of obtaining the data generated by model M in frame E. In general, this requires first generating I/O segment pairs in E_M and then subjecting them to the acceptance processes defined by the input and run control segment specifications. Only in special cases, is it possible to convert this off-line form of data acquisition into more conventional on line-form (see Chapter 16).

Problem 2 explores applicability in the special case of autonomous (input free, Section 4.5) models.

2. Experimental Frames for Isolation of Components

Having developed the formal concepts of derivability and applicability we turn to applications of these concepts: 1) isolation of component via observation of I/O behavior, 2) formalization of modelling assumptions, and 3) criteria for valid simplification of models. In this section we deal with the first of these applications.

An experimental frame specifies a restricted mode of data acquisition and reduction. The data in question derive either from a real system or from a model. The frame may limit access to the data by selecting observational modes relevant to a behavior of interest. It may also control the acquisition of data by passing only that data meeting specified tests. We shall refer to the first role of a frame as its *observation function* and to the second role as its *control function*. The observation function is specified in the choice of output variables and summary processing. The control function relates to the input and run control variables, and the admissible segments specified for them.

Often one attempts to isolate a component of a real system from other components, i.e. one would like to observe the input/output relation (Section 3.1) of the component untainted as far as possible by the rest of the system. A problem arises when, as is usually the case, the accessible measurements contain few or none of the input and output variables associated with the component. Thus for any choice of observation frame (Section 3.1) the data collected in this frame (IORO) cannot be directly attributed to this component. One may however try to scrutinize the IORO so that only input/output segment pairs attributable to the component can be passed. Such scrutiny can be based upon knowledge or hypotheses about the real system and expressed in a control function added to the observation frame so that it becomes an experimental frame.

For example, as illustrated in Figure 13.4, one may constrain the input to the system so that only the component in question is stimulated (via specification of input segments). Choice of frequency band is a common way to do this for linear systems. However, the stimulation may not be confinable to the component in question and signals may be fed back to this component from components that are also excited. In this case, one may attempt to eliminate any input/output pair in which this feedback occurs. Such scrutiny requires access to variables that are sensitive to this feedback. The input

Experimental frames: Operational concepts

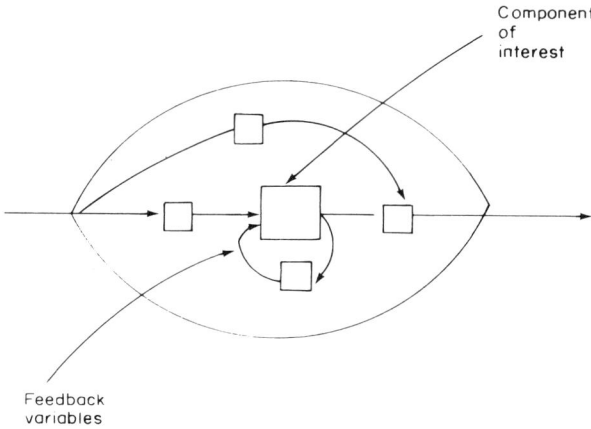

Fig. 13.4. The component isolation problem.

variables of the component connected in the feedback cycle would be the most direct choice. Such feedback sensitive variables would constitute the run control variables of the experimental frame and the admissible segments would be those in which no excitement of these variables is evident. It should be clear that the success of this isolation strategy depends on whether the postulated stimulation and feedback paths do in fact represent those in the real system.

To operationalize these ideas requires that we define an experimental frame E which represents the input/output observations that are possible in experimentation with the real system. Then a second frame E' is defined bearing the input and run control specifications postulated to isolate the component of interest. The criterion for realizability of the desired isolation is that the isolation frame E' is derivable from the accessible interface frame E. Of course, there could be a number of components to be identified from the same data obtained within E, and each would have its own isolation frame derivable from it.

Exercise. For a system and component of your choice, define frames E and E', where E represents possible interfaces with the system and E' is an isolation frame for the component. Show that E' is derivable from E or if it is not, argue that there is insufficient possibility to interface with the system to achieve both a derivable E' as well as one that achieves a meaningful degree of isolation.

3. Formalization of Modelling Assumptions

Recall that validity of a model with respect to a real system was defined with respect to an experimental frame. Validity is a relation defined by the equation:

$$R_{real\ system}/E = R_{model}/E$$

(Section 5.6). Thus we expect of a valid model that it reproduce the I/O behavior of the real system as restricted to the I/O data space defined by the experimental frame E.

Having the full concept of experimental frame now available, we can better appreciate how it can serve to formalize the assumptions under which a model is intended to be valid. These assumptions express restrictions either on the system being modelled or its environment. Restrictions on the environment can be formulated as constraints on the input to the system (hence choice of input variables I and segments Ω_I) and selection of observation mode (choice of output variables). Restrictions on the system can be thought of as defining an "operating space" such that if the "relevant variables" remain within this space then the behaviors of real system and model should agree. A formal definition for this concept is achieved by specifying run control variables C (that constitute the "relevant" variables) and set of control segments Ω_C (the "operating space").

Often the control variables are specified on the basis of a more comprehensive model, the *base model*, of which the model under discussion is a *lumped* version. Please note: this use of base and lumped is only intended as a convenient form of referring to the pair of models. The base model in question need not be the postulated *base* model for a real system (Section 5.2). Under assumptions concerning the behavior of certain base model variables, it might be that validity of the lumped model with respect to the base model can be formally proved. For example, the assumption that the positions and velocities of molecules in an ideal gas (base model) are uniformly distributed is sufficient to prove Boyle's law (lumped model). This assumption can be stated formally by the choice of positions and velocities as control variables and control segments in which uniformity of distribution is maintained. However, it may be much more difficult to prove that the base model remains in the stated control space than to establish the validity of the lumped model under the condition that it does. After formalizing the concept of valid simplification, we shall illustrate how run control specification can capture the uniform distribution assumptions underlying for the random phase-random space simplification procedure.

Experimental frames: Operational concepts 251

4. Experimental Frames for Model Simplification

In Chapter 9, the validity of a simplification was established by demonstrating the existence of a morphism between base and lumped models. Now, with the full concept of experimental frame available we can be more explicit about the validity criteria.

Indeed, for validity of one model with respect to another, just as with respect to a real system, we require agreement of I/O behavior within an experimental frame of interest. Thus we have the following definition:

For models M, M' and experimental frame E, we say that M *is equivalent to* M' *within* E if E is applicable to both M and M', and

$$R_M/E = R_{M'}/E$$

Thus models M and M' are equivalent in frame E if the result of applying the E_M-to-E processing to R_M (the I/O relation of M) agrees exactly with the result of applying the $E_{M'}$-to-E processing to $R_{M'}$. Clearly, the original behaviors R_M and $R_{M'}$ can be quite different and yet appear to be the same when mapped down to the data space D_E.

We say that a lumped model is a *valid simplification* of a base model in frame E if the two models are equivalent in E and the lumped model is simpler than the base model with respect to some measure of complexity (Sections 5.8 and 2.2).

Exercise. Using the definition of validity for a model with respect to a real system (Section 13.3), show that if M is equivalent to M' within E and M is valid for a real system in E then so is M' valid for this system in E. In particular, if a lumped model is a valid simplification of a base model in E, then the base and lumped model are either both valid in E, or both invalid in E.

4.1. Example: Random Phase–Random Space Simplification

We shall now illustrate the ability of experimental frames to express the input/output equivalence desired in a simplification as well as the underlying assumptions required for its validity.

There are two kinds of assumptions that are involved in simplification. *Structural* constraints restrict the class of models on which the simplification can be performed, i.e. define its domain. For example, for the random phase–random space simplification, to be considered in a moment, the class of models is a subclass of the discrete event cell spaces. Such structural constraints are distinguished from the *behavioral* constraints that characterize the operating space of the base model for which valid simplification is

252   4. Multifacetted system modelling

possible. The latter behavioral constraints are formalized by suitable choice of a run control specification for an experimental frame. This frame, E_{simp} is to apply to the base model and accept only those of its state trajectories that satisfy the run control constraints. But E_{simp} is also to apply to the lumped model so that equivalence of the two models within this frame can be considered. Now the lumped model (M) may no longer contain the control variables that are monitored in the base model, indeed they may vanish as a result of the simplification! Thus some means must be provided to enable E_{simp} to apply to M in case it lacks the requisite variables.

Random phase–random space simplification (Section 9.3) is a good example. The various models and frames to be discussed are displayed in Figure 13.5. Recall that it applies to a class of discrete event cell space models. Consider such a base model, B: each cell c has a state $s_c \in S$ and an elapsed time $e_c \in [0, ta(s)]$, as well as spatial co-ordinates (i_c, j_c). To represent this structure in terms of variables, let each cell c have the variables $STATE_c$ with range S, E_c (elapsed time, a non-negative real), L_c (location, a pair of integers).

E_B, the scope frame of this model is a DEVS frame with no input variables

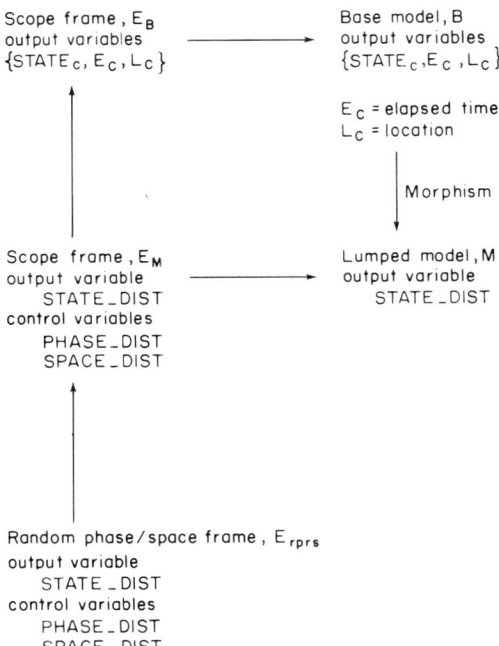

Fig. 13.5. Models and frames.

Experimental frames: Operational concepts 253

(assuming autonomous operation), no control variables, and output variables consisting of the sets $\{STATE_c, E_c, L_c\}$ for each cell c.

Now consider a frame E_{rprs} that represents the random phase–random space simplification. It should be applicable to both base model B and a lumped model M produced by the simplification procedure. Recall that such a lumped model is expressed in differential equation formalism. Thus the time base T of E_{rprs} is the reals (common to both DEVS and DESS) formalisms. It has no input variables. Its output variable O_{rprs} is S__DIST whose range is the set of all distributions of cells over the states in S. A typical value of S__DIST is a vector whose s'th component is N_s, the number of cells in state s. Thus the frame E_{rprs} sets up a data space which consists of segments representing the behavior over time of the distribution of cells in their various states.

The control specification of the frame captures the requirement that the random phase–random space property must hold over an observation interval in order for the output behavior predicted by the lumped model to agree with that of the base model. Thus the run control variables of the frame C_{rprs} consist of the pair of variables PHASE__DIST and SPACE__DIST each having $|S|$ components. The s'th component of PHASE__DIST is intended to represent the set of elapsed times of the cells in state s. Likewise, the s'th component of SPACE__DIST is intended to represent the set of locations of cells in state s. The run control segment restriction is to assert that for each state $s \in S$, the elapsed times and the locations are uniformly distributed throughout an observation interval. Thus Ω_{rprs} can be formulated as the set of all segments over (PHASE__DIST,SPACE__DIST) that satisfy the uniform distribution property at all points in their domains.

Exercise. Provide a complete specification of Ω_{rprs}.

It should be clear that E_{rprs} is applicable to the base model. More specifically, its output and control variables are derivable from the scope frame output variables by means of computations that can be formulated as semantic relations between the set of all variables $\{STATE_c, E_c, L_c\}$ and the distribution variables S__DIST, PHASE__DIST, SPACE__DIST.

Exercise. Define the aggregation mappings for the semantic relations just mentioned. Show explicitly that $E_{rprs} \leq E_B$.

Note however that the lumped model, L produced by the simplification operates in a state space defined by S__DIST. The choice of S__DIST for the output variable O_L of the scope frame of L will allow the output variable O_{rprs} to be derivable from it. But the control variables $C_{rprs} = \{$PHASE__DIST, SPACE__DIST$\}$, are certainly not derivable from S__DIST. To make E_{rprs} derivable from E_L we must give E_L control variables and segments equivalent

254   *4. Multifacetted system modelling*

to those of E_{rprs}. The minimal assignment to do this is to set $C_L = C_{rprs}$ and $\Omega_L = \Omega_{rprs}$. With this assignment, E_{rprs} is derivable from both E_B and E_L and hence applicable to both base and lumped models. Moreover, the discussion in Section 9.3 shows that the models are equivalent in E_{rprs}, hence that L is a valid simplification of B in this frame.

Note that so far it appears that E_{rprs} is identical to E_L. But this need not be the case if E_{rprs} specifies additional control variables that are derivable from the output variables O_L, for example, chosen so as to specify initialization and termination conditions.

Thus our general approach to formulating a scope frame for a model is to incorporate in it a run control specification that reflects the assumptions under which the model is intended to operate. Thus if a model M is to serve as a valid simplification of a base model B in a frame E_{simp} then the run control specification of E_{simp} is added to E_M, the scope frame of M. This specification can be regarded as the *implicit* constraints that the lumped model embodies.

This solution to the problem of applicability differs from that adopted in an earlier development of the theory[31] where the approach of partial applicability was adopted. The present approach has the advantage that only one form of applicability is required and that the distinction between a model and its scope frame is fully exploited. It has the disadvantage that the scope frame definition is relative to another base model. If the latter model, is itself derived from a more elaborate base model, than the run control specification attributed to its scope frame must also be added to that of the lumped model. This is required for equivalence of run control specifications to continue to hold between base and lumped models.

5. Summary

The various concepts that were given operational formulations in this chapter are schematized in Figure 13.6. Starting with the semantic structure of variables and relations, we develop the relation of derivability first, for variables, and then for experimental frames. The derivability of one frame from another sets up a mapping to the I/O data space defined by the first from that defined by the second. Applicability of a frame to a model is defined in terms of the derivability of the frame from the scope frame associated with the model. When a frame is applicable to a model, the I/O behavior generated by the model is mapped into the I/O data space defined by the frame. Two models are equivalent within a frame if their I/O relations viewed within this frame are equal. A valid simplification is one in which the lumped model is

---

[31] See comments in "Other Relevant Literature" at the end of Chapter 12.

Experimental frames: Operational concepts

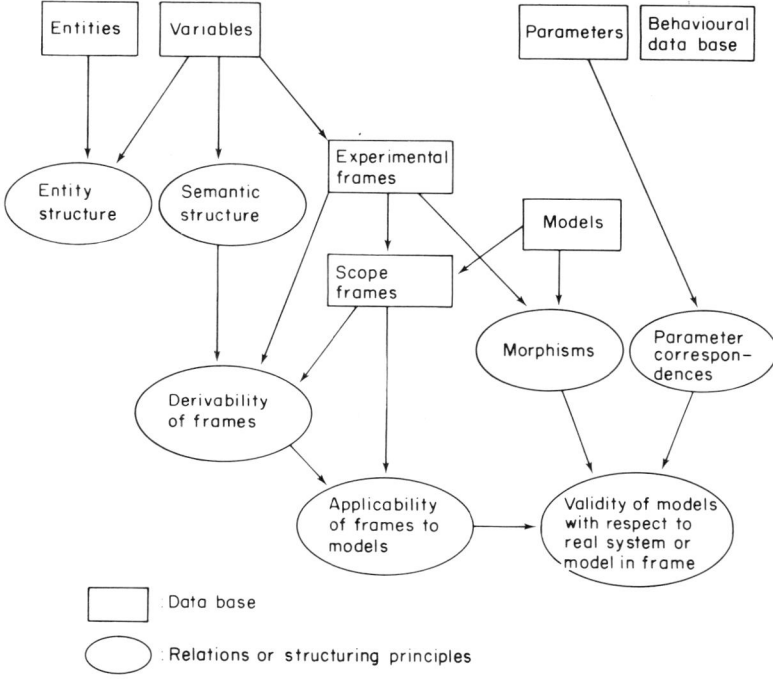

Fig. 13.6.

equivalent to the base model in the frame of interest. Similarly, a model is valid for a real system within a frame if the I/O relations of both are equal as viewed within the frame.[32]

Two applications of these concepts were discussed. The isolation of a component of a real system is formulated as the acquisition of I/O data within an appropriate frame which is derivable from a more global one. The representation of assumptions made during model construction can be formalized in terms of an experimental frame in which validity is sought for the model viewed as a lumped model of a base model. The behavioral assumptions required of the base model are expressed by employing the run control specification part of the frame.

---

[32] We should remember that validity defined in the above ways represents an ideal to be strived for. In practice, the process of validation is a non-trivial activity in the enterprise of modelling and simulation (see comments in "Relevant Literature", Chapter 5).

PROBLEMS

1. Formulate an entity structure for a stretch of highway which has two basic aspects: one representing a decomposition into lanes and the second into segments along the highway; the first aspect applies also to entities of the second. Lanes and segments have variables to measure the arrival and departure of cars, the number of cars currently contained within their boundaries, and the density of cars (number per unit area).
   (a) Describe the semantic structure implied by the above description.
   (b) Point out some of the derivability relations that arise from the closure of the semantic structure under the operations in Section 13.1.1.
   (c) Based on the entity structure, formulate some experimental frames that reflect macroscopic aspects of traffic as well as those interested in more microscopic aspects. Using the semantic structure, develop the derivability relation over these frames (determine which frames are derivable from which other frames).

2. This problem works out how applicability concepts specialize in the case of autonomous models (Section 4.5). Let M be an autonomous model so that I_M can be taken to be the empty set.
   (a) Show that the I/O data space $\mathbf{D}_M$ may be taken to be (Y, T) where Y is the cross product of the output variables, O_M.
   (b) Show that only frames E with no input variables are applicable to M and characterize the E_M-to-E processing for such frames.
   (c) Let B be a model which is not input free, i.e. I_B is not empty. Characterize the conditions under which M can be a valid simplification of B. Give an example in which such a simplification is of interest.

3. Formulate simplification and scope frames for the aggregation simplification procedure of Section 9.2. Note that no control variables are necessary. However, consider relaxing the class of base models to include the case where each component is influenced by a proper subset of the components in a block. For this case, the run control specification can formalize the assumption that the uniformity of influence principle holds in the operating space of interest. Formulate the requirement that all components in a block receive equal stimulation from (other) blocks by proper choice of control variables and segments for the simplification frame.

Chapter 14

MODELLING IN THE LARGE: BOUNDARY EXPANSION, INTEGRATION

So far we have discussed formulating objectives as experimental frames within fixed system boundaries, i.e. with a fixed concept of the system of interest. However, as we have suggested in Section 1.1, the system boundaries under consideration may have to be expanded to accommodate new objectives as they arise. We shall discuss this phenomenon and show how it is reflected in expansion of the system entity structure.

A second topic of this chapter also concerns development of the system entity structure. However unlike the orderly expansion just mentioned, we consider the problem of integrating a large set of models that have not necessarily been developed within the multifacetted modelling methodology. Such a problem arises in attempting to provide a coherent model base for models that have been developed over the years by many different teams adhering to different conventions and dealing with different aspects of a real system.

The chapter closes with a discussion of an important aspect of the integration of models into a model base. This is the propagation of parameter information derived from local calibration of models, first suggested in Section 1.2.5.

1. Objectives and Expanding System Boundaries

Let us return to the bus system problem (Section 1.2) and recall that eventually the review committee may wish to consider wider objectives than those involving the quality of service for university students. We shall see how such wider objectives lead to expanded boundaries and the implications of this entity structure and model construction.

Returning to the task assigned to the review committee let us list the objectives under potential consideration:

1. Passenger Satisfaction
2. Bus Operation
3. Selection of Admission Control Policy
4. Bus System Expansion
5. Future Passenger Distribution

The objectives are listed in increasing order of scope — higher numbered objectives require extending the system boundaries of the lower numbered ones. An entity structure suitable to represent the extended boundaries was shown in Figure 11.2.

Let us consider the new objectives in terms of entities, questions and frames.

1.1. Selection of Admission Control Policy

The committee must decide on policies for controlling the admission of passengers. It might ask questions such as: what are the effects on student passengers of policies which require the bus driver to distinguish students from non students? What revenue can be expected by charging $x.xx admission to non students?

To answer such questions the committee must decide on performance and resource utilization measures for assessing the merits of a policy. We have seen that a performance index such as waiting time can be used to measure passenger satisfaction. Income earned would be a performance variable for assessing commercial payoff.

Experiments might involve investigating alternative student identification schemes and ticket payment policies. The committee may have to deal with a trade-off relation between student inconvenience (e.g. increased time to load the bus) and income earned through a payment policy.

Models constructed for the Admission Control objective would, as illustrated in Figure 10.7, have a parameter such as PAYMENT amount, the admission charged to non students. Experiments would be done at various values of this parameter possibly seeking an optimum in the figure of merit that does the tradeoff between student inconvenience and income measures.

1.2. Bus System Expansion

After considering the previous series of objectives the committee may decide to investigate the possibility of adding additional buses and/or additional stations to the university service. Increasing the number of buses might seem to be necessary, for example, if it appears that the waiting time of students can not be reduced to an acceptable level within the existing constraints. Increasing the number of stations might appear to be attractive for example, if the option of expanding the service to the community is considered.

The kinds of questions that were raised for the existing bus system would continue to apply when expansion is considered. But since now there can be many more combinations to investigate, the experimental frames, and especially, the models of interest may be much more complex. The rational approach to constructing such frames and models is to extend the corresponding existing versions appropriately.

Were the university to expand its service to that of a full fledged company, an appropriate entity structure would be similar to that of Figure 10.12. Such a structure allows buses to ply different routes and passengers to embark and alight at different pairs of stations. Such new possibilities require a finer characterization of a passenger which allows modelling his propensity to leave the bus at a station having embarked at a given one. Each such pair also has a passenger transit time associated with it. But note that measurement of waiting time at a station is not affected by consideration of additional stations.

To specify variables for experimental frames which concern classes of entities we can use the convention developed in Section 11.3. Suppose for example we wish to the frame to simultaneously observe the line length at each station. This is specified by the expression {STATION( ).LINE__LENGTH} which represents the composite variable: (STATION(1).LINE__LENGTH,...,STATION(N).LINE__LENGTH) where N is the current number of STATIONs.

In general, if an entity E has a variable type V, then we may associate a composite variable $\{E( ).V\}$ with the multiple entity, Es. This composite variable represents the simultaneous reading of all the variables V of the components of type E currently in existence.

Exercise. Adapt the passenger satisfaction and bus operation frames for the single bus case to apply to the class of models in which there may be many buses and stations.

1.3. Population Distribution Objectives

At some point in their deliberations the committee may try to forecast the shift in population distribution toward the newer area in which the university is located. The entity structure of Figure 11.2 embodies a decomposition of the city into districts each of which is characterized by population attributes. A simple demographic model may for example divide the city into a university district and a downtown district. Such a model might be a set of differential equations which govern the rates of transfer of population from district to district — purely as functions of the current population sizes. Or more refined models might partition the city more finely or classify the population by age, occupation, training, etc. Indeed a model might take account of the ease of commuting between a district and the city center. With such a model the effect of opening the university bus service to non students could be estimated as we shall see in a moment.

Exercise. Write a frame for the population distribution objective which reflects an interest in the time course of population change in the environs of the university.

The boundary represented by the entity structure of Figure 11.2 clearly encompasses that of the bus system. Thus according to the systems view of Section 3.1, we might expect that the input variables of bus system frames would be output variables of other component models of the larger entity structure.

Indeed, we can use the demographic model to determine rates of arrival at the stations. Such a model structure is illustrated in Figure 14.1a. The demographic and bus system models are now component models of the new composite model (Section 3.1). In addition we have added an interface map which translates population sizes (outputs of the demographic model) into arrival rates which are taken to be inputs to an arrivals generator. An example of such a map might be one that sets arrival rates in some fixed proportion to population size.

It should be recognized that the interface-arrivals generator combination represents a third model component, call it a bus demand model. The committee might begin with the very simple form we have suggested for this model but they might be forced to further refine it. More specifically the above assumption that demand is proportional to population size might be open to question and a more complex relationship, e.g. involving district population over 18, might be sought. Note however, that within a frame such as Passenger Satisfaction such a component model need be refined only to the point where the behavior of variables of interest is no longer affected by further refinements.

Finally let us consider "feeding back" the output of the bus system components to the demographic component. As illustrated in Figure 14.1b, one might have the output of the bus system interfaced to a demographic model which takes account of the ease with which commuting can be done between the university district and downtown.

To be valid in the population distributions frame, such a composite model may need only a simplified version of the bus system, for example one that relates number of buses to ease of commuting.

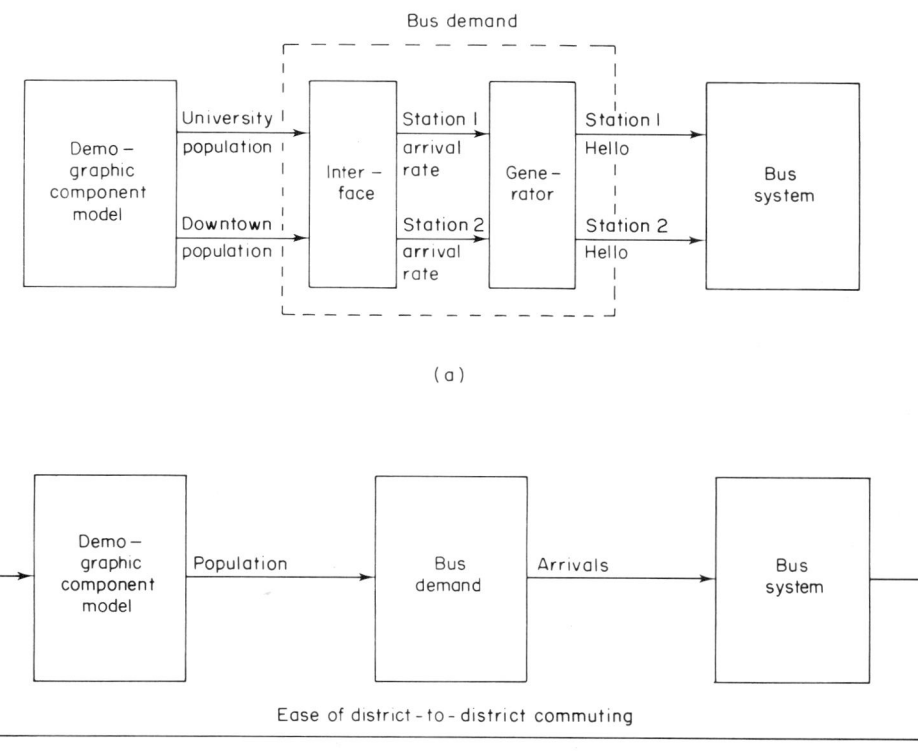

Fig. 14.1. Coupling a demographic component to the buy system model.

To summarize, we see that systems boundaries may grow as new objectives arise in the modelling process. Likewise, the entity structure expands upward to represent more inclusive system boundaries by acquiring new root entities. Software tools for manipulating the entity structure should facilitate this restructuring (see Section 18.3).

2. Entity Structure as an Organizer of Models and Frames

We now formally specify how the entity structure is to organize models and experimental frames. The relations involved are illustrated in Figure 14.2.

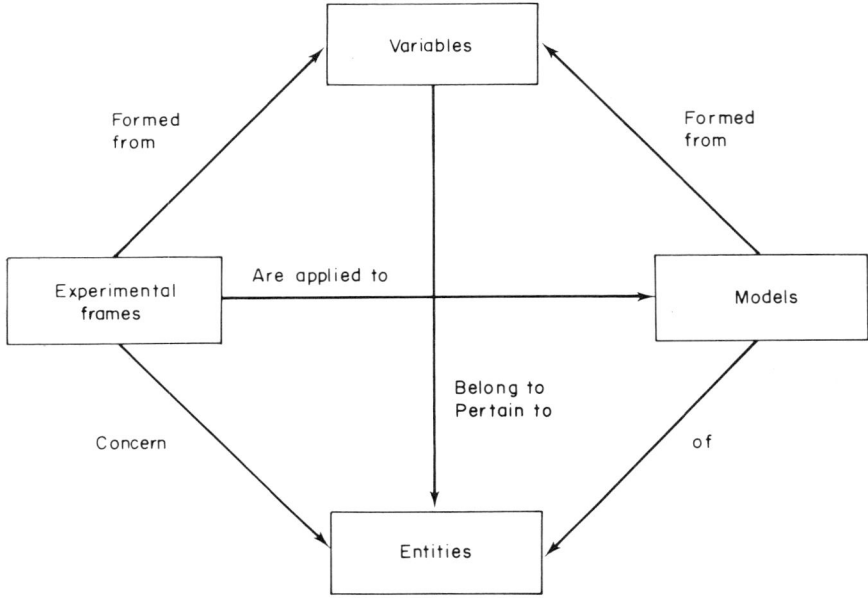

Fig. 14.2.

We see that a model is always a *model of*[33] a particular entity and must employ only variables pertaining to this entity. Similarly, a frame always *concerns* a particular entity and must employ only variables pertaining to this entity. (Recall that the definition of "pertaining" (Section 11.1.2) was crafted to refer to all variables in the substructure of an entity. Thus a model of, or experimental frame concerning, an entity can be hierarchically constructed from components associated with its subentities.)

Thus the entity structure partitions the models and frames into classes associated with entities. Accordingly an entity is a "key" for retrieving models and frames; knowing the entity which interests us, we can cut down the space of all models and frames to just those which are models of, and concern it. Of course since the entity structure itself represents a conceptualization of reality, its structure must be understood by the user in order for it to be an effective aid. This is one place where interactive software for setting up and querying entity structures can play an important role (Section 18.3).

Later chapters will consider how the entity structure is employed to retrieve models and frames from their respective bases in the course of frame

---

[33]Indeed, in general usage, the term "model" used as a noun implies a relation "model of" in which the "model" in question is the left hand element.

and model construction. In what follows, we lay the basis for the organization of a model base using entity structure concepts.

2.1. The Extraction Concept

In the introduction to Chapter 11, we indicated that all models in the model base should have static structures that are extractions of the entity structure. We have now seen an example of an entity structure, (for a CITY including a BUS__SYSTEM, Figure 11.2) and some extractions of it concerning the BUS__SYSTEM (Figures 10.1 and 10.3). These extractions are static structures of models prepared for the university committee of Section 1.1 that were amalgamated in a process we shall soon discuss.

However we have not formally defined what an extraction is. Let us now do so. Let MS be a model static structure and ES an entity structure. Then MS is an *extraction* (or *extract* for short) of ES if the following holds:

(1) The root component C of MS is synonymous with (has the same label as) some entity E in ES.
(2) If C is non-atomic,[34] let its first level decomposition components be C_1, $C_2,...,C_m$. Then E must also be non-atomic and have an aspect A with subentities $E_1,E_2,...,E_n$. Moreover the C_is must be synonymous with a subset of the E_js.
(3) If C is non-atomic then its variables are synonymous with a subset of variables belonging to E and A. Otherwise, (i.e. when C is atomic) its variables are synonymous with a subset of variables pertaining to E.
(4) The subdecomposition trees rooted by the C_is are each extractions of ES.

Note that it may be convenient to allow an entity or variable to have more than one name. Such names are called *synonyms*. For example a long descriptive name, a short abbreviation and an English equivalent might be synonyms. The classes of such synonyms must all be disjoint so "synonymous with" is an equivalence relation and may substitute for "has the same label" in the above definition.

Thus an extraction of an entity structure is obtained by starting with some entity, selecting one of its aspects and selecting a subset of the subentities of this aspect to form the first level components of the model structure. One continues in this way until stopping at any entity one wishes. Such an entity becomes an atomic component. One may select any variable for a non-atomic model component from those belonging to the corresponding entity and the selected aspect. For an atomic component, one may select from all

---

[34] A node is *atomic* if it is a leaf node in a decomposition tree. It is thus, *non-atomic* if it is in the interior of the tree (Section 2.2).

the variables pertaining to the corresponding entity. All names given to the extraction must be synonyms of entity structure counterparts.

The set of all extractions of an entity structure characterizes the set of all models static structures compatible with it. Conversely, given a set of model structures one can seek an entity structure of which they are all extractions. Such an entity structure, if it exists, would be called an *amalgamation* (or *amalgam*) of the model structures. Indeed the problem of amalgamation was the one we raised originally in Chapter 11.

2.2. Entity Structure Amalgamation: Model Integration

In some domains of large scale system modelling such as energy, groundwater and ecosystems there are already many models, in the hundreds or even thousands [1]. Efforts are being undertaken to catalog and document these models to form libraries accessible to interested users. Such efforts come under our rubric "modelling in the large". They are difficult not merely for the amount of manual labor involved — just the sheer amount of manpower required to input information about each model would be significant. But more fundamentally, they are difficult because of the variety of approaches, often idiosyncratic, taken by the various modellers or modelling teams.

With the concepts we now have available we can understand, and deal with, some of the dimensions of this problem of model unification and integration. We can analyse each model, and its builder(s) if possible, asking questions like:
(1) Of which system entity is this a model (what are the system boundaries implicit in the model)?
(2) What is the decomposition tree of the model (what are its components and subcomponents)?
(3) What are the variables attached to each component (how are the model's variables organized)?

If, as is to be expected, the models were not built top down in our framework, there may be no final answers to the questions. Indeed, we may have to iterate the amalgamation process, we shall describe, back to this primary model structuring phase. But let us suppose that we have answered the above questions for each model. Then seeking an amalgamation would be possible and would, if achieved, be a basis for further integration and unification.

Let us remark also that even if modelling is proceeding "top down" from the entity structure, there are decisions in its development which are analogous to those which must be made in amalgamation. We have to remember that the entity structure is a growing "organism" responding to new questions by accommodating new entities, variables, experimental frames and models. Each modification raises questions like those involved in amalgamation: where should a certain variable be attached? Does a new aspect have to be defined?, etc.

The phases of the amalgamation are:
(1) setting down the item information
(2) constructing an entity structure from item information
(3) testing the entity structure for validity
(4) reducing the entity structure to a smaller equivalent.

We shall discuss each phase in turn.

Setting down the item information

An entity structure is completely specified by giving its set of item names (and their synonyms) and for each item i, the pair (S_i, V_i) where S_i is its set of sub-items and V_i is the set of variable types belonging to it.

To satisfy the alternating mode axion, if i is an entity, then S_i is a set of aspects, and vice versa.

In the case of amalgamation, we are given the set of entities and for each, one or more model static structures. We assume that this set is complete in the sense that each node in a static structure is synonymous with some entity in the set and proceed as follows:

To each entity e we assign a set A_e of aspects such that each aspect $a \in A_e$ corresponds one-one with a model static structure for e; the set V_e of variables belonging to e is empty. To each aspect $a \in A_e$, we assign the entity set E_a which corresponds to the first level components of the model structure named by a; the variables V_a are the variables attached to the root component (synonymous with e) of this same model structure.

The items $\{e\} \cup A_e$, entity pairs (A_e, V_e), and aspect pairs (E_a, Va) constitute item information, which clearly is alternating in mode.

Constructing an entity structure from item information

Call an item *maximal* if it never appears as a sub-item, i.e. it is in no S_i for any item i in the given set I. In a valid entity structure, there should be exactly one such maximal item — the root item — of mode entity. However, one can build a structure in any event as follows: Create nodes labelled by the maximal item names. For each of these nodes i create successor nodes labelled by its subitem set S_i. (An item with empty subitem set S_i is atomic and produces an atomic or leaf node.) For each of the nodes i so labelled, create successor nodes labelled by its set S_i. Continue in this way until every just-created node is atomic. To each node attach the variables V_i of the item i which labels it.

Testing the entity structure for validity

The entity structure created as above from item information satisfies the uniformity axiom. This is because each time a node with some label n_i is created it leads to the same attachment of variables V_i and the same labelling of successor nodes S_i. By induction, the uniformity axiom can be easily proved. The alternating node axiom can be tested directly from the item information as we have just indicated. If there are more than one maximal items then at most three fictitious nodes with no variables can be added so that an entity rooted tree results.

Exercise. Show that depending on the composition of nodes of the maximal items (all aspects, all entities or mixed) one, two or three items can be added to create a tree rooted by an entity. Which case characterizes the item information obtained from amalgamation?

There remains only the test for strict hierarchy. Since it involves paths, it must be done on the entity structure *per se* or better its pre-structure. Each time a new node is created, the path to it from the root to it must be checked for a node labelled by the same item. If this check fails (to uncover a superior item occurrence) then the node may be attached and the entity structure remains valid. Otherwise, this item information must be rejected as being inconsistent with the prior information.

In the case of amalgamation, assuming strict hierarchy of each model structure, strict hierarchy of the entity can be violated if, and only if, there is a pair of entities e and f such that f is subordinate to e in some model structure and e is subordinate to f in some other structure.

Reducing the entity structure to a smaller equivalent

Two entity structures will be said to be *equivalent* if they give rise to the same set of extractions. An entity structure is *smaller* than another if it contains fewer nodes or fewer variable attachments.

The following are some size reducing equivalence preserving procedures.

1. Removing redundant aspects. Let two aspects a_1 and a_2 be brothers 'i.e. sub-aspects of the same entity). If $E_{a_1} \subset E_{a_2}$ and $V_{a_1} \subset V_{a_2}$ (variables) then a_1 and all its substructures can be eliminated. This is so since any extraction employing a_1 can be obtained employing a_2.

2. Re-assigning variables from aspects to entities. Let V be a variable and e an entity, such that V belongs to *each* aspect $a \in A_e$. Then V may be removed from each V_a, $a \in A_e$ and be added to V_e. Thus the variables belonging to an entity are common to each of its aspects.

Call an entity structure *reduced* if there are no redundant aspects in the sense of (1) above nor re-assignable variables as in (2). Every entity structure can be brought to reduced equivalent by applying the procedures (1) and (2).

Exercise. (Open Question) Prove or disprove: The reduced entity structure equivalent to a given structure is unique.

Exercise. What effect is there on the set of extractions if two or more brother occurrences of the same entity E are replaced by a multiple decomposition triple $Es - a_{melt} - E$?

Applying the above procedures to a set of model static structures we arrive at the:

Theorem. Amalgamation of Metal Static Structures

Let **S** be a set of consistent model static structures, where by "consistency" here we mean that there is no pair of components e and f such that e is subordinate to f in some structure and f is subordinate to e in another. Then there is a reduced entity structure called an *amalgamation* of **S**, of which each static structure in **S** is an extraction.

This theorem provides a first step toward integration of sets of models. It assumes that the static structure can be gleaned from the given models, a process that might prove to be extremely difficult in cases where little or no documentation exists on the structural considerations that went into the model design. Also, the amalgamation of static structures into an entity structure is the first step toward the decomposing of the models into modules that can be freely coupled along the lines set up by the entity structure. Nevertheless, such a first step is necessary precursor to tackling the more difficult problems of model integration.

3. Propagation of Parameter Information

An important aspect of model integration is that brought about by the interrelation of their parameters. Such a parameter often has a physical or theoretical significance that transcends its use in a particular model. Thus many models may share the "same" parameter. For example, the earth's gravitational parameter g may appear in many mechanical models. Where ever such a share parameter is used its value assignment is the same (otherwise, we would not regard it as the same parameter in different models). A second means in which inter-relation of parameters occurs is via the transformation and simplification of models. We have seen that such transformations induce correspondences on the parameter spaces of pairs of models (Section 6.4.3 and 9.5). Such correspondences place constraints on the assignments that can be jointly made to the respective parameter sets. Collectively, such parameter correspondences set up a network of relations which provides a valuable means of judging the mutual consistency of models.

More formally, suppose that we have a collection of parameterized models **M**. A typical model is denoted $M(p)$, where p is an assignment to the set of parameters of the model P_M, i.e. $p \in P_M$.range. We suppose that the parameters of all models are all distinct, i.e. the sets $\{P_M, M \in \mathbf{M}\}$ are all mutually disjoint. Suppose also that there exists a set of relations **R** on these parameters. Such a collection of parameters and relations forms a *parameter structure* whose formal properties are the same as the semantic structure of variables introduced in Section 13.1.1. That is, we close the parameter structure under the operations listed in Section 13.1.1 (composite variables, identity relations, extension to composites, composition of relations).

But what are the relations that are given at the beginning? These are the ones arising out of sharing of parameters and model transformation as mentioned above. In the first case, suppose that models M and M' employ parameters P and P', respectively. If P and P' are really different names for the same parameter then we have the relation $R \subset P.\text{range} \times P'.\text{range}$ where $(p,p') \in R$, if and only if, $p = p'$.

In the second case, let **P** be a parameter correspondence between models M and M' induced by a procedure that transforms M into M' (for example, a simplification procedure as in Section 9.5). Then **P** relating P_M to P'_M is a relation in the parameter structure.

Now let **P** be a relation relating the parameter sets of models M and M' (either given, or arising from closure of the parameter structure). We say that **P** is *justified in experimental frame* E if the following is true:

$$R_{M(p)}/E = R_{M(p')}/E$$

for each pair $(p,p') \in \mathbf{P}$.

That is, the relation is justified in E if each pair of models that it puts into correspondence are equivalent in E.

In particular, for a simplification procedure, the parameter correspondence that it induces is justified in frame E if the procedure always produces valid simplifications (Section 9.4).

From this perspective, relations arising from parameter sharing are to be viewed as justified in all experimental frames. When relations are combined in closure operations, the frames in which the output relation is justified are those which are derivable from all of the frames justifying the input relations.

To see this, first note that if a relation involving M and M' is justified in frame E, then it is also justified in any frame E'' derivable from E.

Exercise. Prove this assertion. Hint: show that

$$R_{M(p)}/E = R_{M(p')}/E \text{ and } E'' \leq E$$
$$\text{implies } R_{M(p)}/E'' < R_{M(p')}/E''$$

(see the exercise in Section 13.1.8 and definitions in Sections 13.1, 9–10).

Then, for example if **P** is justified in E and **P'** is justified in E', their composition is justified in E'' where $E'' \leq E$ and $E'' \leq E'$. (In case we are considering a lattice of frames, the maximal frame in which the composition holds can be characterized as the greatest lower bound of E and E' (see Section 15.5).)

The parameter structure thus developed places constraints on the values that can be assigned jointly to the parameters of all the models in **M**. If we regard each model as a source of empirical knowledge about the parameters in its set, then the parameter structure propagates this knowledge throughout the model base. Such knowledge input from different sources is subject to the consistency requirements imposed by the parameter structure.

How is a model function as a source of empirical knowledge concerning its parameters? The process of *calibration* or *parameter identification* is that in which parameter assignments are sought which result in acceptable agreement between model and real system behavior ("acceptable" according to chosen criteria). Empirical knowledge is gained in such a process in the sense that only certain parameter assignments are consistent with empirical observations.

We shall formulate model calibration in an ideal form in which exact agreement can be obtained in matching model and real system data (see Section 5.6). Let there be a fixed real system under discussion and let M be a model and p an assignment to its parameter set. We say that p is *acceptable* in frame E if $M(p)$ is valid for the real system in frame E, i.e.

$$R_{real\ system}/E = R_{M(p)}/E$$

The calibration process seeks as its most desired outcome that exactly one parameter assignment from the set P_M is acceptable in E. The usual result however, is that no assignments are acceptable or that many assignments are acceptable. The empirical information derived from a calibration activity is thus to associate with model M a set of parameter assignments acceptable in some frame E (a possibly empty subset of P_M.range).

Now we switch from the local calibration process to assess its global consequences. Consider the set of models whose parameter sets are linked with that of M through some relation in the parameter structure that is justified in E, i.e. the "E-neighborhood" of M. The parameter assignments that can be consistently made to these models are constrained through the relations to the set associated with M.

For example, if M is a lumped model which is valid in E for some base model B. Then the parameter correspondence set up by the associated simplification procedure constrains the parameter assignments that can be made to B after M has been calibrated. Indeed, this represents a practical common situation, since it may be much easier to calibrate a lumped model, with its smaller parameter space, than a base model.

Explicitly, the constraint linking M and B can be stated in the form:

if (p,p') is in the parameter correspondence justified in E then p' is acceptable for M in E if, and only if, p is acceptable for B in E

Exercise. Prove this statement.

Exercise. Let the parameter correspondence be given by a mapping f from P_B.range into P_M.range (for example, that given in Section 9.5). Consider the two cases: (a) $p' \in f(P_B.\text{range})$, (b) $p' \notin f(P_B.\text{range})$ and suppose that p' is the only acceptable assignment for M in E. What is the set of acceptable assignments for B in E? Consider the same question in the case that the acceptable set for M in E is empty.

In general, we can imagine the process of parameter identification for a model base linked by a parameter structure to proceed as follows: At any time t, each model has a set of acceptable parameter assignments indexed by experimental frames.[35] At some time t' later, a calibration experiment is done for some model M in some frame E that results in the narrowing of the acceptable set for M in E. This information (reduction in uncertainty) is propagated to the E-neighbors of M through the parameter structure. One result may be that the acceptable set of M or one of its E-neighbors becomes empty. In this case, such a model can not be validly employed in E. Or it may happen that the acceptable set of a model becomes a singleton, in which case, there is now no uncertainty about the appropriate parameter values to assign it to when it is used for decision making.

In this approach, we can plan parameter identification experiments not only on the basis of which model is in greatest need of calibration but also on which experiment will result in the greatest overall uncertainty reduction in the model base (see Section 18.2).

4. Summary

The concept of entity structure was introduced in order to organize variables, models and experimental frames around the system entities they are relevant to. The entity structure represents the choices of system boundaries and decompositions implicit in an existing model collection. New modelling objectives may force an expansion in the boundaries containing

---

[35] Acceptability here is time dependent: p is acceptable at time t for M in E if there is agreement between the I/O pairs observed in the real system and those generated by $M(p)$ in E until time t.

the real system of interest and this is reflected in the upward extension of the entity structure. New objectives may also force the addition of new aspects (reflecting the consideration of new decompositions) and entities (components in these decompositions). Moreover the structure may grow downward as entities that were previously atomic are given substructures (giving the possibility to consider decompositions of previously indecomposable components).

An amalgamation procedure was given for unifying and integrating a collection of models into a coherent model base. This procedure constructs an entity structure from which the original model static structures are extractable. The set theoretic characterization of the entity structure developed provides a basis for computer assistance in formulating and manipulating entity structures.

Concepts for viewing the integration of models in a model base in terms of their validation and verification were discussed. The consistency checking that is possible in such a model base offers a much broader base for assessment of model credibility than does validation of models in isolation.

References

[1] Gass, S. I. (1980). *Validation and Assessment Issues of Energy Modelling*, National Board of Standards, Special Publication Number 569. Provides an appreciation of the state of affairs in energy modelling. See especially the articles by Cherry, and by Hudson and Jorgenson.

Chapter 15

MULTIFACETTED MODEL BUILDING METHODOLOGY

The system entity structure encodes the set of decompositions and components that have been identified up to some time for a particular system. The set of all models of, and the set of all frames concerning, an entity are constrained by the entity and its substructure. Thus we have seen how model and frame bases may be organized by the entity structure (Chapter 14).

In this chapter we assume that the entity structure, the frame and model bases already exist and use them as a basis for guiding model and frame specification. Hierarchical model construction as illustrated in Section 8.7 constitutes the basic form of specification. A formal concept of the composition tree is developed as the formalism for hierarchical construction. The tree can be employed to represent model construction via both synthesis and aggregation.

Having the composition tree concept for hierarchical model construction, we develop a methodology that operates on the entity structure to produce such a composition tree. The methodology is guided by experimental frames and exploits the available knowledge represented in the model base. The basic methodology is depicted as follows:

entity structure
$\quad \rightarrow$ pruned entity structure
$\quad\quad \rightarrow$ composition tree

It is presented here for reference to as it will undergo considerable elaboration.

1. Pruning the Entity Structure

We discuss a process called pruning, in which a substructure of the entity structure is extracted for use as the skeleton for model or experimental frame specification. The extraction, called a *pruned entity structure*, is an entity structure with the following further restriction: *unique aspect*: every entity either has no aspects or a single aspect.

An entity with no aspects is said to be *atomic*. All non-atomic entities in a pruned entity structure have exactly one aspect which represents the unique decomposition selected for this entity. Thus descending the structure from the root to leaves we trace out a single hierarchy of decompositions for the root entity which culminates in the atoms. We shall later describe how this hierarchy will be elaborated into model or experimental frame specifications.

In the pruning process, we must consider each occurrence of an entity or aspect individually. For example we may wish to select one means of decomposition for one entity occurrence and another for a second. Thus if the result is still to be an entity structure, we must be allowed to change the names of item occurrences if they are pruned differently so as to retain the uniformity property. To simplify matters in the following pruning procedure we shall assume that each occurrence of an item is given a unique name.

Let ES be an entity structure having an entity occurrence called ENT. We describe a recursive procedure Prune (ES,ENT,E,d) that constructs a tree T_{ENT} of depth d. The name given to the node in T_{ENT} that corresponds to ENT is EN.

Prune(ES,ENT,EN,d):: let EN label the root of T_{ENT}
    if $d=0$ or ENT has trivial substructure in ES[36]
        then T_{ENT} is the single entity EN;
        select zero, one or more variables
        pertaining to ENT in ES and
        attach them to E in T_{ENT}.
  else
        select zero, one or more variables
        belonging to ENT in ES and
        attach them to EN in T_{ENT}
        select one of the aspects A of ENT in ES
        call it $A.j$ (add 1 to j) and
        add it as the aspect of EN in T_{ENT}
        select one or more entities E of A in ES
        as the entities $E.j'$ of $A.j$ in T_{ENT},
        for each such entity E
        employ procedure Prune(ES,$E,E.j,d-1$)
        to construct T_E and
        attach it to $A.j$ in T_{ENT}.end

---

[36]ENT has trivial substructure in ES if it has no subentities whatever, i.e. it is either atomic or all of its aspects are atomic.

Here j is a global variable which is initialized to zero and counts the number of item occurrences encountered. It is employed to uniquely label the item occurrences. The procedure is initialized as Prune(ES,ENT,ENT.1,d).

Thus starting at ENT one proceeds down the entity structure such that each time an entity is selected a unique aspect is assigned to it. The variables attached to the nodes of the pruned ES are selected from those pertaining to the corresponding entities of the original entity structure. The following choices must be made:

*the entity at which pruning starts (the choice of system boundary (Chapter 14) and hence the system to be modelled)

*the aspect selected for each entity (the choice of decomposition of a component)

*the variables selected for each entity (the descriptive variables of a component)

*the depth d (the maximum number of decompositions that will be undertaken)

The objectives that motivate the model construction should guide the making of these decisions as is now shown.

1.1. Objectives Driven Pruning

We indicated that the modelling objectives guide the selection of initial entity, ENT through the determination of system boundaries (Chapter 14).

Chapter 12 discussed the objectives-driven methodology in which experimental frames are formulated to express modelling objectives. This methodology interacts with the pruning process, but for simplicity let us suppose that all steps in the formulation of experimental frames have been carried out first and the results are available to constrain the pruning process. Later we shall return to the influence of the entity structure on the frames formulation process.

As an outcome of the objectives-driven methodology we have a set of variables (interest and mediating) that express quantities related to the objectives. These constitute the variables that will be incorporated into an experimental frame, E. We refer to them as the *frame variables* $\mathbf{V}_E$.

On the other hand the entity structure makes available a set of variables which we refer to as the *entity structure variables*, $\mathbf{V}_{ES}$. A pruned entity structure T_{ENT} makes available a set of variables, called the *pruned structure variables*, $\mathbf{V}_{ENT}$.

Pruning the entity structure with respect to an experimental frame, E should satisfy the following constraints:

(*) the frame variables should be available in the pruned structure:

$$\mathbf{V}_E \subset \mathbf{V}_{ENT}$$

(**) the pruned entity structure should be minimal in the set of all pruned structures that have property (*), i.e. no further pruning of the entity structure T_{ENT} should have property (*).

This suggests that starting with ENT, the pruning process should proceed down the entity structure, making choices of aspects and entities that to which frame variables belong. It should stop as soon as all frame variables accounted for, i.e. conditions (*) and (**) are satisfied. Viewed this way, the depth parameter of the pruning procedure is not fixed beforehand but is determined by the requirements of the modelling objectives.

We call this approach *objectives guided pruning* and note that it provides a rationale for continuing, and terminating, system decomposition in terms of modelling objectives. Each time a new aspect is selected, i.e. a new level of decomposition is undertaken, access is obtained to a new set of variables (those belonging to the entities at the next level). The refinement stops when having access to these additional variables no longer contributes to variables employed by the experimental frame of interest. This pruned entity structure is *necessary* in the sense of being as small as possible, yet meeting modelling objectives. However it may not be *sufficient* in the sense that it may require expansion to meet the needs of model construction as we shall soon see.

1.2. Example: Patch Structured Ecosystems

A central question of interest to theoretical ecologists concerns the conditions under which species can coexist indefinitely in the same ecosystem. For example, the rule of competitive exclusion states that, of two species competing for the same resources, the one with the greater efficiency will be the only one surviving in the long run. Yet there are many apparent counter-examples to this rule. Space is one key factor in conditions for mutual coexistence, and ecologists have found the decomposition of an ecosystem into "patches" a helpful one in this regard. *Patches* are regions that are relatively hospitable to the species in question and that are isolated from each other by inhospitable terrain. Patches can for example be trees in a forest for termite populations, or lakes linked by rivers for fish, etc. Population growth is assumed to take place only in the patches. Patches are loosely coupled through migration of individuals, which while small, is still significant. Indeed, a patchy ecosystem may permit coexistence that is impossible in a homogeneous environment since while a species may be going extinct in any one patch, it may just be beginning a growth phase in another. Thus patch structured ecosystems as wholes may have radically different properties from their homogeneous counterparts.

A system entity structure for a family of models is displayed in Figure 15.1. Two fundamental decomposition modes (aspects) are represented. The first is

276   4. *Multifacetted system modelling*

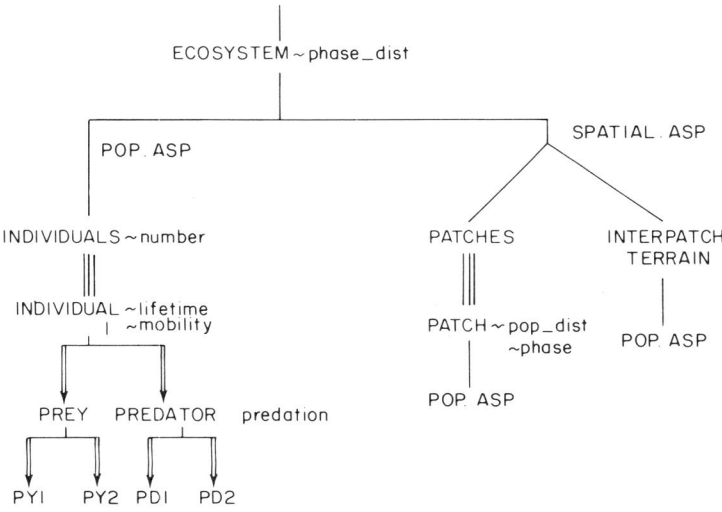

Fig. 15.1. Entity structure for ecosystem example.

a decomposition into populations of individuals each of which may be of type predator or prey. The effect of replacing the specialization operator expressing this possibility is to create two homogeneous populations consisting of predator and prey individuals repectively (Section 11.2). Since we may have more than one prey or predator species, the specialization relation is continued one more level.

Exercise. Draw the pure entity structure obtained by replacing all occurrences of the specialization relation.

The second aspect represents a decomposition into patches and interpatch terrain. A patch being a hospitable region in which populations of preys and predators interact. Patches are separated from other patches by the inhospitable terrain. The population aspect is shown applied to individual patches and terrain so that the kinds of species and their characteristics that can be represented in aggregate form for the ecosystem as a whole, can also be selected for each patch, and the interpatch terrain, as well.

Several pruned entity structures are displayed in Figure 15.2 along with names of corresponding experimental frames. One may be interested in resolving a patch to the level of species as in Figure 15.2a or only in coarsened form in which only colonization phases can be distinguished as in Figure 15.2b. Similarly, the ecosystem can be resolved to the population level ignoring any patch structure (Figure 15.2c). It can also be decomposed into

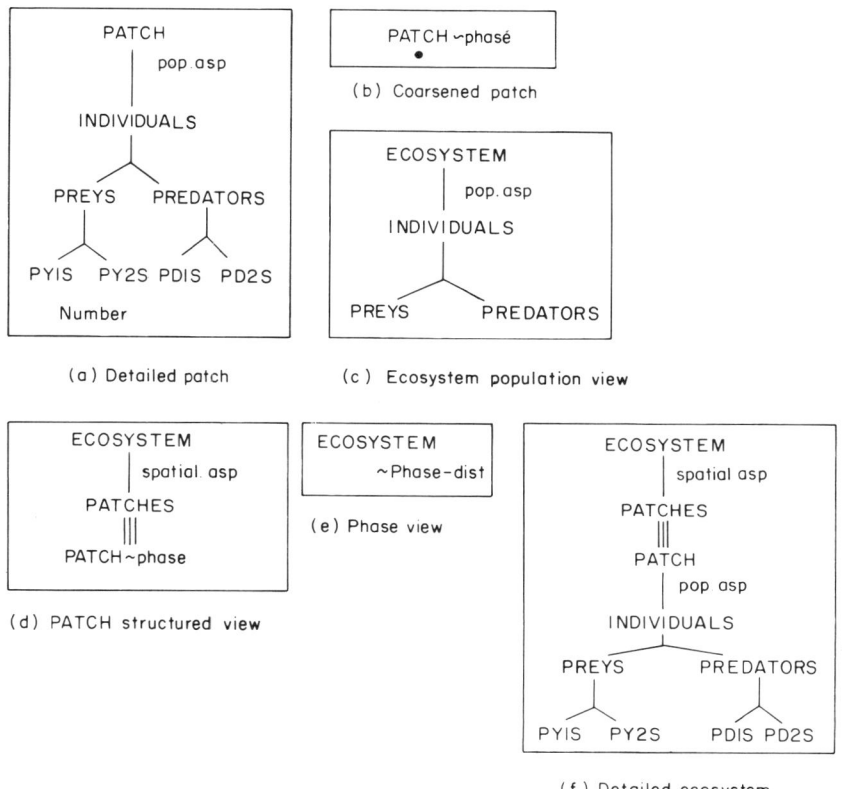

Fig. 15.2. Pruned entity structure and associated frame titles.

coarsened patches (Figure 15.2d), and a derived phase view (Figure 15.2e), or into patches that themselves are decomposed into species (Figure 15.2f). Although it is not shown, we can also focus on individuals, employing biological concepts to further decompose them, in order to question and model the biological determinant of their ecological behavior. Having illustrated the many prunings generally possible for an entity structure, we turn to the use of the pruned entity structure for model construction.

2. Concepts for Model Coupling

This is a good point to review, and elaborate upon, the concepts of multicomponent model specification formally introduced in Chapters 3 and 8.

A multicomponent model consists of a set of *components* (which are themselves models) coupled together so that they interact. Components have *inputs* and *outputs* through which the interaction is mediated. A *coupling* scheme specifies how the outputs of one component are fed to the inputs of another. A component has *state transition and output functions* that determine how inputs are processed and what outputs are produced. Models are dynamic, i.e. exist over a time base. The *state* of a component at any time instant is a summary of its past history up to that instant. When a component processes inputs, the internal effect is a change in its state. Discrete event models may be characterized as passive or active. A passive model can undergo state transitions and generate outputs only in response to external events. An active model can in addition execute state transitions and generate outputs in the absence of external events. Inputs and outputs are represented concretely by sets of *input and output variables*, respectively, collectively called *interface variables*. Likewise, the state set concretely represented by *state variables* (to be more fully discussed in Section 16.2.2).

Usually, one considers a component to be an instance of a *model class*. All instances of such a class have the same static and dynamic specification. In a *parameterized* model class, a set of *parameters* is associated with each instance. The particular values assigned to the parameters of a component determine the nature of its specification.

Depending on the questions being asked, especially the time periods of interest, variables of a real world entity may be represented by model counterparts as either state variables or parameters. Over relatively short periods such variables may be considered as constants with values determined when the component came into existence. In this case, their model counterparts are appropriately treated as parameters. Over relatively long periods, if these variables are subject to change, they may be treated as state variables (see Section 16.2.3).

While coupling schemes have been discussed at the abstract level (Sections 3.1, 8.2), let us consider a particular concrete form that applies when the input and output sets are structured by input and output variables respectively.

The scheme, to be fully formalized later, is a set of pairs or *links* of the form:

$$<\text{output variable}>_i \rightarrow <\text{input variable}>_j$$

indicating that a particular output variable of component i is to be fed to a particular input variable of component j. One can visualize the scheme in terms of a digraph (Section 2.2) whose points are labelled by the components and such that there is an arrow from point i to point j if there is a corresponding link involving i and j in the coupling scheme.

Various *types of linkage* can be identified, e.g. control, communication and transfer of material or energy. Accordingly, the coupling scheme can be put together as a union of coupling schemes for each type of linkage existing in the model. Each such *sub-coupling scheme* may be represented by a separate digraph. Also these subdigraphs may be unified into single multigraph in which the arrows are labelled by the type of linkage. This approach makes it possible to use properties of digraphs, and measures of their complexity to characterize the nature and complexity of component interaction (Section 2.2).

Examples of interactions and linkage types at various levels are:

*** interactions between individuals:
mating, reproduction, capture/kill of prey

*** interaction between patches:
migration

*** interaction between species
competition, predation, co-operation

A model constructed by linking component models may itself be employed as a component in a higher level model. To do so requires first that its input, state and output variables be identified. This is straightforward: any component input variable that has not been linked by the coupling to an output is an input variable of the model; likewise any output variable of a component that has not been linked to an input variable is an output variable of the model. The state variable set of the model is the union of the state variables sets of each of the components. Hierarchical model construction results when this process of employing already constructed models as components in higher level models is repeated.

2.1. Preparing the Pruned Entity Structure for Model Construction

The pruned entity structure that is developed by objectives guided pruning of Section 15.1.1 represents a minimal entity structure that has all the variables required by the experimental frame. These are variables of direct interest in relation to the objectives or required to mediate the observation of such variables. Many more variables may have to be employed by a model intended to accommodate the frame. (Recall that a model M accommodates a frame E just in case E is applicable to M.) Such variables are necessary to express the transition and output functions of the model.

Exercise. Reread Section 10.1 concentrating on the relation between the static and dynamic structures of a model.

However, the process of constructing a model is a creative activity involving human understanding not only of the objectives to be met (formalized in the experimental frame) but also of the nature of the system being modelled. At this point we shall assume that the pruned entity structure has been expanded to incorporate the variables that are required for model construction. This will enable us to skip ahead to the next stage in which the pruned entity is converted into a hierarchical model specification. Having this appreciation what must be done in model construction we will return to consider the question of the creative leap from experimental frame to accommodating model.

After its expansion, the pruned entity structure still does not indicate what roles the variables attached to it will play in the model to be specified. We shall refer to the process of specifying such roles as *role designation*. It consists of two sub processes: *orientation* (selection of interface variables) and *internal role designation* (selection of state variables and parameters).

Role designation is summarized as follows:
  Orientation (selection of interface variables):
    input variables (values determined externally to the model)
    output variables (model computed values, available externally)
  Internal Role Designation specification of:
    state variables: (model computed values, initialize and restart model)
    parameters (values select particular model from class of possibilities)

In general, a variable can be given at most one such designation except in the case of state and output: a variable may be both a state and an output variable. Also a variable may play different roles in different models. For example, a variable may serve as a parameter in one model and as a state variable in another (as pointed out in Section 15.2).

Having established input/output orientations, one is in a position to couple components together. A second process called *coupling constraint* places restrictions on the kinds of coupling schemes than can be employed subsequently. Note that role designation applies to entities (for which component models are to be constructed). It is natural to associate the coupling constraints with *aspects* since these stand above the entities whose models will be coupled together.

Coupling constraint: (constraints on the kinds of coupling schemes that can be associated with an aspect). Examples of such types of constraints:
  spatial: components are located in space and immediate interaction is restricted to components near to each other; if the interaction is uniform in nature, it may be specified by the template of a cellular space (Chapter 6).
  type of flow: input and output variables relating to flow of communication, material, energy, etc. can be interconnected only within the same flow type.

series/parallel: components can only be connected so that no feed back loops result.

Figures 15.4–6 display the pruned entity structures of Figure 15.2 that have been given role designations and coupling constraints appropriate to some ecosystem models to be discussed.

3. The Composition Tree Formalism

The pruned entity structure, as it emerges from the preceding processing, sets out a family of hierarchically constructed models which are consistent with it. We now develop a formalism for uniquely specifying such models, called the *composition tree*. A composition tree is like a decomposition tree (Section 10.1) except that it contains much more information relating to the dynamic structure of the component models and their linkage via coupling.

As depicted in Figure 15.3, a composition tree is a tree with certain kinds of objects attached to its nodes. Attached to its leaf nodes are system specifications. These are the atomic model components which will be coupled together in hierarchical fashion. Every interior node n has three attached items: a system specification S_n, a coupling scheme C_n, and a correspondence H_n. The idea is that the coupling scheme C_n is to be used to couple the system specifications assigned to the successors. The result is a new system S'_n, which if the tree is valid, will be morphically related to S_n using the correspondence H_n. In particular the result of coupling the first level components will morphically relate to the root system specification.

The formalism will be parameterized by the three kinds of objects just mentioned. Thus in any particular application, a class of system specifications, a class of coupling schemes and a class of morphisms must be specified. The classes of system specifications and coupling schemes are left as free parameters so as to facilitate specialization to a wide variety of modelling formalisms. A natural choice for the class of morphisms is the class of isomorphisms applicable to the chosen class of system specifications. In this case the composition tree represents a hierarchical construction process which starts with the atomic components and results in a system isomorphic to that specified at the root. However, by allowing the extra freedom in choice of morphism, we also capture the case of model construction using aggregated components, an important approach to model construction (Chapter 9).

Formally, a *composition tree* over a class of system specifications, **S**, coupling schemes, **C**, and morphisms, **M** is a structure

$$\text{tree}(\mathbf{S},\mathbf{C},\mathbf{M}) = <T,m>$$

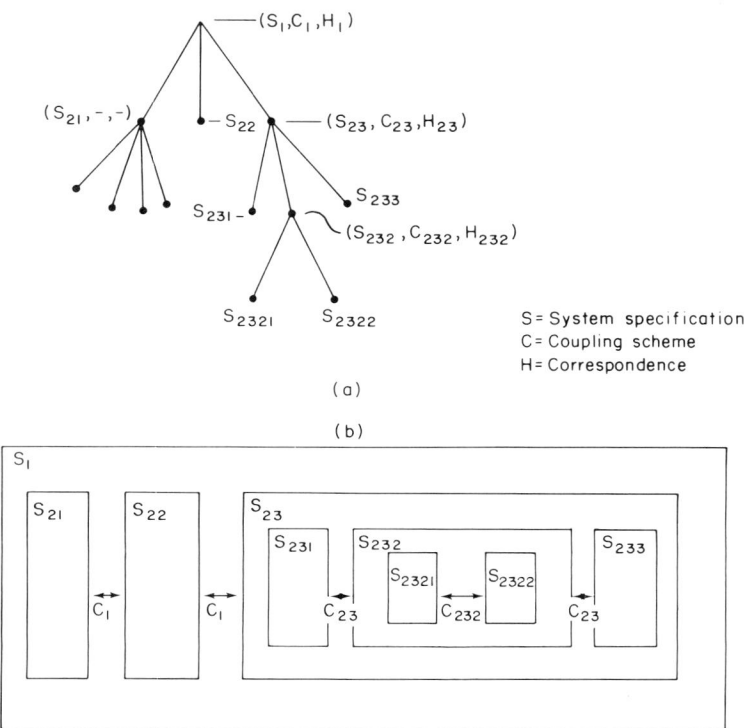

Fig. 15.3. A composition tree showing some of its node assignments and a representation of the hierarchically specified model it represents.

where

T is a finite tree (Section 2.1)

m is a mapping, the *node labelling* of T

subject to the constraints:

m: interior__nodes$(T) \to \mathbf{S} \times \mathbf{C} \times \mathbf{H}$[37]

m: leaves$(T) \to \mathbf{S}$

[m assigns to each interior node (including the root) a triple consisting of a system specification, a coupling scheme (for coupling the system specifications at the successors of the node) and a correspondence for comparing the system specification with the resultant of the coupling of its components; the

---

[37]$\mathbf{H}$ is the set of correspondences underlying the morphisms in $\mathbf{M}$. For example, a state-state correspondence becomes a homomorphism if it preserves the state transition and output functions.

leaves receive only system specifications which are *atomic* and not subject to decomposition. The atomic systems are the ones which are the ultimate components of the composition.]

Let n be an interior node and let $m(n) = (S_n, C_n, H_n)$. Let S'_n be the resultant of applying the coupling scheme C_n to the set of systems labelled by the successors of node n, i.e.

$$S'_n = \text{Resultant}(\{S_a | a \text{ is a successor of } n\}, C_n)$$

Then we say that node n is *valid*, if the correspondence H_n induces a morphism from S'_n to S_n in class **M**. In particular, that the resultant S'_n exists is a prerequisite for validity. As just introduced Resultant is a partial mapping which associates a system in **S** with a pair $<\{S_a\}, C>$ specifying a set of systems in **S** and coupling scheme C. The Resultant must be given an algorithmic definition appropriate for a class **S** of interest. A prerequisite for existence of a resultant is the consistency of the coupling C with the set of system specifications it is intended to couple. Such a consistent pair $<\{S_a\}, C>$ constitutes a Coupling of Systems specification at level 5 (Section 3.1).

The composition tree is *valid* if each of its interior nodes is valid. The *resultant* of a composition tree is the system assigned to the root. A valid composition provides a hierarchical specification for the resultant. As the atomic components are the ultimate constituents of this construction, it is natural to suppose that under certain circumstances, the resultant can also be specified directly as a coupling of the atomic components. We call such a composition, the *flattened composition* associated with the composition tree.

The composition tree can be used to formalize hierarchical model construction with or without aggregation. Unaggregated construction, the more familiar in non-modelling contexts, is brought about by restricting the morphism class **M** to isomorphisms. Construction of a composite model using aggregated components is obtained when the morphism class includes proper morphisms. A morphism is *proper* if it is not an isomorphism (Chapter 9, Section 13.4).

Interpretation of the composition tree formalism in the contexts of coupled system specification, multicomponent DEVS, and I/O coupling schemes (Section 15.2) are detailed in Appendix 1.

Hierarchical model construction using aggregation can be readily formalized by retaining the preceding concepts and allowing some of the morphisms to be proper ones. To see this, let n be an interior node, let $m(n) = (S_n, C_n, H_n)$, and let S'_n be the resultant of the coupling scheme C_n applied to the component systems at the successor nodes of n. Then S_n is a *valid simplification* of S'_n if the correspondence H_n induces a morphism from S'_n to S_n and S_n is simpler than S'_n with respect to some complexity measure (Sections 5.8 and 2.2). Usually, but not always [3] (Chapter 16), the morphism involved is proper in this case. If S_n is a valid simplification of S'_n then, we also say that S_n is a valid *aggregation* at node n of the subtree hanging below it.

3.1. Mapping a Pruned Entity Structure into a Composition Tree

At this point we have a formalism for specifying hierarchical models (the composition tree) which must be generated from the pruned entity structure. There are two basic approaches to generating such a hierarchical model specification: so-called *top down* and *bottom up*. (It should be noted that the actual process that humans seem to follow is an iteration between the two.) The *bottom up* methodology constructs or selects "off the shelf" model components for the atomic entities consistent with the entity orientation. It then proceeds up the pruned structure, selecting coupling schemes consistent with the aspect constraints encountered at each stage. In *top down*, one starts with a model specification for the root entity which is consistent with the entity orientation and proceeds downward, selecting components for the next level entities and a coupling scheme consistent with the aspect coupling constraints.

Hierarchical *construction using aggregation* is a bottom up process which is more complex. This is because a simplification process also takes place in which the resultant of the coupling of systems at a node is simplified to produce the system at the node which will itself be employed at the next higher level. Thus in this case, one starts with the atomic components and proceeds to couple them, and simplify the resultant, assigning it to the next higher level node. This upward process proceeds until all components are coupled in one system assigned to the root.

3.2. Hierarchical Model Construction: The Ecosystem Example Continued

We continue with the ecosystem example (Section 15.1.2) to illustrate the generation of composition trees from pruned entity structures. The models specified by these trees are discussed informally and many details of ecological substances are omitted. The interested reader is referred to [1,4] for further information.

A variety of models can be constructed on the basis of prunings of the given entity structure. The overall ecosystem can be represented without regard to patch structure focussing only on aggregate population numbers. Or each patch can be modelled as an I/O system and the patch models can be coupled together to form a disaggregated model of the ecosystem. Correspondingly there are experimental frames that apply to the ecosystem as a whole (to which the aggregated and disaggregated models apply), those that apply to individual patches or to individual populations. The constructed models may also be represented in a variety of formalisms (differential equation, DEVS, etc.) or their subformalisms (cell space, modular, etc.). We shall enumerate some of the models demonstrating the pruned entity structure and the composition tree that underlies the structure of each.

Model M1 (Figure 15.4)

M1 models individual motion in the interpatch terrain. This is discrete event model that samples a time interval for an individual to travel and the location at which it will be at the end of the interval. The model is based on a random walk hypothesis summarized by diffusion like distribution. A lifetime is also sampled for the individual, which if shorter than the travel time, causes the death of the individual. Since, in the interpatch terrain, individuals are assumed not to interact, the coupling constraints restrict composition to the simple parallel form.

Model M2

M2 models the interaction of species populations on a patch. Its state variables are the population numbers and differential equations describe the rates of change of these numbers in terms of such interactions as predation (conversion of preys to predators) and birth/death rates (effects of a species population on itself). Immigration of individuals of each species is represented by suitable input variables of the model and its time behavior is

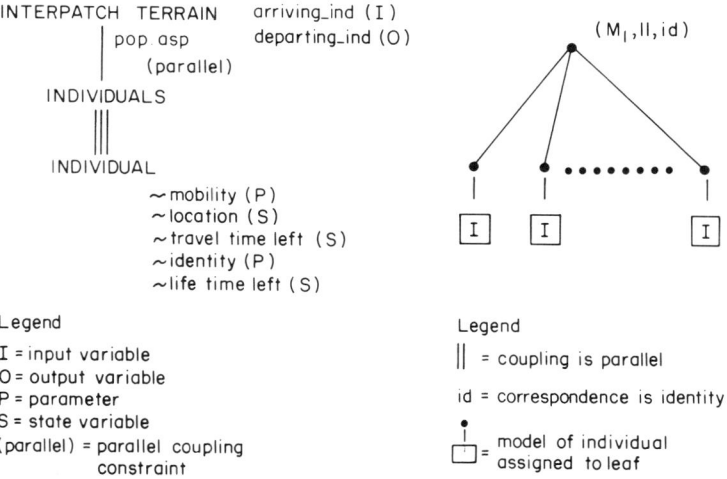

Fig. 15.4. Developed entity structure and composition tree for model M_1.

286   *4. Multifacetted system modelling*

represented by discrete event segments. The effect of an arrival is to increment the appropriate population by unity. Emigration constitutes the output variables and is also discrete event in form, migrants being extracted from the population due to overcrowding or lack of food (triggered when certain partition boundaries of the state space are crossed).

Although, it is not shown as such, Model *M2* can be constructed by aggregation of individuals partitioned into their species blocks. The correspondence from the base model (coupling of individuals) to the lumped model *M2* is the natural counting of block elements. This correspondence may be justified as a system morphism under the uniformity conditions of Section 9.2. Uniformity of components within blocks is obvious; uniformity of influence occurs if the populations maintain a homogeneous distribution in space and individuals cannot distinguish other individuals any finer than their species membership.

Model M3

M3 is a discrete event model of a patch derived from model *M2* by the transformation procedure of Section 9.4. The population number state space is partitioned into blocks and tables are developed for the time advance, next state, and external transition function of the DEVS via simulation of *M2*. For example, consider starting with states in which there are no predators and the prey population is non-zero. Tables are constructed for the time it takes the prey to reach the population level at which migration is initiated, its population size at that time, and the effect of an immigrant's arrival during this phase are tabulated. Model *M3* retains the same input/output interface of *M2* so that it can replace *M2* as a component in coupled systems.

Model M4

M4 is a discrete event model of a patch which identifies only a small number of patch states such as empty, prey only, etc. (see example in Section 6.2.1). *M4* may be placed into correspondence with *M2* (or *M3*) by identifying these states with a partition of the population numbers state space. A system morphism could be established were such a partition to have the congruence property (Section 9.1), a condition that is unlikely in most realistic models. However, approximate congruency may achieved with judicious partitioning and an approximate validity may result sufficient for gaining qualitative insight.

Models M5, M6, M7 (Figure 15.5)

Models *M5*, *M6* and *M7* are models of the ecosystem derived by coupling together as components, patch models of types *M2*, *M3* and *M4* respectively. In the first two cases, the component models have input/output interfaces

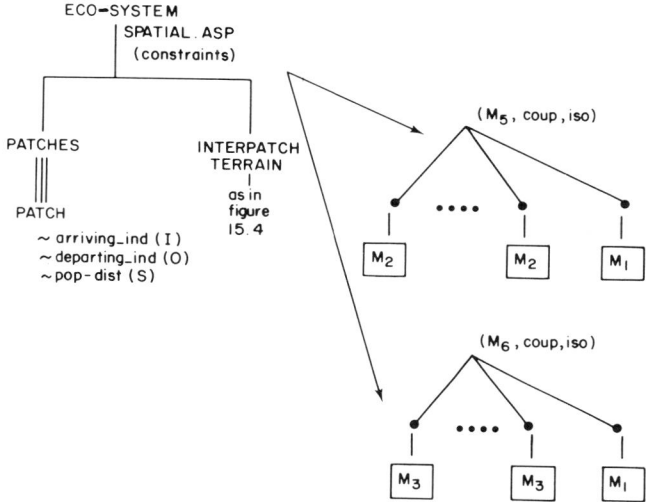

Fig. 15.5. Developed entity structures and composition trees for models M_5, M_6.

based on individual migration. Thus they are coupled by linking emigration outputs to immigration inputs through migration model *M1*. Thus the output (emigrant of a patch) is fed to a copy of *M1* to decide whether it survives the ordeal of migration, and if so to which patch and with what delay it is to be input. Since *M2* and *M1* are differential equation and discrete event models, respectively, the coupling of components of such types is a so-called *mixed formalism* system specification. Once its discrete event version *M3* replaces *M2* in the coupling of systems, the specification becomes a pure multicomponent DEVS model in modular form. In the last case, in which the patch model is *M4*, migration is represented through neighborhoods and probabilities of conversion of cells from one state to another, for example, from empty to prey growth phase. The resulting model is a discrete event cellular space model of the form described in Section 9.3.

Model M8 (Figure 15.6)

Model *M8* is a differential equation model derived from the discrete event cellular space model *M7* using the random phase–random space simplification procedure of Section 9.3. Thus *M8* has as stated space PHASE__ DISTRIBUTION, the vector whose components indicate the number of cells in the space in the various phases, e.g. the number of empty cells, number of prey only cells, etc.

Figure 15.7 summarizes this discussion by presenting the collections of

288   4. Multifacetted system modelling

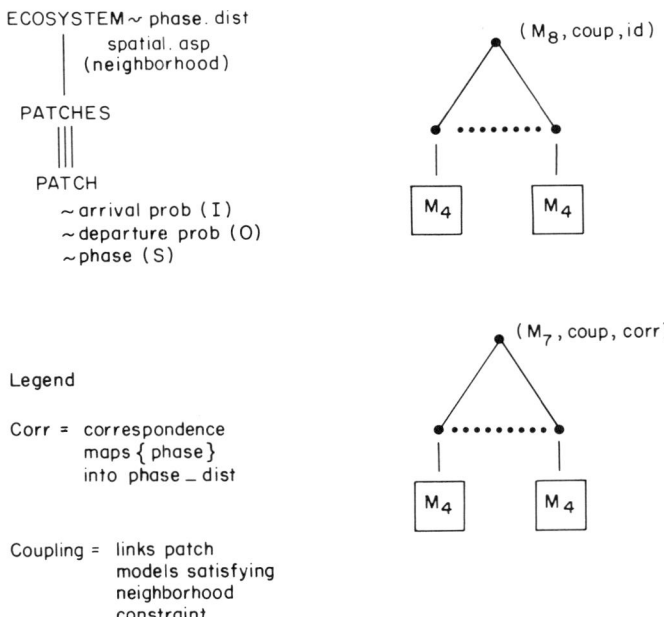

```
ECOSYSTEM ~ phase. dist
         | spatial. asp
         | (neighborhood)
PATCHES
  |||
  PATCH
    ~ arrival prob (I)
    ~ departure prob (O)
    ~ phase (S)


Legend

Corr = correspondence
       maps { phase }
       into phase _ dist

Coupling = links patch
           models satisfying
           neighborhood
           constraint
```

Fig. 15.6. Developed entity structures and composition trees for models M_8, M_7.

frames and models organized by the derivability, applicability and simplification relations.

Exercise. Verify the relations displayed in Figure 15.7. Also deduce all other relationships that can be inferred from those displayed using the fact:

$$E' \leq E \text{ and } E \text{ applicable to } M$$

implies E' applicable to M.

4. Iterative Methodology for Model Construction

The approach to model construction so far discussed can be summarized as follows:
1. operate on the entity structure:
   1.1 identity the entity of interest, ENT
       (choice of system boundaries)
   1.2 formulate an experimental frame, E
       (using objective-driven methodology)

Multifacetted model building methodology   289

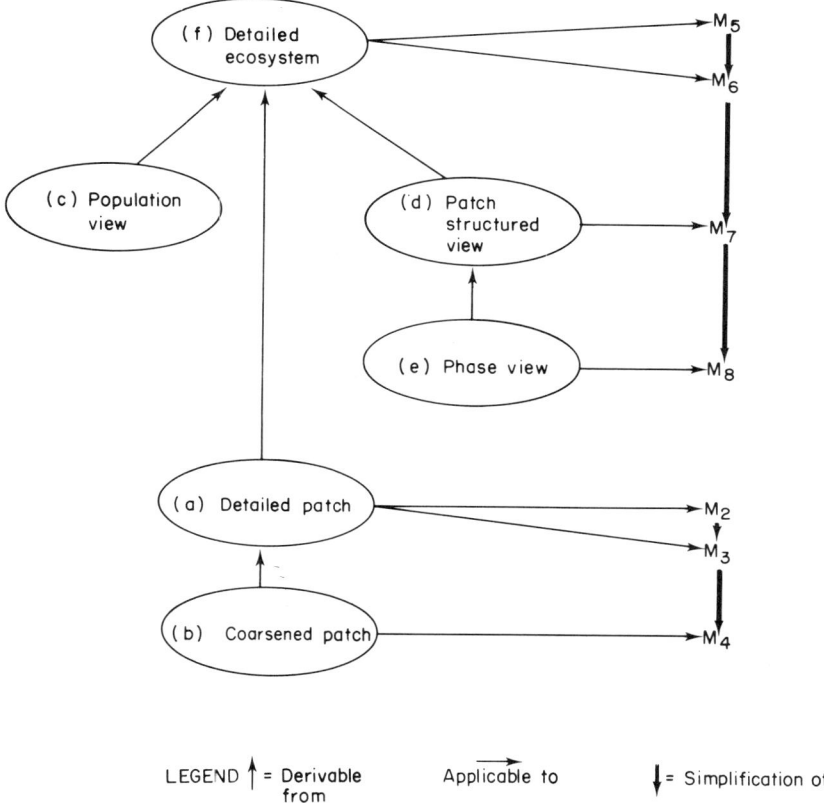

Fig. 15.7. Derivability, applicability and simplification relations for ecosystem example.

    1.3 produce a pruned entity structure, $T_{\text{ENT}, E}$
        (using objectives-guided pruning)
    1.4 expand $T_{\text{ENT}, E}$ to include variables for
        expressing model state and output function
2. map $T_{\text{ENT}, E}$ into a composition tree $<T,m>_{\text{ENT}, E}$
    2.1 orient interface variables
    2.2 designate state variables and parameters
    2.3 specify coupling constraints
    2.4 assign component models and coupling schemes
        meeting constraints of 2.1–2.3
        (using a top down or bottom up approach)

It is time to correct a fiction that we have assumed for simplicity until now. This is that the order in which the above steps are carried out is as portrayed above. In actual fact, the process is more likely to involve shuttling back and forth between entity structure, experimental frame, and model construction operations. For example, a variable desired for an experimental frame may not exist in the entity structure which will thus have to be modified to accommodate it. Also, carrying out later steps in the above order, may remind the modeller of an omission made in an earlier step or indeed may clarify what was unclear earlier. In this case, looping back to earlier steps will be appropriate. Thus while all the steps must be performed no fixed order can be assigned before hand on their execution. This difference between the *logical order* in which steps should be performed according to a methodology, and the *actual* iteration of steps, leads us to think in terms of a flexible set of *activities* rather than a fixed sequencing of steps. We take up this point of view in Chapter 18.

We have delayed discussion of the transition from the objectives-driven mode to the model construction mode represented above by step 1.4. One answer is that in applying the modelling formalisms of Part 2 and our understanding of the real system we are led to employ suitable variables to express the transition and output functions of the model. We iterate back and forth between the substeps of step 2 where we discover what variables are needed and step 1.4 where the pruned entity structure is suitably expanded.

This answer would have to suffice were we not to have a model base available to access in search of models that could help in the construction process. However, by taking the model base into account we can delve more deeply into the processes that are involved in trying to apply existing knowledge (expressed the formal, publicly accessible form of model and experimental frame bases) to meet new modelling objectives.

5. Paradigms for Frame Based Model Retrieval and Synthesis

Consider the following problem: As in Figure 13.1, we have sets of experimental frames **E** and models **M** stored in respective bases. Suppose that we have constructed a frame E and wish to construct a model M to accommodate E. How should we employ the knowledge available to us, i.e. **E** and **M** to assist in the construction of M?

In the formulation of this problem we have abstracted away the role of the entity structure, which we shall re-introduce later. Further, we shall assume the following closure properties:
 1. for each $E \in \mathbf{E}$ there is model $M \in \mathbf{M}$ to which it applies
 2. the scope frame, E_M of each $M \in \mathbf{M}$ is in **E**.

Both these postulates are reasonable. (1) states that we have already constructed models accommodating each of the frames that have been defined. (2) states that we have added the scope frame of each constructed model to the frame base.

Making more restrictive assumptions, we consider a set of experimental frames which forms a lattice **L** relative to the derivability relation.

A lattice structure provides two operations that produce the least upper and greatest lower bounds respectively of all pairs of frames. Consider the set of all frames $U(E,E')$ that dominate frames E and E', i.e. both E and E' are derivable from each frame in $U(E,E')$. Then the *least upper bound*,

$lub(E,E')$ is the unique frame F in $U(E,E')$ which is derivable from all other frames in $U(E,E')$.

Thus $lub(E,E')$ represents the smallest frame that dominates both E and E'. We can think of it as the smallest frame that has the experimentation scope of both E and E'.

The *greatest lower bound* is defined in a dual manner. If $L(E,E')$ is the set of frames derivable from both E and E', then

$glb(E,E')$ is the frame F in $L(E,E')$ from which all frames in $L(E,E')$ are derivable.

Thus $glb(E,E')$ is the largest frame which is derivable from both E and E'. We can think of it as the frame that represents what is common to the experimental conditions expressed in both E and E'.

In addition a lattice has unique maximal and minimal elements 1 and 0, respectively. We assume that frames in **E** belong to **L** so that the *glb* and *lub* operations can be performed on them. We need not suppose that **E** has been closed under these operations, i.e. that **E** is a sublattice of **L**.

The existence of *lub* and *glb* operations place very strong requirements on the derivability relation. The working out of useful classes of frames which have such lattice properties is still an open research problem (see Problem 1). However, there is a class which is readily shown to have the requisite properties.

Exercise. Show that the following class of frames forms a lattice structure: the set of all frames of the form $E = <I,O>$, where I,O are sets of variables (input, output, resp.) and $<I',O'> \leq <I,O>$ just in case, each primed set is included in the corresponding unprimed set (e.g. $I' \subset I$). Hint: take unions for *lub* and intersections for *glb*.

The class of frames, specifying only input and output variables, can be viewed as an abstraction of the class of all frames in which one applies a "forgetful" mapping to wipe out all the other structures that frames can have. As with any abstraction, when reflected back to the original class, results obtained using it may be valid in some domain, may be good approximations in a wider domain, and may be invalid elsewhere. However, the lattice structure of the class makes it a useful abstraction (Section 2.3).

The *glb* operation supplies us with a criterion for relevance of frames. A frame F is *relevant* to a frame E if

$$glb(E,F) \neq 0$$

i.e. E and F have something non-trivial in common.

Call the set of all frames in **E** that are relevant to E, $Rel(E,\mathbf{E})$. The set of models:

$$Rel(E,\mathbf{M}) = \{M \mid M \in \mathbf{M}, M \text{ accommodates some } F \in Rel(E,\mathbf{E})\}$$

constitutes all models in **M** that are relevant to E.

Exercise. Show that

$$Rel(E,\mathbf{M}) = \{M \mid M \in \mathbf{M}, glb(E,E_M) \neq 0\}$$

where E_M is the scope frame of M. Hint: use postulate 2.

Exercise. Express $Rel(E,\mathbf{E})$ and $Rel(E,\mathbf{M})$ in the case of the class of I/O frames.

Let us formulate conditions under which this set of models provides a set of components which can be coupled together to accommodate E.

When E is already in **E** or is derivable from some F in **E**, then by postulate 1, and the transitivity of derivability, there is a model M in **M** that accommodates E. In this case, no coupling is required.

In lieu of such a fortunate occurrence, suppose that there is no frame in **E** from which E is derivable, but that the relevant frames form a *complete* set in the following sense:

$$E \leq lub\{F \mid F \in Rel(E,\mathbf{E})\}$$

that is, E is derivable from the smallest frame formed by putting together the relevant existing frames.

Exercise. Show that the completeness criterion is equivalent to:

$$E \leq lub\{E_M \mid M \in Rel(E,\mathbf{M})\}$$

that is, E is derivable from the smallest frame formed by putting together the scope frames of the relevant models.

Exercise. Express the criterion for completeness in the case of the class of I/O frames.

Then if we couple models in $Rel(E,\mathbf{M})$ we stand a good chance of constructing ones that accommodate E. In particular we require a subset of models in $Rel(E,\mathbf{M})$ and a coupling scheme whose resultant M_R accommodates E.

Exercise. Show that in the case of I/O frames, if the completeness criterion is satisfied by the models relevant to a frame E, then there exist many Wymore coupling schemes (Appendix 1) for which E is applicable to the resultant M_R.

Any such model however is likely to have much structure irrelevant to E in it and we would want to apply simplification procedures (Chapter 9) to it to produce a valid simplification in frame E (Section 13.4).

Model construction using aggregation (Section 13.3) offers another approach to the simplification problem. In this approach we first simplify each of the component models to accommodate only the part of E that it is relevant to. That is for each model $M \in Rel(E,\mathbf{M})$, construct a valid simplification of it in frame $glb(E,E_M)$. Then couple the simplified models together to construct a resultant. In general, the resultant is not guaranteed to accommodate E. What is required is that the simplified models continue to form a complete set for E. However, completeness is assured in the case of distributed lattices, and in particular in the case of the class of I/O frames.

Exercise. Show that the set of simplified models is still complete with respect to E if $\mathbf{L}$ is a distributive lattice, i.e.

$$lub(glb(E,F),glb(E,G)) = glb(E,lub(F,G))$$

for all frames E,F,G in $\mathbf{L}$. In particular the lattice of I/O frames is distributive.

If, as is likely, the set of relevant models, whether in simplified or original form, is not complete for E, then these models must be elaborated and/or augmented with additional components constructed from scratch. Thus ultimately the model builder will no doubt have to turn to his creative intuition and understanding. However, the irreducible creative part may be much less extensive than would have been the case without the support of the frame and model based system. Moreover, more energy can be focussed on this residual creative activity rather than being spread out over the whole of the model building process. Of course, this supposes that algorithmic activities such as discovery of relevant frames and models, and model simplification are largely carried out by the computer. To build computer systems capable of realizing this goal is the challenge presented by multi-facetted methodology.

We shall summarize frame directed retrieval of models by incorporating it within the entity structure based activities formulated in Section 15.4. Note that frame directed activities are sandwiched between those related to the entity structure and those related to composition tree. We suppose that after discovering the relevant models, and going through a simplification process, the model structures are amalgamated among themselves and with the pruned entity structure derived from the frame (Section 14.2). *This amalgamation represents both what is necessary to meet the modelling objectives and what of relevance is available in the model base to meet these objectives.* Again, we stress that the order presented is a logical one and there is likely to be considerable iteration between the entity-based, frame-directed, and model-construction activities.

1. operate on the entity structure:
    1.1 identity the entity of interest, ENT
    (choice of system boundaries)
    1.2 formulate an experimental frame, E
    (using objective-driven methodology)
    1.3 produce a pruned entity structure, $T_{ENT,E}$
    (using objectives-guided pruning)
2. Access the experimental frame and model bases
    2.1 discover frames and models relevant to E
    2.2 simplify relevant models relative to E
    2.3 expand $T_{ENT,E}$ to include amalgamation of models (using amalgamation (Section 14.2))
    2.4 add additional variables required for necessary model elaboration
3. map $T_{ENT,E}$ into a composition tree $<T,m>_{ENT,E}$
    3.1 orient interface variables
    3.2 designate state variables and parameters
    3.3 specify coupling constraints
    3.4 assign component models and coupling schemes meeting constraints of 3.1–3.3
    (using a top down or bottom up approach)

6. Summary

This chapter ends our discussion of multifacetted system modelling with a presentation of its methodology for model building. All of the concepts developed in previous chapters have been marshalled to support this methodology. Objectives-driven methodology leads to development of experimental frames that in turn, serve to guide pruning of the entity structure, and access to the model base, ultimately, resulting in the specification of a composition tree for model synthesis.

Our discussion of this methodology is readily seen to have only scratched the surface. Much more remains to be learned empirically through experimentation with computerized versions of the methodology and this will interact with further conceptual and theoretical developments. In the last part of the book, we shall describe and analyse existing software and hardware systems as potential steps toward support of multifacetted methodology. However, before doing this, we turn to the simulation side of multifacetted methodology to examine how the hierarchical and modular models specified in a composition tree can be implemented on simulators and subjected to experimentation.

PROBLEMS

1. Referring to the ecosystem system example of this chapter, formulate a novel experimental frame and "simulate" following through the methodology summarized in Section 15.5 to specify a composition tree for the model.
2. Extend the concepts of commonality, union, relevance and completeness (developed under the convenient assumption of lattice structure) in two ways: a) to the class of all frames under the derivability relation of Section 13.1.7 or b) relax the lattice assumptions partially and find useful classes of frames that satisfy the relaxed assumptions.

Appendix 1. Interpretations of Composition Tree Formalism

Hierarchical construction with Wymore's Coupling Recipe

We consider Wymore's [2] coupling recipe concept, that was informally introduced in Section 15.2. Let the class of system specifications **S** be the set of *ported input/output specifications*. S in **S** is given as a pair $<input\ ports, output\ ports>$ where *input ports* and *output ports* are finite sets of ports, a *port* being a pair $<name\_spec, range\_spec>$. The interpretation of a port specification is that name_spec is its unique identifier and range_spec is the set of values that it can take on. In fact, except for its nautical associations, the concept of port is "isomorphic" to the concept of variable, defined in Section 11.1. Thus we are considering system specifications at the Observation Frame Level (Chapter 3) except that the input and output sets are structured. For the balance of this section we employ the port, rather than the variable, terminology.

The class of coupling schemes **C** is defined as follows. Let a *link* be a pair of ports ($<p_1,r_1> \rightarrow <p_2,r_2>$) satisfying the requirement that the range set r_1 is included in the range set r_2. The interpretation of a link is that the output port p_1 is to be connected to the input port p_2. A *coupling scheme* is a finite set of links such that no port appears more than once.

A *coupling recipe* is a pair consisting of a finite set of ported I/O specifications and a coupling scheme. The coupling recipe is *consistent* if for each link ($<p_1,r_1> \to <p_2,r_2>$) in the scheme, $<p_1,r_1>$ is one of the output ports of the component I/O specifications, and $<p_2,r_2>$ is one of the input ports. Thus a coupling recipe is consistent if the links of the coupling scheme refer to existing ports in the set of I/O specifications.

The *resultant* of a consistent coupling recipe always exists and is defined as follows. An (input or output) port of a component I/O specification is *free* if it does not appear in any of the links of the coupling scheme. The resultant of a coupling recipe is the I/O specification having as input ports the free input ports, and as output ports, the free output ports.

To set up the appropriate morphisms for Wymore coupling consider the following definitions. Two sets of ports are *isomorphic* if there is a one-one correspondence between them such that corresponding ports have identical range specifications. Two ported I/O specifications S_1 and S_2 are *isomorphic* if their input and output port sets are (respectively) isomorphic.

We shall require that the resultant of applying the coupling scheme at a node to the I/O specifications at its successors be isomorphic to the I/O specification at the node. Thus the appropriate morphism concept for the class of ported I/O specifications is the class of isomorphisms just defined. The underlying correspondence that must be associated with a node is the pair of one-one correspondences relating the input and output ports of the specification at the node to the free input and output ports (respectively) of the resultant.

To summarize: a composition tree based on Wymore's coupling concept is a structure

$$\text{Tree}(\mathbf{S},\mathbf{C},\mathbf{M}) = <T,m>$$

where

$\mathbf{S}$ = the class of ported I/O specifications
$\mathbf{C}$ = the class of Wymore coupling schemes
$\mathbf{M}$ = equality morphisms

Let m assign to an interior node n an I/O specification, S_n and a coupling scheme C_n. A coupling recipe at node n is formed from the I/O specifications at the successors of n and the coupling recipe C_n. If the coupling recipe is consistent, then we associate its resultant S'_n to the node n. Then node n is valid if $S'_n = S_n$. The composition tree is valid, if each interior node is valid.

Exercise. The flattened composition is readily defined using Wymore's coupling concepts. Show that it is the coupling recipe whose set of components is the set of atomic I/O specifications and whose coupling scheme is the set of links in the coupling schemes of all the interior nodes.

Hierarchical Coupled System Specification

As the second example, consider the case of constructing models from atomic components that are specified at the Structured System Level (level 4 of the specification hierarchy Chapter 3). A consistent coupling scheme applied to such components constitutes a Coupling of Systems specification at level 5. The resultant of such a coupling, when it exists, is just the association of a structured system with it (the level transition $5 \to 4$.) If we require that the resultant of coupling at a node is isomorphic to the system at the node, then the appropriate morphism class is the Structured System Isomorphism.

Thus a composition tree for this case is a structure:

$$\text{Tree}(S,C,M) = <T,m>$$

where

  S = Structured System Specifications
  $C = \{C | C = <D, \{I_\alpha\}, \{Z_\alpha\}>\}$
  M = Structured System Isomorphisms.

Note that a coupling scheme C consists of an indexing set D, an indexed family of subsets of D (potential influencers), and an indexed family of functions (potential interface maps). For an interior node n, let $m(n) = (S_n, C_n, H_n)$. The coupling scheme C_n is *consistent* if the index set D_n is the set of successor nodes of n and each interface map Z_α maps the composite of the output sets of its influencers I_α into its input set. Thus if C_n is consistent it forms a Coupling of Systems when combined with the set of structured systems at the successors of n. The mapping H_n is a one-one correspondence from the states of S'_n (the resultant of the Coupling of Systems) to those of S_n. The node n is valid if this one-one correspondence preserves the transition and output functions so as to become an isomorphism.

Hierarchical Modular DEVS Specification

Hierarchical model construction within a particular modelling formalism is readily specified with the composition tree concept for any formalism that is closed under composition (Chapter 3). Of special interest here is the DEVS formalism, shown to be closed under composition in Section 8.2. All that must be done is to specialize the interpretations given in the previous section for arbitrary structured systems to the DEVS subclass.

A *Hierarchical DEVS* is a structure

$$\text{Tree}(S,C,M) = <T,m>$$

where
  S = the set of DEVS models
  $C = \{C \mid C = <D, \{I_\alpha\}, \{Z_{\alpha,\beta}\}>\}$
  M = the set of DEVS isomorphisms ([3], Chapter 10).

A coupling scheme C consists of an indexing set, an indexed family of subsets of D (the potential influencee sets), and an indexed family of functions (the potential output translation maps). It is straightforward to provide the appropriate definition of consistency for such schemes and for validity of nodes and the composition tree.

References

[1] Hogeweg, P. (1981). "Two Predators and One Prey in a Patchy Environment: An Application of MICMAC Modelling" *J. Theor. Biol.* **93**, 411–432.
[2] Wymore, W. (1980). *A Mathematical Theory of Systems Design*, Tech. Rept, College of Engineering, University of Arizona, Tucson.
[3] Zeigler, B. P. (1976). *Theory of Modelling and Simulation*, Wiley, NY.
[4] Zeigler, B. P. (1976). "Multi-level, Multi-formalism Modelling: An Ecosystem Example". In *Theoretical Systems Ecology*, (ed, E. Halfon), Academic Press, NY.

Other Relevant Literature
Zeigler, B. P. (1978). "Structuring the Organization of Partial Models", *Int. J. Gen. Systems* **4**, 81–88.
Zeigler, B. P. (1983). "Structures for Model-Based Simulation Systems". In *Simulation and Model-Based Methodology: An Integrative View* (eds, T. I. Oren *et al.*), Springer-Verlag, NY.

PART 5
MULTIFACETTED SYSTEM ARCHITECTURES

Chapter 16

MODULARITY IN MODELS AND EXPERIMENTAL FRAMES

The model/frame separation principle was introduced in Section 12.3.2 as a general guideline for the modular specification of a model and the frames intended to be applicable to it. This chapter develops the programming technology to support the realization of simulation schemes specified by model/frame/execution control triples. We distinguish between *on-line* experimentation, in which experimental analysis is performed during the running of a model, and *off-line* experimentation, in which results of experimentation are analysed *after* simulation of the model.

For on-line experimentation, we shall show how a construct called a change detector, based on the monitored variable feature of SIMSCRIPT 11.5, can be employed to link together code derived from model and experimental frame specifications in such a way that the integrity of each is not affected by the other. The modularity afforded in this way provides for independent modification of model and frame segments. Moreover, such model/frame assemblies can be disassembled into the original code segments and re-used in other combinations. The flexibility that results is fundamental to a software system that supports multifacetted modelling and simulation methodology.

A crucial part of on line experimentation is the third component of a simulation scheme, the execution control module. We shall emphasize the design of such a module to provide advanced assistance to the user in interactive experimentation with models. The concepts of state variable sets and of stratification of variables and parameters according to their likelihood of change will provide the basis for the design approach.

The chapter closes with a discussion of off-line experimentation as the most general form of viewing the behavior of a model or real system under various experimental frames.

1. Modular Realization of Model/Experimental Frame Couplings

Recall that an experimental frame is a structure

$$< I, C, O, \Omega_I, \Omega_C, SU >$$

specifying sets of input variables I, run control variables C, and output variables O; admissible input segments, Ω_I and run control segments, Ω_C; and summary processing SU (Section 12.2).

When interpreted in structural form, an experimental frame specifies three systems connected to the model as shown in Figure 12.3. The Input system is most commonly, a generator of the input segments Ω_I, but could also be an acceptor, which selects the input segments from a wider class. The output system is a transducer which observes model input/output segment pairs and performs the statistical and other processing specified by the SU component of the frame. The Run Control system is an acceptor which observes a model control variable segment and indicates acceptance or rejection of an experiment according to whether or not the segment belongs to the admissible class, Ω_C.

Linkage of frame derived components to the model requires that certain interface requirements be satisfied. Satisfaction of these requirements is embodied in the "applicability" relation between frames and models (Section 13.1.10). The most direct form of applicability requires the presence of the input, output and control variables specified by the frame as variables in the model. In this section we restrict our discussion to this direct form of applicability and to its realization in on-line experimental form.

Coupling of the Input, Output and Run Control DEVS systems to the model, itself a multicomponent DEVS, results in a hierarchical multicomponent DEVS as defined in Section 8.2. Consequently, it can be realized with the constructs presented in Section 12.5. More specifically, note that the Input system is an influencer of the component models with free input variables. Thus its output translation map should send signals to the external transition functions of these components. Likewise, the output transducer and run control acceptor are influencees of the model components whose variables they read. So signals carrying the updated values of these variables should be generated whenever changes occur to these variables.

However, this approach works to violate the goal of separation of model and experimental frame in both conception and realization. This is so because the output specifications of the component models would be required to play two distinct roles: one, to compute inputs to the other model components as specified by the model coupling scheme, and two, to compute inputs to the transducer and control systems, as specified by the experimental frame. This duality of function is not in keeping with the goal of maintaining a clean separation between model and experimental frame.

302 5. *Multifacetted system architectures*

A second solution to the coupling of the experimental frame systems to those of the model is that represented by the monitored variable concept of SIMSCRIPT 11.5 [2]. A variable which is declared as *monitored on the left* must be supplied with a so-called left routine of the same name. Whenever the variable is about to be assigned a value, the corresponding routine is automatically called to intercept it. The intercepted value may be observed, and altered, by this and other routines before being returned to the original point of control.

1.1. Realization of Model/Experimental Frame Modules in SIMSCRIPT 11.5

Consider an example of the use of the monitored variable to realize programs that exhibit separation between model and experimental frame components.

The example is that of a simple grocery store for which the DEVS multicomponent structure is shown in Figure 16.1. Customers arrive at the store, do some shopping, wait to be served at the checkout, and depart (the

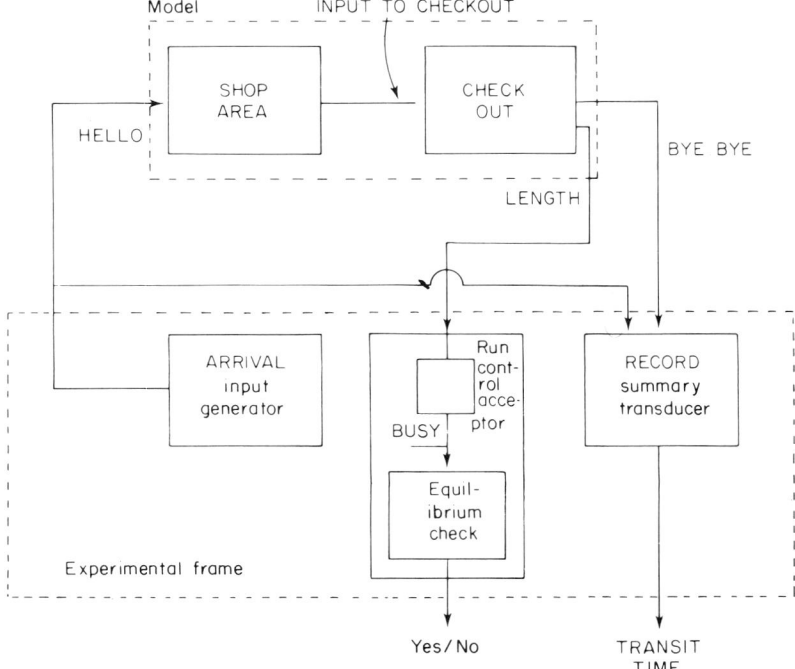

Fig. 16.1. DEVS model and experimental frame for grocery store.

original model was expressed in next event formalism in [4] pp. 175–188.

An extract of the SIMSCRIPT program for realizing the model and an applicable experimental frame is shown in Figure 16.2. The transition function of the model is realized in SIMSCRIPT with the two events SHOPAREA and CHECKOUT belonging to components of the same name. The SHOPAREA and CHECKOUT events occur when a customer has finished shopping and when a customer has finished checking out, respectively.

```
PREAMBLE
    :
    :
DEFINE BUSY AS A VARIABLE MONITORED ON THE LEFT
DEFINE UTILIZATION AS A REAL VARIABLE MONITORED ON THE LEFT
DEFINE HELLO AS A VARIABLE MONITORED ON THE LEFT
DEFINE BYE.BYE AS A VARIABLE MONITORED ON THE LEFT
END


MAIN
"****************   MODEL DESCRIPTION   ****************

"::::::::::::::internal transition function::::::::::
EVENT SHOPAREA SAVING THE EVENT NOTICE
REMOVE THE FIRST CUSTOMER FROM SHOPPERSLIST
CALL INPUT.TO.CHECKOUT(NAME(CUSTOMER))
IF SHOPPERSLIST IS NOT EMPTY "THEN
SCHEDULE THIS SHOPAREA IN TIME(F.SHOPPERSLIST)
ALWAYS RETURN
END


EVENT CHECKOUT SAVING THE EVENT NOTICE
REMOVE THE FIRST CUSTOMER FROM THE LINE
LET LENGTH = N.LINE
LET BYE.BYE = NAME(CUSTOMER)
DESTROY THE CUSTOMER
IF LINE IS NOT EMPTY "THEN
SCHEDULE THIS CHECKOUT IN EXPONENTIAL.F(SERVETIME,3) UNITS
ALWAYS RETURN
RETURN
END


ROUTINE INPUT.TO.CHECKOUT(NAME)
DEFINE NAME AS A VARIABLE
IF LINE IS EMPTY "THEN
SCHEDULE A CHECKOUT IN EXPONENTIAL.F(SERVETIME,3) UNITS
ALWAYS
FILE NAME IN LINE
LET LENGTH = N.LINE
RETURN
END
```

```
::::::::::::external transition function::::::::::::
ROUTINE EFFECT.OF.ARRIVAL
CREATE A CUSTOMER CALLED NEW.CUST
LET NAME(NEW.CUST) = NEW.CUST
LET TIME.LEFT(NEW.CUST) = EXPONENTIAL.F(SHOPTIME,2)
LET HELLO = NAME(NEW.CUST)
IF SHOPPERSLIST IS EMPTY "THEN
SCHEDULE A SHOPAREA IN TIME(NEW.CUST)
FILE NEW.CUST IN SHOPPERSLIST
RETURN
OTHERWISE
FOR EACH CUST IN SHOPPERSLIST
    SUBTRACT ELAPSED.TIME FROM TIME.LEFT(CUST)
IF TIME.LEFT(NEW.CUST) < TIME.LEFT(F.SHOPPERSLIST) "THEN
CANCEL THE SHOPAREA
DESTROY THE SHOPAREA
SCHEDULE A SHOPAREA AT TIME(NEW.CUST)
ALWAYS
FILE NEW.CUST IN SHOPPERSLIST
RETURN
END

   FUNCTION ELAPSED.TIME

DEFINE LTIME AS A SAVED,REAL VARIABLE
DEFINE ET AS A REAL VARIABLE
LET ET = TIME.V - LTIME
LET LTIME = TIME.V
RETURN WITH ET
END
"********** EXPERIMENTAL FRAME SECTION ***********
EVENT ARRIVAL SAVING THE EVENT NOTICE
CALL EFFECT.OF.ARRIVAL
RESCHEDULE THIS ARRIVAL IN EXPONENTIAL.F(INTAR,4) UNITS
RETURN
END
"_____
ROUTINE BUSY.COMPUTATION(LENGTH)
DEFINE LENGTH AS A VARIABLE
IF LENGTH > 0 "THEN
LET BUSY = 1 RETURN
OTHERWISE LET BUSY = 0
RETURN
END
"_____
ROUTINE UTIL.COMPUTATION(BUSY)
DEFINE BUSY AS A VARIABLE
DEFINE LTIME AS A SAVED,REAL VARIABLE
DEFINE ELAPSED.TIME AS A REAL VARIABLE
DEFINE TOTAL.ELAPSED.TIME AS A SAVED,REAL VARIABLE
DEFINE INTEGRAL AS A SAVED,REAL VARIABLE
LET ELAPSED.TIME = TIME.V - LTIME
ADD ELAPSED.TIME TO TOTAL.ELAPSED.TIME
LET LTIME = TIME.V ADD BUSY*(ELAPSED.TIME) TO INTEGRAL
IF TOTAL.ELAPSED.TIME IS NOT ZERO
    CALL EQUILIBRIUM.TEST(INTEGRAL/TOTAL.ELAPSED.TIME)
```

```
ALWAYS
RETURN
END
"_____
ROUTINE RECORD(NAME)
DEFINE NAME AS A VARIABLE
CREATE AN ENTRY
LET ENAME(ENTRY) = NAME
LET ETIME(ENTRY) = TIME.V
FILE ENTRY IN ELIST
RETURN
END
"_____
ROUTINE TRANSIT.COMPUTATION(NAME)
DEFINE NAME AS A VARIABLE
FOR EACH ENT IN ELIST WITH ENAME(ENT) = NAME
FIND THE FIRST CASE
IF FOUND "THEN
PRINT 1 LINE WITH ENAME(ENT),TIME.V-ETIME(ENT) THUS
CUSTOMER ********* SPENT TOTAL TIME ******.******* IN THE STORE
REMOVE ENT FROM ELIST
DESTROY THE ENTRY CALLED ENT
ALWAYS
RETURN
"_____
ROUTINE EQUILIBRIUM.CHECK(UTILIZATION)
DEFINE UTILIZATION AS A REAL VARIABLE
DEFINE PREVIOUS AS A SAVED,REAL VARIABLE
IF ABS.F(UTILIZATION – PREVIOUS) > 0.01 "THEN
PRINT 1 LINES THUS
***********EQUILIBRIUM VIOLATED***********
STOP STOP
ALWAYS
LET PREVIOUS = UTILIZATION
ALWAYS
RETURN
END
"**************** MONITORING SECTION *****************
"_____
LEFT ROUTINE LENGTH
DEFINE TEMP AS A VARIABLE
ENTER WITH TEMP
CALL BUSY.COMPUTATION(TEMP)
ALWAYS MOVE FROM TEMP
RETURN
END
"_____
LEFT ROUTINE BUSY
DEFINE TEMP AS A VARIABLE
ENTER WITH TEMP
CALL UTIL.COMPUTATION(TEMP)
ALWAYS MOVE FROM TEMP
RETURN
END
"_____
```

```
LEFT ROUTINE HELLO
DEFINE TEMP AS A VARIABLE
ENTER WITH TEMP
CALL RECORD(TEMP)
ALWAYS MOVE FROM TEMP
RETURN
END
"----------------------------------------------
LEFT ROUTINE BYE.BYE
DEFINE TEMP AS A VARIABLE
ENTER WITH TEMP
CALL TRANSIT.COMPUTATION(TEMP)
ALWAYS MOVE FROM TEMP
RETURN
END
```

Fig. 16.2. Extract of SIMSCRIPT realization of grocery store model in equilibrium transit time experimental frame.

Note that the SHOPAREA and CHECKOUT components are realized as separate modules with the ROUTINE INPUT.TO.CHECKOUT being called from the SHOPAREA event to provide the coupling that sends a customer from the SHOPAREA to the CHECKOUT. (We shall examine the modular realization of model components in Chapter 17.)

The input variable to the model is HELLO that indicates the arrival of a customer to the store. The output variables available for experimental frame inspection include LENGTH (of the CHECKOUT.LINE) and BYE.BYE (indicating the departure of customers from the store).

The experimental frame realized in the program manifests an interest in the transit time of customers under equilibrium conditions (Figure 16.1). In the SIMSCRIPT program, transit time calculation is realized in a transducer that is implemented by the ROUTINE TRANSIT.COMPUTATION. This routine is called whenever a customer arrives and departs the store. The invocation however is not direct but is mediated by the LEFT ROUTINEs for HELLO and BYE.BYE. As indicated above, these routines are called automatically whenever the respective variables are about to be given new values, viz, when HELLO and BYE.BYE receive the names of an arriving and departing customer, respectively. Thus customers do not do their own transit time computation in accord with the discussion of the model/frame separation principle of Section 12.3.2.

Note that the input segments of the frame are realized by a Poisson generator that is implemented by the EVENT ARRIVAL. This event calls EFFECT.OF.ARRIVAL which is the event realizing the external transition function of the model, passing to it the name of an arriving customer.

EFFECT.OF.ARRIVAL sets HELLO to this name and inserts the customer in the SHOPAREA.

The run control portion of the frame employs LENGTH as the control variable. Time segments of this variable are subjected to successive transducers which compute the variable BUSY (1 if there is a customer being served, 0 otherwise) and perform an average over time to obtain CHECKOUT.UTILIZATION. This statistic is fed to an acceptor that does a check for constancy, the criterion for equilibrium employed here. (A very rough test is shown for illustration. The modularity employed allows to include tests of arbitrary sophistication within the acceptor and implement them in its realization, ROUTINE EQUILIBRIUM.CHECK.)

Note that the routines implementing these transducers and acceptors may employ saved variables (that retain their values between successive calls) and so are properly characterized as I/O systems with state spaces (Section 3.1).

To summarize, this example illustrates how the left monitored variable of SIMSCRIPT 11.5 may be employed to implement the model/experimental frame separation principle (Section 12.3.2) in a straightforward manner.

1.2. The Change Detector

Two shortcomings of the SIMSCRIPT monitored variable become apparent. First, it allows write (as well as read) access to the variable. This is both unnecessary as well as dangerous, since the Output and Control modules should not interfere with the model behavior. Second, it can not distinguish between a change which is significant at the model level from a change at the program implementation level. Let us explain this last idea and then propose a new simulation language construct to be called a change detector.

First recall the definition of a composite variable as a collection of model variables considered as a unit (Section 10.2.4). For example, the queue lengths in a network might be considered jointly as a composite variable in the form of a multi-element vector. While it is true for a simple variable, it is even more apparent for a composite variable, that it may undergo a sequence of changes while the simulation clock is not advanced, only the final value assumed being significant. For example, if a customer leaves one queue and immediately joins a second one, both queues change simultaneously in the model. In a sequential simulation however, this global change in the composite queue variable would be executed as a decrement in the first queue followed by an increment in the second. A monitor of the first queue that called a routine to read the composite queue variable would get a false reading. Moreover, there may be multiple additions to a queue from different sources, so that it is generally not possible to impose a fixed order in which such simultaneous changes are to take place.

Exercise. Explain what would go wrong if ROUTINE BUSY.COMPUTATION of Figure 16.2 were to be replaced by the following:

    ROUTINE BUSY.COMPUTATION(LENGTH)
    DEFINE LENGTH AS A VARIABLE
    LET BUSY = 0
    IF LENGTH > 0 "THEN
       LET BUSY = 1
    ALWAYS
    RETURN
    END

Hint: Note that BUSY is a global variable that is monitored on the left.

A *change detector* for a simple or composite variable detects any change in the value of the variable; moreover, it waits until all changes have taken place at any instant in time before transferring control to a routine of the same name that is allowed only read access to the variable.

We shall use the term *change detectable variable* to refer to a simple variable or a composite variable declared by the user (or other program generator) as requiring a change detector.

Such a construct is realizable in a discrete event simulation language as follows: For every change detectable variable declared by the user, the compiler sets aside a storage area containing a change indicator bit and the addresses of the variable or variables composing the composite. Code is added so that every assignment statement involving any variable is followed by the setting of the change indicator bit of all the change detectable variables of which it is a component. (The indicator bit of a detectable variable will not be affected by subsequent changes in its component variables until reset.) After all imminent events have been processed for a given instant, the simulation executive scans the indicator bits and transfers control, in turn, to the routines associated with the indicator bits that have been set. Execution of the routine causes the resetting of the indicator bit.

To enforce the read access restrictions, the compiler checks that no variable appearing in assignment statements of routines called by change detector routine is a model variable.

The change detector and its associated routine provide a clean and correct means of linking the frame derived modules to the model component modules. The linkage is done as in the case of left monitored variables, i.e. for each of the run control and output sub-modules influenced by a detectable variable V place calls in its detector routine to these modules. However unlike the monitored variable, the value of V passed to these routines is the final value assumed by V for the current clock instant.

An example of a multiple server queueing network is shown in Figure 16.3.

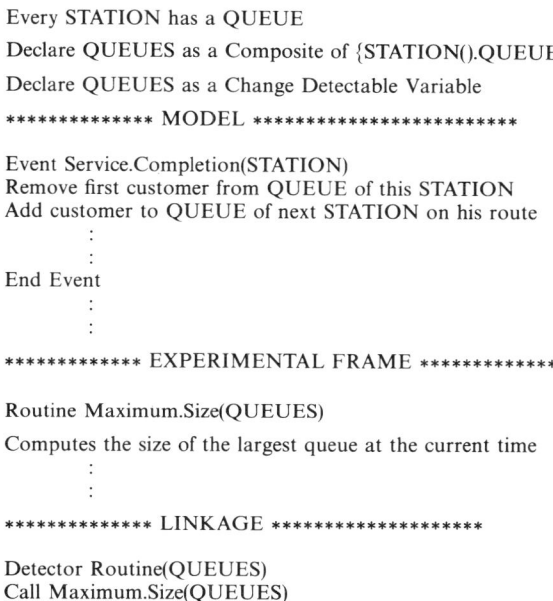

Every STATION has a QUEUE
Declare QUEUES as a Composite of {STATION().QUEUE}
Declare QUEUES as a Change Detectable Variable
************* MODEL *************************

Event Service.Completion(STATION)
Remove first customer from QUEUE of this STATION
Add customer to QUEUE of next STATION on his route
  :
  :
End Event
  :
  :
************* EXPERIMENTAL FRAME *************

Routine Maximum.Size(QUEUES)
Computes the size of the largest queue at the current time
  :
  :
************* LINKAGE *********************

Detector Routine(QUEUES)
Call Maximum.Size(QUEUES)
End Detector

Fig. 16.3. Change detector in multiple server queuing implementation.

2. State Variables and Parameters in Execution Control

Recall that the third component, after model and experimental frame specifications, to a simulation program is the *execution control* segment (Section 12.4.2). This segment is responsible for selecting the simulation runs that are to be executed on the model within the given frame. To set in motion such an experiment, the execution control module must set the model into an initial state, select an input segment and initiate the execution of the code expressing the model and experimental frame realizations. Actually, if the admissible input segment set is realized by a generator then the selection of an input segment is determined by the choice of initial state. Thus in this case the execution control module is concerned with selecting initial states for both model and frame components.

As shown in Figure 16.4, a second important task of the module is to set the parameter values of both model and frame components, thus selecting a specific model and frame pair from a class of possibilities. In this section we shall discuss the design of execution control modules intended to provide

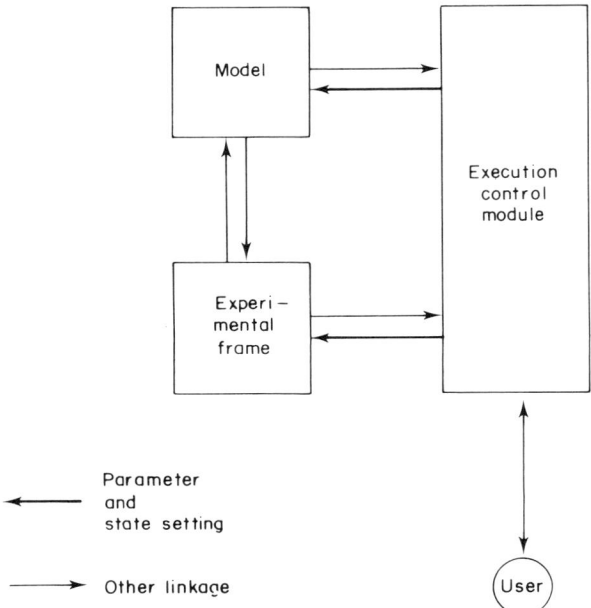

Fig. 16.4. The modules of a simulation diagram.

advanced assistance in the conduct of experimentation. To do so we draw upon concepts of state variables and of stratification of variables and parameters.

2.1. The Concept of a State Variable Set

Recall from Section 3.1 that the "state" of a system is to capture all of its past history necessary to compute what it will do in the future. The state set and transition function are part of the system specification postulated at the I/O Systems level. So if one starts with a system formalized at this level or higher there is no uncertainty in what these objects are. However, if one starts from a lower level specification, from an informally described model, or a simulation program text, then the question of how best to represent it as an I/O system specification arises. Actually, the choice of state set is a critical one, whether or not, a complete formalization as an I/O system is undertaken.

In Section 3.1 we saw that the essential property of the state space was that a state and input segment uniquely determine the state at the end of the input segment. Such is the case in general when the system has input variables. However, as we have seen, once the model/frame pair have been realized as a coupling of systems, the resultant is a *closed* system, having no input variables (unless explicit inputs from the user or other devices during simulation runs are allowed).

Let us then consider an informally specified closed system. The concept of state for it then reduces to property that the state at any time uniquely determines the state and output at any future time. We now formulate this abstract property in more concrete terms.

The static structure (Section 10.1) of a closed system specification structure is *complete* with respect to an informal description of dynamic interaction if the values of all the descriptive variables at any time uniquely determine their values at any later time. We also say that the model is *well described* in this case. Actually, a well described model description cannot be ambiguous, inconsistent or incomplete so it represents a goal which we strive for during model construction.

A *set of state variables* is a subset of descriptive variables such that the values of these variables alone uniquely determine the values of all the descriptive variables at any future time. Thus it is easy to see that a static structure is complete if, and only if, it has a set of state variables!

Some remarks regarding state variables:

The choice of state variables may not be unique

Information contained in the descriptive variables may be redundant and reconstructable in more than one way. For example, any two of the STORAGE variables CAPACITY, CONTENTS and REMAINDER determine the third. So any two could be chosen as state variables.

A set of state variables may not be minimal

A set of state variables is said to be *minimal* if it contains no proper subset which is also a set of state variables. In a well described model the set of all variables is a set of state variables but because of the inclusion of output, auxiliary variables, etc., it need not be minimal.

Usually at the beginning of model formulation state variables are lacking

Otherwise said, at the beginning of model formulation the static structure is usually not complete, and one proceeds to complete it by looking for necessary state variables.

Deterministic treatment of random variables

To specify a random variable requires that one gives a mapping on the unit interval [0,1], (we also allow this mapping to depend on other variables). To sample a value of the random variable, the mapping is applied to the current value of an associated random-generator seed and reported as the sampled value. A new value of seed is then generated by the generator transition rule from the old one. Thus each such seed is a state variable of the model. In this way stochastic models are formulated as deterministic ones.

2.2. State Variables in the Design of Simulation Control Modules

We turn now to the importance of state variables for the control of simulation runs. In particular we refer to the execution control module of the model/frame realization discussed in Section 12.4.2.

Among other tasks, the execution control is responsible for the initialization of simulation runs, i.e. is a computation of a model state trajectory in an experimental frame over some observation interval. To specify a unique state trajectory the members of a minimal set of state variables must each be assigned a value in their respective ranges. We call this operation *setting the simulator into a particular state*.

Thus to *initiate* a run, the execution control must set the simulator into a state desired by the user. To *repeat* a run the simulator must be set into the same state as it was at the beginning of the original run (same state means same state variable settings).

What must be done to *continue* a run after it has been terminated so as to produce the same trajectory as would have been produced if it had not been interrupted? The simulator must be set into the same state that it was in *at the time of termination*.

Exercise. Reread Section 3.1 and compare the definitions of state and transition function in the abstract with the concrete interpretation given just now.

Let us now see what happens when a minimal set of state variables is not used to initialize a run. There are two cases: overdetermination and underdetermination.

In *overdetermination* a set of state variables, e.g. the whole set of variables, is used which includes a minimal set. In this case, the variables not in the minimal set must be set consistently with the settings of the minimal set.

In *underdetermination* a set properly contained within a state variable set is used. In this case some of the state variables are not determined so that a family, not a single, trajectory is specified.

A third problem arises even with a minimal state variable set — that of *intervariable relations*. Such relations constrain the values which may be simultaneously assigned to state variables. That they may be present even though the state variable set is minimal indicates that in general minimality is not the same as independence.

Each of the above problems is of interest in the design of a user-friendly execution control modules, which by and large do not exist today. For example, intervariable relations are especially likely to arise when one restricts the states of interest to those accessible from certain ground states, such as all queues empty, all facilities free, etc. Most simulation languages allow the user to start from only such ground states. To start from any other accessible state the user must first run the simulator into such a state, which may be a difficult and costly approach.

In a more flexible approach, the computer may be given knowledge of the range sets of the variables, the minimal state variable sets and the intervariable relations. However, there are limits to the knowledge that it can employ here since problems concerning characterization of accessible states are often recursively unsolvable. Nevertheless in a cooperative interaction the computer can provide greater assistance in selection of initial states and in continuation from previous states.

2.3. Interactive Parameter Exploration

Models and experimental frames are usually defined parametrically, that is, in terms of families rather than as singles. In a parametric model, a set of variables, called *parameters*, span the class of models in the sense that specifying a value for each of the parameters determines a unique model from the class. Similarly, the parameters of a frame provide degrees of freedom for its specification (Section 12.4). When a model/frame pair is realized as a coupling of systems, the parameters of each component must be set to specify a unique simulation run. Unlike state variables they cannot change during a run since this would amount to changing the model in mid stream.

In the modelling context a parameter provides a degree of freedom in matching model and real system data (Section 5.4) and a search is conducted for parameter assignments that maximize measures of goodness of fit between these two data sets. In a design or decision making context, commonly a set of alternatives is parameterized and the set is explored via simulation runs. Often such explorations are controlled automatically by a search procedure that selects a new point or points in the parameter space based on the evaluated goodness of fit, performance and utilization indexes of previously sampled points. However, at crucial stages of such studies the modeller or designer may wish to understand the effects of parameter variation by direct hand manipulation. In this case, a user-friendly interface for interactive parameter exploration is in order. The design of such an interface may be based on the following concept of stratification of variables and parameters.

Depending on the questions being asked, especially the time periods of interest, variables of a real world entity may be represented by model counterparts as either state variables or parameters. Over relatively short periods such variables may be considered as constants with values determined when the component came into existence. In this case, their model counterparts are appropriately treated as parameters. Over relatively long periods, if these variables are subject to change, they may be treated as state variables. In this case, of course, the internal mechanism of the component must dictate how the changes are brought about. To control the complexity of a component, its variables may be stratified into *structural levels*. The variables at a given level act as parameters for all lower levels and as state variables for all higher levels. Thus variables at the lowest level are state variables while those at the highest level are parameters. Such stratification should be correlated with volatility and time scale: the greater the frequency with which a variable is likely to change the lower the level that it should occupy in the stratification.

Often, the structural levels are associated with levels of adaptive control. Variables are presumed to represent settable controls of decision making components and higher level components can determine the set points of lower levels based on performance criteria.

A similar concept applies to the parameters of a model/frame pair. They may be stratified according to volatility in the sense that the lower the level of a parameter, the less is it likely to be changed by the user or search procedure from one simulation run to the next. One may choose the rank of a parameter to reflect its level in an adaptive control scheme as above. Another possibility is that higher level parameters represent more extensive modifications than do lower level ones. Or it may just be that the user wishes to explore the space in which certain parameters of interest are to be frequently altered while others are to remain constant or nearly so.

An interesting possibility arises if we allow the interface to modify the state variables of the model/frame realization as well. In this case, the user may alter the state resulting at the termination of one simulation run to serve as the initial state for the next one. (An important special case is the continuation of a run from the state it terminated in.) Accordingly, one should also rank the state variables by volatility criteria just as with parameters.

A state machine design of an interface based on the volatility concept is shown in Figure 16.5. The central idea is that after a simulation experiment, control is returned to the interface which presents the lowest level variables and parameters to the user for possible modification, If modifications are limited to this level, then control may return to the simulation executive for running the next experiment. If not then successively higher levels are made accessible to alteration.

In Figure 16.5, the state machine has n states corresponding to n levels of

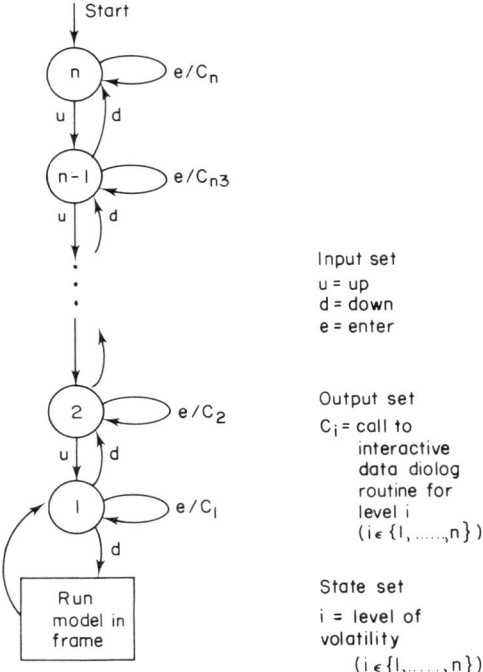

Fig. 16.5. State machine design of interactive interface.

volatility. At each state, inputs of u and d will cause state transitions to the next higher and lower level states, respectively. The input of e will call a routine to mediate the interactive assignment of values at the level represented by the state. This routine should be a scheme for displaying parameters and their values, accepting modifications, echoing, and accepting further changes. It should be given the range sets of the parameters and state variables and be able to check input values for inclusion in the range. In state 1 an input of d will call the execution control module to run an experiment. Control returns to state 1 when the experiment is complete.

In an advanced design, the stratification itself is open to modification during run time rather than being frozen at compile time. Thus the user may lower the ranks of variables that he wishes to change frequently or the system itself may keep frequency of use counts and automatically alter rankings to align the stratification with these counts.

3. Off-line Realization of Experimental Frame Application

As we have seen, on-line experimentation offers a powerful and immediate form of interactive model experimentation. In such experimentation, the processes of model behavior generation and of behavior analysis take place simultaneously. There are however reasons to separate these processes in time in what is called *off-line* experimentation.

In the off-line approach, the input/output segment pairs generated by simulation of a model are stored in a *behavior data base* to be available for subsequent analysis [3]. This makes it possible to apply different analyses to the same data set without having to regenerate it each time, as would be required in on-line experimentation.

Off-line processing can be readily pressed into service for the application of experimental frames to models, and more generally for the mapping of the data space of one experimental frame to that of another. Recall from Section 13.1.8, that the derivability relation from a frame E to a second frame E', sets up the E-to-E' processing for mapping the $\mathbf{D}_E$ to $\mathbf{D}'_E$. Such processing first takes an I/O segment pair $(\omega, \rho) \in \mathbf{D}$ into a triple of segments (ω', ρ', v') in the input, output and run control spaces of E', respectively. Then the pair (ω', ρ') is accepted for $\mathbf{D}'_E$ if both ω' and v' are admissable in frame E' ($\omega \in \Omega'_I$ and $v' \in \Omega'_C$).

Now recall that applicability of a frame E to a model M is defined in terms of the derivability of E from the scope frame E_M and the execution of such application is nothing but the E_M-to-E processing applied to the data space of E_M (Section 13.1.10). Thus in off-line experimentation with a model M under a frame E, we apply the E_M-to-E processing to the I/O data stored for M in the data base (in the data space associated with E_M).

Realization of the frame-to-frame processing in procedural form can be broken up into the following component processes:

segment-to-segment mapping induced by variable derivability
This requires provision for managing the semantic structure of the variables attached to entity structure (Section 11.3) and generating procedures to carry out the segment-to-segment mappings induced by the derivability of variables existing in the semantic structures.

segment acceptance testing
The high level language approach can be employed to specifying acceptors for admissible input and run control segments (Section 4.5). The specifications can be translated into procedural form such as illustrated in Section 16.2. In this case, however, the input to such acceptors is in the form of

discrete event segments rather than in the form of calls from a change detector mechanism.

summary mappings

The summary mappings can be implemented by transducers realized as in Section 16.1.1. In addition, one can exploit any standard statistical processing (such as available in SPSS [1]) and data set operations that are provided by the data base management system used as a host (see Section 18.3).

4. Summary

The realization of the model/experimental frame separation principle in procedural form was demonstrated. In on-line experimentation, model and frame specifications are realized as independent modules and coupled together by means of a change detector construct patterned after the left monitored variable of SIMSCRIPT. As opposed to the more standard coupling approach, the change detector scheme does not require introducing any explicit coupling information into the code derived from the model specification. Model and frame code thus remain separate and can be reassembled in different combinations.

Once model and frame components are coupled the resultant system can put under the supervision of an execution control module. The design of such a module intended for advanced computer assistance of interactive experimentation was discussed employing the concepts of state variables and of stratification of variables and parameters based upon volatility criteria.

The model/frame separation approach is taken one more step in off-line experimentation. Here on-line experimentation generates input/output data of a model in its scope frame and stores it in suitable data base form. Application of an experimental frame to the model is then accomplished by performing the associated frame-to-frame processing on the stored model data.

References

[1] Hull, C. H. and N. H. Nie (1979). *SPSS Update*, McGraw Hill, NY.
[2] Kiviat, P. J., R. Villanueva and H. M. Markowitz (1975). "SIMSCRIPT 11.5 Programming Language" (ed, E. C. Russell), CACI.
[3] Standridge, C. R. and A. A. Pritsker, (1980). "Using Data Base Capabilities in Simulation". In *Proceedings of Simulation '80*, Interlaken, Switzerland.
[4] Zeigler, B. P. (1976). *Theory of Modelling and Simulation*, Wiley, NY.

Chapter 17

SIMULATION OF MODULAR AND HIERARCHICAL MODELS

The multifacetted modelling methodology that has been developed in previous chapters is based on the delineation between modelling and programming. Ultimately, this distinction must be justified in the sense that the forms of model/experimental frame specification we developed must be shown to be implementable in realistic software form. This chapter takes up this task. It formulates a criterion of correct simulation of discrete event specified systems (DEVS) and demonstrates the existence of an abstract simulator that satisfies this criterion. A coupling of such simulators, each responsible for a DEVS component, is then shown to correctly simulate a multicomponent DEVS model in modular form. These abstract simulation structures are shown to be realizable in current simulation language concepts. Although the addition of features that directly cater to these structures would make their realization much more convenient, the clarity of structure and modularity achieved is significant none-the-less. Simulators of hierarchical modular system specifications are then developed in a form naturally paralleling the recursion underlying such specifications. Finally, we consider the attractive possibility that simulators of multicomponent system specifications might execute on interconnected processors in parallel. Realization of such asynchronous distributed simulation systems is shown to follow quite directly from the concepts of abstract simulators developed earlier in the chapter.

1. Abstract Simulator of a DEVS

An interpretation of the dynamics specified by the DEVS M is given by considering a simulator S_M for M. As shown in Figure 17.1, the simulator has two state variables, one for the sequential state of M, the second for the *time of the last event* t_L. S_M also has three storage cells whose values are determined by its state variables and its input. The first holds the *time of the next*

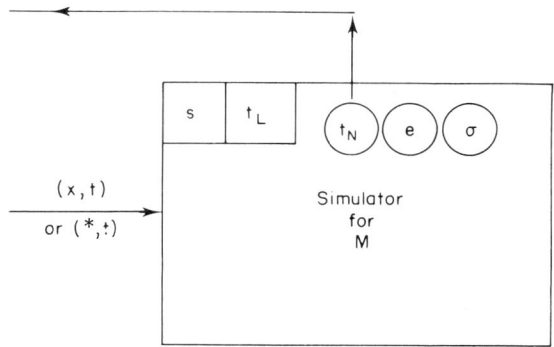

SIMULATOR FOR DEVS M

when receive an input (x,t):

    done := false

if $t_L \leq t \leq t_N$ then

    $e := t - t_L$

    $s := \delta_{ex}(s,e,x)$

    $t_L := t$

    $t_N := t_L + ta(s)$

    done := true

when receive an input $(*,t)$:

    done := false

if $t = t_N$ then

    $s := \delta_\phi(s)$

    $t_L := t$

    $t_N := t_L + ta(s)$

    done := true

Fig. 17.1. Simulator of an abstract DEVS.

scheduled internal event, t_N, where

$$t_N = t_L + ta(s) \tag{1}$$

If it is given the global model time t, S_M can compute the time elapsed since the last event e, where

$$e = t - t_L \tag{2}$$

and the *time left to the next event* σ, where

$$\sigma = t_N - t = ta(s) - e \tag{3}$$

The simulator communicates with the external world via an output port through which it reports t_N and an input port through which it receives signals informing it of external events and providing synchronization for internal events. A signal of the first kind takes the form (x,t) where x belongs to the external event set of the model and t is the global time at which the event named by x is supposed to occur. A signal of the second kind is of the form $(*,t)$ where if proper synchronization has been maintained, t is equal to t_N, the model's next event time reported at the output port. When receiving either signal the simulator checks for consistent synchronization and then executes a model state transition, employing the internal or external transition function as illustrated in Figure 17.1.

An external event notice (x,t) requires that t lie in the interval bounded by the times of the last and next events:

$$t_L \leq t \leq t_N \tag{4}$$

or equivalently, in view of Equation 1,

$$0 \leq e \leq ta(s) \tag{5}$$

In other words, an external event arrives within the allowed bounds given by Equation 4, if, and only if, the model state represented by the simulator is one of the legal total states Q, as indicated by Equation 5. We shall employ this equivalence in formulating the criterion for simulator correctness in a moment.

Thus we see that the sets and functions specified by a DEVS are just those required for a discrete event simulator to:

*schedule interval events
*execute transitions due to such events
*execute transitions caused by external events.

Note that the DEVS description of a model is time invariant (Section 4.1), while the simulator is synchronized relative to an absolute global time base. This difference clearly separates the model formulation from its computer implementation.

To prove the correctness of a DEVS simulator requires the statement of a simulation relation involving model and simulator (Section 5.3). For our purposes the following will serve:

We shall model a DEVS simulator as a passive DEVS. Certainly this applies to the simulator just given since it is activated only by external synchronizing signals. Let such a simulator S_M have sequential state set $\mathbf{S}$ and let the DEVS M have sequential state set S. We say that S_M *correctly simulates* M if for each $t \in T$ (time base) there is a correspondence h_t from $\mathbf{S}$ onto the total states $Q = \{(s,e) | s \in S, e \in [0, ta(s)]\}$ of M which is a homomorphism in the following sense:

(a) if h_t maps a state of **S** to $(s,e) \in Q$, then it maps its successor under (x,t) to the state $(\delta_{ex}(s,e,x),0) \in Q$.
(b) if h_t maps a state of **S** to $(s,ta(s)) \in Q$, then it maps its successor under $(*,t)$ to the state $(\delta_\phi(s),0)$.

This concept of correctness requires that the simulator be able to replicate both the external and internal state transition of the model when it receives the respective co-ordinating signals. Note however that it does not assure the proper arrival of the latter signals, since this must take place external to the simulator.

It is readily shown that the simulator defined above performs a correct simulation in the given sense.

Theorem 1
The simulator S_M in Figure 17.1 correctly simulates the DEVS M.

Proof. The required correspondence h_t maps the simulator state (s,t_L) to the model total state (s,e) where $e = t - t_L$. For a signal (x,t), if $e \in [0,ta(s)]$ then $t \in [t_L, t_N]$ (using Equation 1). Following the simulator operation in Figure 17.2, we see that it enters state $(\delta_{ex}(s,e,x),t)$ which corresponds to $(\delta_{ex}(s,e,x),0)$ under h_t as required. The operation under signal $(*,t)$ can be similarly proven to be correct.

1.1. Abstract Simulator a Multi-component DEVS

Having developed a workable simulation concept for DEVS models at the I/O Systems level, our next step is to develop such a concept for DEVS systems specified at the Coupling of Systems level. In other words, we shall build simulators for multicomponent DEVS models in modular form by coupling together simulators for each of the component DEVS.

A simulator for a multicomponent DEVS (Section 8.2) is shown in Figure 17.2. Each model component M_i is handled by its own simulator S_i, which is assumed to correctly simulate it. In addition, a global co-ordinator is responsible for synchronizing the component simulators and handling external events. The co-ordinator sees to it that the selected imminent component is the one that receives the $(*,t)$ signal with $t = t_N$, its next internal event time. This component sends the (x,t) signal to each of its influencees with x properly computed using the output translation map for each. When an external event notice (x,t) arrives, the co-ordinator channels the (x,t) signal to each of the component simulators.

Note that the input/output interface of the multi-component simulator remains of the same form as a component simulator: it makes available a time-of-next-event signal t_N as output and receives the signals (x,t) and $(*,t)$ as

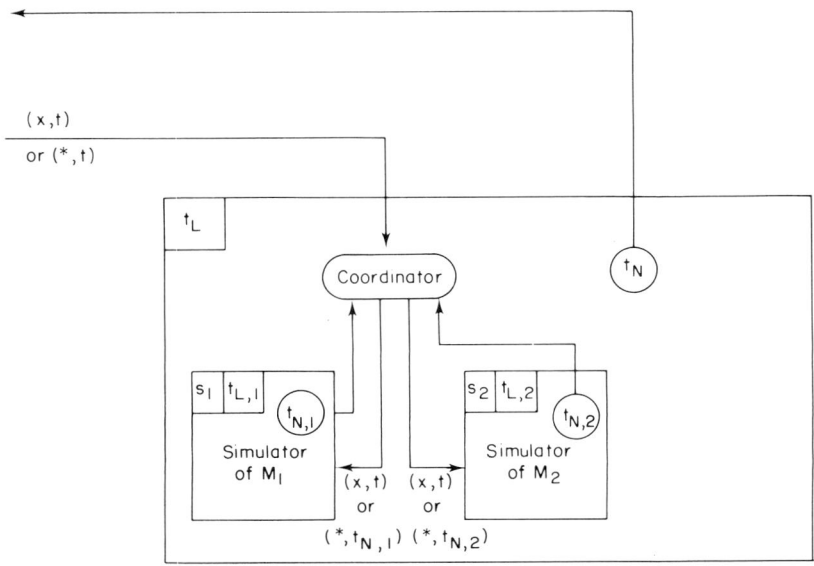

CO-ORDINATOR

when receive an input (x,t):
    done := false

if $t_L \leq t \leq t_N$ then

    send input (x_i,t) to each component simulator i

    wait until all simulators done

    $t_L := t$

    $t_N :=$ minimum of component t_Ns

    done := true

when receive an input $(*,t)$:
    done := false

if $t = t_N$ then

    find the imminent simulators (those with minimum t_N)

    SELECT one, i^*, and send the input $(*,t)$ to it
    send the signals $(x_i^*, _j, t)$ to each of its influences j
    wait until this simulator i^* and each of its
        influencees are done
    (all other simulators are unaffected)

    $t_L := t$

    $t_N :=$ minimum of component t_Ns

    done := true

Fig. 17.2. Simulator of multicomponent DEVS.

input. This makes possible the self recursion required for hierarchical construction to be discussed in Section 17.3.

The simulator of Figure 17.2 can be proven correct under the criterion of Section 17.1.

Theorem 2

The simulator of Figure 17.2 correctly simulates M_{DEVN}, the DEVS which defines the result of coupling the multi-component model DEVN (Section 8.2).

Proof. A typical state of the simulator takes the form $((s_1, t_{L,1}),...,(s_n, t_{L,n}), t_L)$. Define a correspondence h_t which at time t takes maps this state to the model total state (s, e) where $s = ((s_1, e_1),...,(s_n, e_n), e)$, $e_i = t_L - t_{L,i}$ for $i = 1,..,n$ and $e = t - t_L$. Note that t_L, the time of the last event handled by the co-ordinator, equals $t_{L,i}$, the time of the last event handled by simulator i, if the latter was one of the simulators involved in the last event computation (so $e_i = 0$); for any simulator not so involved $t_{L,i}$ is less than t_L (so $e_i > 0$).

For signal (x, t) received by the co-ordinator, it can be verified that $e \in [0, ta(s)]$ implies that $t \in [t_L, t_N]$ by employing the definition of $ta(s)$ for M_{DEVN}. Under these conditions, the signal (x_i, t) is sent to each simulator i. In state $(s_i, t_{L,i})$ this simulator represents its component model total state $(s_i, t - t_{L,i})$ under the correspondence employed in Theorem 1. Assuming correctness of this simulator, it computes the transition $\delta_{ex,i}(s_i, t - t_{L,i}, x_i)$ which equals that defined for M_{DEVN} when we note that $e_i + e = t - t_{L,i}$.

For signal $(*, t)$ received by the co-ordinator, it can similarly be shown that $e = ta(s)$ implies that $t = t_N$. Under these conditions one shows that the imminent simulators are those of the imminent components, IMMINENT(s) and that the component selected by the co-ordinator is i^* = SELECT(IMMINENT(s)). It can then be checked that the transition computed by the simulator for i^* and the transitions computed by each of the simulators of its influencees are correct with respect to lines (1) and (2) of M_{DEVN}'s internal transition function (Section 8.2). Finally, any simulator j not affected by these actions will represent the state of component j given by line (3), $(s_j, e_j + ta(s))$ recalling that t_L is updated to $t = t_N = t_L + ta(s)$.

2. Simulation Languages Realization of the Modular DEVS

The previous sections have developed the abstract basis for simulation of multicomponent DEVS systems. It remains to show how such a basis can be implemented within state-of-the-art simulation languages. As a prior comment however, we should note that such implementation demonstrates the feasibility of the concepts but does not establish that this is the most desirable approach. One alternative simulation scheme for multicomponent DEVS models for example, is the implementation of languages similar to those developed in Sections 8.3 and 8.7. Nevertheless, since the major simulation languages are likely to be the only languages to be generally available for some time, it is important to be able to employ them for modular and hierarchical model realization, albeit in a form that is not as convenient as it might be. (Chapter 18 discusses proposals for future languages built on system theoretic concepts.)

A strategy for implementing multicomponent DEVS models within conventional discrete event simulation languages is illustrated in Figure 17.3. In addition to standard event scheduling constructs, we make use of modern programming constructs which support the definition of multi-functional modules which restrict access to their internal variables [6]. SIMULA as extended by Unger [7] provides these constructs. Although in other languages, such as SIMSCRIPT 11.5 the implementation strategy still provides a clean model representation, the additional convenience and safety afforded by the module concept cannot be obtained.

The mapping from model description to implemented form is close to a one-one correspondence. Each model component is represented by a module which corresponds to the simulator of Figure 17.1. Module and simulator state variables are in exact correspondence. The module is comprised essentially of an event routine that implements the internal part of the model's transition function and (an ordinary) routine for its external transition function. The latter routine is to be called with specified input value from a point external to the module (how this is done will be discussed in a moment). The former routine reschedules itself using the time advance function. Serving in an auxiliary role are routines for the handling of elapsed time. Note that the current time is a global variable so that it is available whenever an internal or external transition event routine is to be executed. The event list executive sees to it that the event notice for the internal transition is posted for time t_N and that its event routine is called when imminent.

The coupling of modules follows the lines described in the formal DEVS multicomponent specification. Recall that each model component has a set of

MODULE implementing a (component) model i

s, sequential state:local,saved
                  visible to: OUTPUT i

t_L, time of last event:local, saved

t, current time:global

::::::::::::::::::::::::::::::::::

EVENT INTERNAL TRANSITION
    call OUTPUT i

  execute internal transition function:
    $s := \delta_\phi(s)$
  RESCHEDULE this event in $ta(s)$ time units

END EVENT

::::::::::::::::::::::::::::::

ROUTINE EXTERNAL TRANSITION given INPUT VALUE, x

  execute external transition function:
  $s := \delta_{ex}(s, \text{ELAPSED TIME}, x)$
  RESCHEDULE internal transition even in $ta(s)$ units

END ROUTINE

::::::::::::::::::::::::::::::

FUNCTION ELAPSED TIME
    return $(t - t_L)$
END FUNCTION

::::::::::::::::::::::::::::::

ROUTINE RESCHEDULE given TIME LEFT, σ

t_N, time of next event: local
           visible to event list executive

    $t_L := t$
    $t_N := t + \sigma$

  call event list executive
    (to reposition event notice
    for internal transition event)

END ROUTINE

::::::::::::::::::::::::::::::

ROUTINE OUTPUT i

read access: s, sequential state of Module i

for each influencee j, of i.
  compute the translation mapping:
  $x_{i,j} := Z_{i,j}(s)$
  call EXTERNAL TRANSITION of j with INPUT $= x_{i,j}$

END ROUTINE

Fig. 17.3. Discrete event language realization of DEVS component.

influencees, and for each an output translation map. At the onset of an internal transition event, the translation maps are used to send a signal from the active component to its influencees. This concept is implemented in Figure 17.3. As can be seen, a call to an associated output routine appears in the module body so that it is invoked before the state is updated. The appropriate calls to influencees are placed within this output routine.

Note that if the coupling of a system is changed (while the components are not), then the only changes required are in the output routines.

An example of a program implementation of multicomponent DEVS using the foregoing modularity concepts was given in Figure 16.2 of Chapter 16. There the grocery store is a coupling of two component DEVS, the SHOPAREA and the CHECKOUT. An output translation map sends a customer from the SHOPAREA to the CHECKOUT by calling the latter's external transition function implemented as the INPUT.TO.CHECKOUT routine.

A feature that greatly enhances the convenience of this module concept is the ability to reset the model state (i.e. both the sequential state and the elapsed time components) to a desired setting. Moreover, the ability to initialize the state at the beginning of a run follows as a consequence of the reset capability. To do the resetting, the simulation executive has to be able to reschedule the module's event notice to maintain consistency with the new state. If set into model state (s,e), the internal event should be rescheduled in time $\sigma = ta(s) - e$. Conventional discrete event simulation languages do not provide reset capability. One may speculate that this is because the proper concept of model state was not understood by the designers.

3. Hierarchical DEVS Models and Simulators

Recall from Section 8.7 and Appendix 1 of Chapter 15, that the DEVS formalism provides a fundamental characterization of hierarchical construction with which to approach the design, and prove the correctness of, simulators intended to support top down modelling. Due to closure of the DEVS formalism under composition, the coupling of modular DEVS components is itself a modular DEVS. This fact implies that DEVS systems may be constructed recursively from DEVS components, each being itself a coupling of DEVS components. Thus top down specification, as well as bottom up reconstitution, are justified in the DEVS formalisms.

Correspondingly, the simulation of a hierarchically constructed DEVS may be realized by a hierarchically constructed simulator. This follows by examining the simulator of Figure 17.2. Each of the component simulators operates correctly if it is given the specification of a DEVS. It does not matter in principle whether this DEVS is itself the result of a coupling of more elementary DEVS components. Thus the simulation algorithm given in Figure 17.2 is self recursive. More explicitly, if a component DEVS is itself a composite, then the component simulator which handles it will consist of a co-ordinator and component simulators coupled in the manner of Figure 17.2.

The above discussion can be summarized by saying that a hierarchical, modular specification of a DEVS, in the form of a composition tree (Appendix 1, Chapter 15), can be readily mapped onto a hierarchical, modular simulator that correctly simulates it. Figure 17.4 demonstrates a two-level hierarchically constructed DEVS and its corresponding hierarchical simulator. This simulator may be compared to a one level simulator obtained by flattening the hierarchy, i.e. considering the isomorphic DEVS derived by directly coupling the lowest level DEVS components. It is then evident that the same minimization and selection of imminent component occurs in both architectures, except that it is done stage wise in the hierarchical simulator. While an in-depth study has not been carried out, there seems to be no apparent loss in efficiency, and perhaps some gain, in the performance possible from a hierarchical simulation architecture, especially using distributed processing (Section 17.4). However, the main motivation for such an architecture is the conceptual benefits it affords in permitting the one-one implementation of hierarchically constructed DEVS models.

Simulation languages for hierarchical, modular discrete event simulation are being developed [4]. The system theory based language, GEST is discussed in Section 18.3.

4. Distributed, Asynchronous DEVS Simulation

Since discrete event models may represent systems with varying degrees of parallelism it is natural to seek simulators which distribute the execution of a model among more than one computer (See Section 18.3). Particularly attractive is the possibility of simulating very large scale modularly constructed models, the components of which might be located in computers at dispersed sites. An analogous application is to the linkage of simulated model components with real system components in real time operation [2]. Since processors responsible for different computations may proceed at unpredictable speeds, no assumption can be made concerning relative timing, i.e. they must be assumed to be asynchronous. Several approaches to the synchronization of asynchronous processors for discrete event simulation have appeared [1, 3, 5] and are reviewed in [5]. They all suffer from the same flaw: no distinction is made between the simulator and the model being simulated.

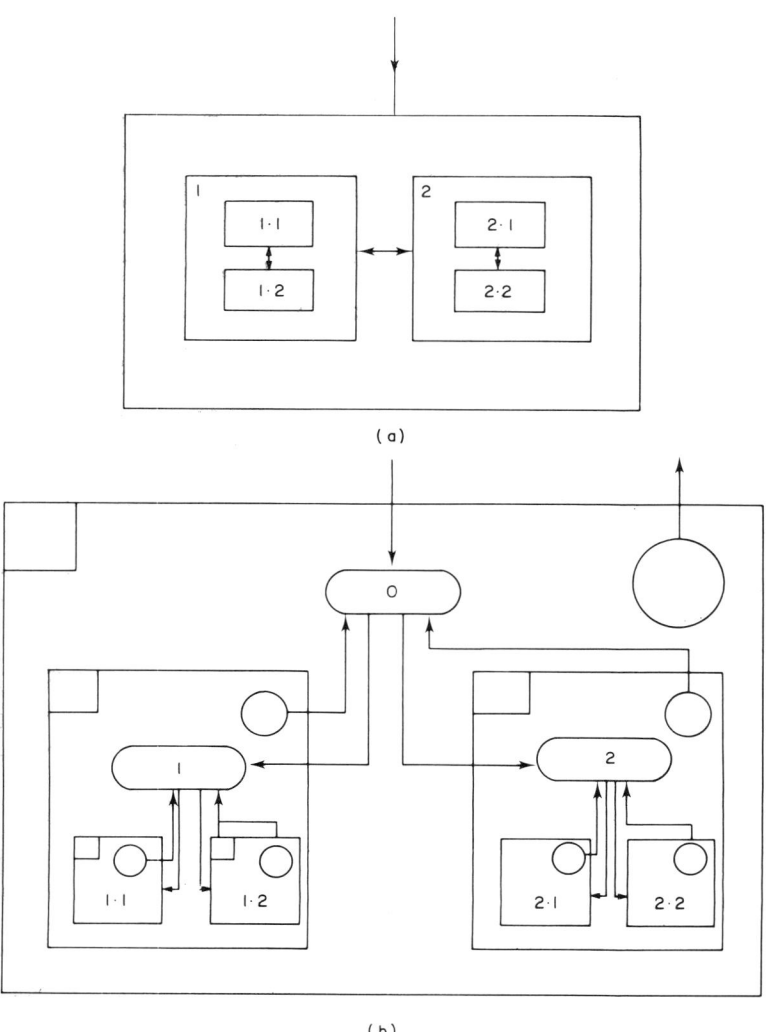

Fig. 17.4. Hierarchical model (a) and simulator (b).

Thus it is difficult to judge for which class of models is the proposed simulation strategy correct. In contrast, the approach we shall be taking makes an explicit distinction between the model and simulator.

An architecture for distributed simulation of multicomponent DEVS models is derived by assigning each simulator and the co-ordinator in Figure 17.2 to a distinct processor. Although no assumption about the timing of computations by the processors is made, the algorithm given to the co-ordinator in Figure 17.2 will assure that it properly synchronizes the component processors in the case of an input free system model. This is because the co-ordinator must wait until all activated processors have finished computing their component model transitions before it goes on the next internal event. However, in the case of a non-input-free model, no guarantee can be given that an external input signal will not be received while processing of an internal event is still going on. The only way to achieve synchronization of such interrupts is to expand the model to include all sources of external input, e.g. by adjoining the experimental frame input generator. Once this is done, the new model has only internal transitions and is therefore synchronizable as just discussed. This expansion can be done hierarchically, as we have shown in Section 16.1.

The parallelism achieved in this scheme is that logically permitted by the DEVS transition definition viz the selected component and its influencees may be simultaneously active. Thus on average this distributed simulation will run $I+1$ times faster than a sequential simulation where I is the average influencee set size (disregarding overhead). The average utilization of processors is therefore $(I+1)/N$ where N is the number of model components. The speed-up achievable by this distributed scheme is thus highly dependent on the influencee set sizes.

Several questions are raised by the preceding scheme. Is it possible to remove the global co-ordinator? Can the speed-up afforded by distributed processings be improved? The answer to both these questions is yes under certain conditions, no in general. It can be shown that the global co-ordinator can be replaced in networks of regular structure (such as lattices, rings, or trees) where synchronization signals may be relied upon to traverse the network [4]. We illustrate this result in Figure 17.5 for a one-dimensional uniform array of asynchronous processors.

The model simulated is a modular discrete event cell space (Section 8.5). Each processor handles a component simulator as before, the model being assumed to be input free. The processor at the left hand end initiates each global transition cycle by sending a signal to the right which accumulates the smallest next event time as it travels. A processor which effects a reduction of the accumulated minimum, marks itself as imminent. Upon reaching the right hand end, a high level signal is initiated travelling rightward. The first marked processor that it reaches is considered selected. It downgrades the signal so that no other imminent processors can be enabled. The selected processor is activated and activates its influencees. Upon the completion of the activated transition computations, the processor causes the return signal to continue its leftward journey to the right hand end.

We note that processors must be able to distinguish left from right hand

330   5. Multifacetted system architectures

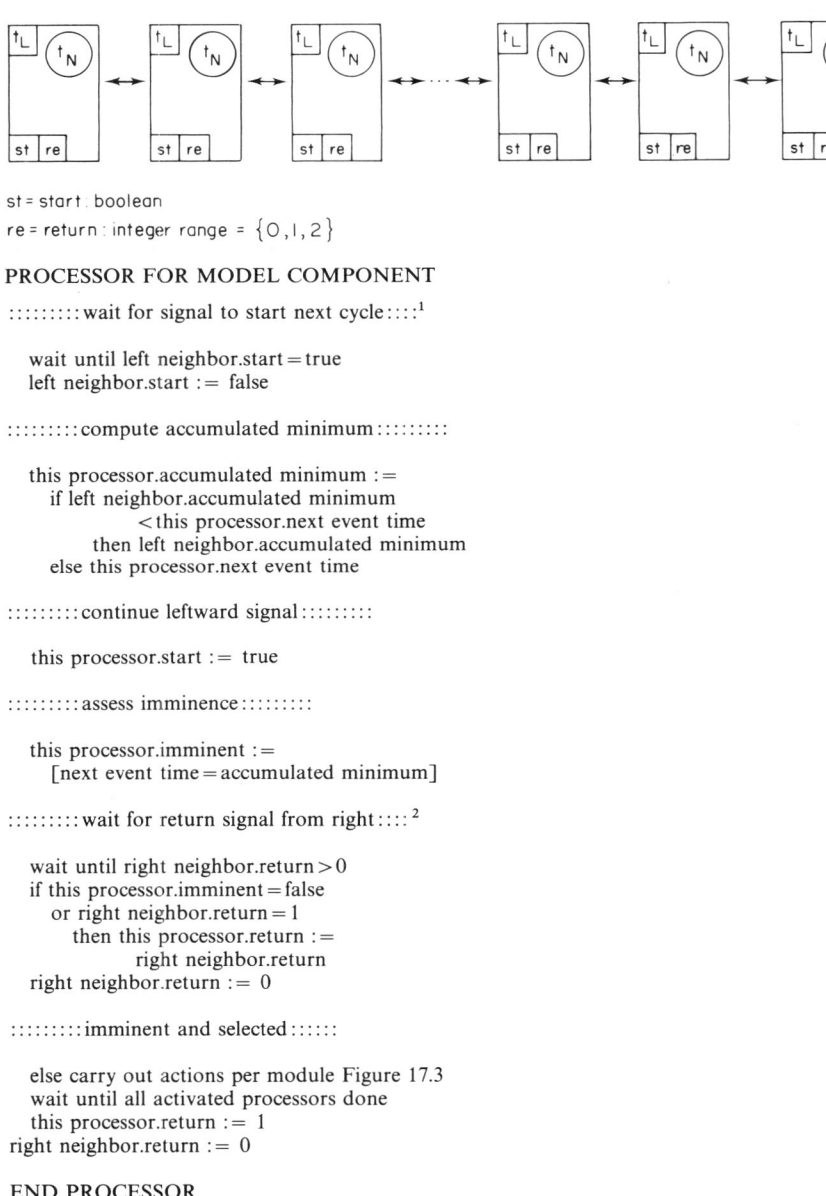

st = start : boolean
re = return : integer range = $\{0, 1, 2\}$

PROCESSOR FOR MODEL COMPONENT

::::::::: wait for signal to start next cycle ::::[1]

    wait until left neighbor.start = true
    left neighbor.start := false

::::::::: compute accumulated minimum :::::::::

    this processor.accumulated minimum :=
       if left neighbor.accumulated minimum
           < this processor.next event time
        then left neighbor.accumulated minimum
       else this processor.next event time

::::::::: continue leftward signal :::::::::

    this processor.start := true

::::::::: assess imminence :::::::::

    this processor.imminent :=
      [next event time = accumulated minimum]

::::::::: wait for return signal from right ::::[2]

    wait until right neighbor.return > 0
    if this processor.imminent = false
      or right neighbor.return = 1
        then this processor.return :=
            right neighbor.return
    right neighbor.return := 0

::::::::: imminent and selected ::::::

    else carry out actions per module Figure 17.3
    wait until all activated processors done
    this processor.return := 1
right neighbor.return := 0

END PROCESSOR

:::::::::::::::::::::::::::::::::::::

1. The left end processor sets its own start to true here.
2. The right end processor set its return to 2 here.

Fig. 17.5. One-dimensional array of asynchronous simulators.

ports. Also existence of right and left hand ends of the array is the basis for achieving co-ordination without a global component. However, finiteness of array is not necessary. Indeed, the array may be allowed to grow at either end by activating passive processors. Being an end processor can be recognized by detecting the presence of a neighbor in the passive state.

This simulation when initiated from a global state in which all processors are in their ground states except the left end processor which has its start = true, will return to such a state after having completed the correct model global state transition. Thus the observer need only look at the left hand processor to know when a model global state transition has been completed.

This property (for every global state transition of the model there is a moment when the next model global state is represented by a global state of the simulator) is a rather strong one to require for correct simulation. If we give it up, we give up the assurance that an observer could inspect all the component simulators at the right time to see a global model state. On the other hand, there may be no observer capable of simultaneously inspecting all the simulators, for example if they are widely dispersed in space. Thus a more realistic criterion is that each local simulator correctly reproduce the global state trajectory as it is projected on the co-ordinate it is responsible for.

In other words, we require that the local state trajectory, computed by a simulator for its DEVS component model, is the same that this model would experience as a component of the simulated DEVS network. However, we do *not* require that the times at which events in these state trajectories are computed bear any fixed relation to each other.

The local correctness criterion just introduced has the advantage that a simulator may be able to move ahead in computing its components state trajectory without being slowed down to co-ordinate with other simulators. However, a simulator cannot move beyond a point in the state trajectory at which inputs from other componenents that would affect its course are not yet available. More precisely, it can do so, but it must be prepared to backtrack to this point and start again once an inconsistency has been detected.

A class of models for which distributed simulation satisfying local correctness is possible without backtracking is the cellular automata (Section 6.1). (Actually, the properties employed in the following approach make no use of the geometry of the space so that it applies more generally to discrete time coupled systems.) Figure 17.6 gives the algorithm for the simulator of a cell. Each model cell is handled by its own simulator. These simulators are connected so that communication is possible with simulators of neighbors as well as inverse neighbors. Each simulator keeps track of its cell's current and next state and also its cell's time step. The simulator waits until all the simulators of its neighbors have reached the same time step that it

1. waituntil all neighbor simulators in phase
            1 or 2 with their clocks at t
            (where t is your clock time)
2. compute $s' :=$ transition function applied to
            neighbor states
3. waituntil all inverse neighbor simulators
            are in phase 1 with clocks at $t+1$
            or in phase 2 with clock at t
4. update: let $s := s'$ and $t := t+1$
    go to 1.

Fig. 17.6. Algorithm for simulator of a cell.

has and then computes its cell's next state based on the information supplied by the neighbor simulators. Since it does erase the current cell state in this action, the latter is still available to the simulators of the inverses neighbors for computing their next states. Once all the inverse neighbor simulators have utilized this information, the simulator is free to update the cell's current state by transferring the computed next state to it while advancing its local clock by unity.

It can be shown that such simulation strategy satisfies the local correctness criterion in the sense that if a simulator has the triple (s,s',t) stored in memory, and all simulators started with $t=0$, then s is the state that its cell would be in at time t.

While it is not immediately apparent, a moment's thought reveals that simulators can move ahead of others up to the following limitation: For any pair of cells, let d be the distance separating them (d is the minimum number of time steps that it takes for information to travel from one to the other). Then the local clocks of this pair of cells may differ by as much as d in the scheme of Figure 17.6. Thus the further two cells are apart in space, the greater can be the disparity in the time step that their simulators are currently working on.

Distributed simulation schemes for cellular automata based on the global correctness criterion can be devised as well and are studied in Problem 1.

Having characterized modular and hierarchical structures for multicomponent DEVS models (Chapters 8 and 15), it is straightforward to associate simulators which accept model specifications in such forms. Simulators of this kind were put forward for both conventional simulation languages as well as for asynchronous, distributed architectures. Correctness of simulation is shown by establishing the existence of a suitable morphism relation between simulator and DEVS model.

These concrete forms of representation justify the approach adopted early in this work whereby model specification is abstracted from program implementation. The separation principle for models and experimental frames was justified at the same time since the realization of model/frame pairs was shown to result in a hierarchical modular system specification that can be simulated with approaches developed here.

Possibilities for exploiting asynchronous microcomputer networks for multifacetted modelling and simulation are taken up again in Section 18.3.

PROBLEMS

1. Consider a one-dimensional cellular automaton array with a finite number of cells. Let each cell be handled by its own simulator with communication links existing between adjacent simulators. Devise a synchronization scheme that satisfies the global correctness criterion, i.e. it is possible for an observer to identify the global state of the automaton array by inspecting the global state of the simulator array. An additional property that is possible to achieve is full parallelism of the simulators. This means that the time taken to compute a global model state transition is the maximum of the times required by each simulator to compute its local transition, rather than the sum of these times.

Hint. Arrange for a rightward signal to initiate local transition function computation and for a leftward signal to initiate local state update. Make sure that all simulators are started together (assuming 0 signal propagation delay) but that updating does not occur prematurely.

References

[1] Bryant, R, (1979). "Simulation on a Distributed System". *IEEE CH4445*.
[2] Cellier, F. E. (1981). *Combined Continuous/Discrete Real-time Simulation*.
[3] Chandy, K. M. and J. Mistra (1981). "Asynchronous Distributed Simulation via a Sequence of Parallel Computations". *C.A.C.M.* **24**, 198–205.
[4] Decker, H. D. and J. Geissler (1983). "Modelling and Simulating Nets of Agencies with BORIS". In *Model Acceptability* (ed, H. Wedde), Pergamon Press, NY.
[5] Concepcion, A. and B. P. Zeigler (1983). *Distributed Simulation of Distributed System Models* CS Dept. Tech. Rept., Wayne State University, Detroit, MI 48202.
[6] Peachock, J. K., Wong, J. W. and E. Manning (1979). "Distributed Simulation Using a Network of Processors". Computer Networks 3, 44–56.
[7] Wulf, W. A. (1977). "Languages and Structures Programs". In *Current Trends in Programming Methodology*, Vol. 1. (ed, R. T. Yeh), Prentice Hall, NJ.
[8] Unger, B. W. (1978), "Programming Languages for Computer Systems Simulation". *Simulation* **30**, No.4.

Chapter 18

SYSTEM ARCHITECTURES FOR MUTLIFACETTED MODELLING

1. Architectural Framework for Support of Multifacetted Methodology

In a scheme for support of multifacetted methodology, *users* interact with the computer system through *interfaces* that enable them to initiate or engage in *activities*. The sequencing of activities may be partly fixed and partly open to users' control. An activity is executed by one or more *processors* (in conjunction with a user) and acts upon one or more *data bases*. In executing an activity, information may be stored in the bases and is accessible to users through interfaces.

Among the activities involved in modelling and simulation are those coming under the following categories [1]:

Model generation/referencing:– the generation of models (i.e. system descriptions) either by construction from scratch or by employing models retrieved from a model repository as components to be coupled together.

Model processing:– the manipulation of model texts, (e.g. to produce documentation, to check consistency, etc.) and the generation of model behavior, of which simulation, restricted to mean computerized experimentation with models is a predominant form.

Behavior processing:– the analysis and display of behavior in static (e.g. equilibrium), dynamic, (i.e. concerning state trajectories) or structural, (i.e. concerning changes in model structure) modes.

Real system experimentation:– the gathering and organized storage of behavioral data from the real system or any of its components of interest.

Model quality assurance:– the verification of the simulation program or device as correctly implementing the intended model, the validation of the model as an adequate representation of the real system, and the testing of other relationships in which models participate.

We shall refer to the models, experimental frames, etc., manipulated by these activities as *objects*. Objects constitute the contents of some of the data bases, called *object bases*. Programs are required to construct and manipulate

such objects. These programs, called *tools*, help carry out the following basic kinds of object related activities:
construction and destruction of objects
querying and displaying the state of an object
operating on the state of an object
transforming one or more objects into a new one
testing relationships between objects
storing and retrieving objects in object bases.

Examples of objects and associated activities that have been discussed for the multifacetted methodology are displayed in Table 18.1. As can be seen, the term "data base" refers to an organized form of storage of "data" that facilitates its retrieval when required by an activity. Thus such data may refer to object descriptions such as model and frame specifications, and primary data such as real system or model input/output segment pairs. Also, the tools for manipulating objects may also be stored in *tool bases* so that they may readily be identified and retrieved when required.

Many types of users may be interacting with a computer-based modelling system: modellers, experimentalists, data analysts, engineers, managers,

Table 18.1. Objects and activities in multifacetted methodology.

Objects	Activities
variables	definition
	association to entities
	coarsening
	refining
	establishing semantic relations
entities	extracting from models
	composing
	decomposing
parameters	reconciling
	establishing correspondences
	identification
	propagation
experimental frames	associating with entities
	coarsening
	refining
	testing derivability
	testing applicability to models
models	associating with entities
	coarsening
	refining
	testing validity in frames

system designers — persons whose skills are specialized in one of the categories of activities listed above. The interfaces designed for user-system communication should also be specialized to the nature of the activity be carried out. Communication goes in two directions: user-to-computer in the form of commands and data, and computer-to-user in the form of displays. Commands may be transmitted in several forms: as texts containing instructions and data as in a conventional program, or through interactive dialogs. Such dialogs may be entirely under the control of the user or may be driven by the computer. In the first case there will be a *command language* that user must keep in mind in which he formulates commands. In the second case, the applicable commands are displayed in questionnaire or menu form. Account should be taken of the fact that users may have many command languages (appropriate to different activities) to master. Uniformity in syntax and style, and consistency in meaning, reduce the memory burden and the chance of wrongly generalizing from one language to another. An example of uniformity and consistency: every interface language should have a command of the same form and meaning: help.

Concerning display of information, increasingly sophisticated graphics facilities are becoming available. They should be employed to present perspectives of an object that foster a user's grasp of its structure. This is especially true for abstract objects that may not have obvious geometrical representations. Inventing and employing such representations may greatly facilitate users' abilities to deal with the abstractions required by the methodology. An example of such geometric representations is the tree structure labelled with various symbols to portray an entity structure (e.g. Figure 15.1).

2. Contexts For Support of Multifacetted Methodology

While the concepts that we have been developing are intended to be applicable in a wide variety of application contexts, each context is likely to elicit an emphasis on some subset of the overall scheme. The design of computer based systems for multifacetted modelling should consider the environment in which the system will be used so as to match its capabilities to the needs of users in this environment. The system should be designed so that the minimal set of facilities characterizing the application context can be implemented first and an operational system brought into being. However, the natural tendency of any system, once operational, will be to create a demand for extension to greater multifacetted capability. Thus the design should set up a hierarchy of levels of extensions graded according to an estimate of the future evolution of user interests. The design should incorporate points of flexibility so that successive levels do not require reconstruction of the system anew but can be realized by suitable modifications of the core system.

With this approach in mind we shall consider some representative contexts from the point of view of the features of multifacetted methodology that they emphasize.

Long-term development of models for a specific real system

The first context we consider is that in which a study of a real system is being done over a protracted period of time (measured in decades). For example, a particular ecosystem such as a lake or forest is under investigation with data being gathered, and models constructed, by several research teams. This context is characterized by its emphasis on *knowledge acquisition*, i.e. the gathering of empirical data and its codification through the construction and validation of models. The following structures are therefore emphasized: system entity structure, behavioral data base, model base, experimental frame base, parameter base. A well planned approach to data gathering would lay out the experimental frames in which experiments will take place. Data gathered would be stored within the behavioral data base according to system entity that the data describes and the experimental frame in which it was gathered. Using derivability principles, data gathered in one frame can be processed to meet the specifications of others. With this in mind a research program would plan the experimentation so that unintended redundancy in experimentation is eliminated, while desired redundancy that enables consistency checking of data sets gathered in different frames is consciously incorporated.

Also important in knowledge acquisition is the encoding of empirical data in the form of parameter values for models derived by model calibration or measured directly in the real system. A *parameter base* enables a model to access information providing up-to-date estimates of the values of its parameters (Section 14.3). This sharing of knowledge facilitates the model calibration process since constraints are placed on the range of those parameters for which information is available. This enables searching to be concentrated on parameters whose values are currently unconstrained. Successful, or unsuccessful, matching of model and real system data, should feed back to the confidence with which a parameter value may be accepted as known. Tools for testing of model data against real system data, searching of parameter spaces, and propagation of parameter information are thus crucial in a knowledge acquisition environment.

Long-term development of models for a class of similar systems

Science always strives to generalize its findings beyond the particular space/time point in which they were ascertained to hold. Thus knowledge acquisition for a particular real system is likely to proceed in parallel with similar projects concerning systems of the same type. In this case attempts will be made to put comparable structures into correspondence and to seek general relationships applicable across all systems in the same family. For example, several lakes may be under investigation, with the lakes sharing many aquatic and ecological properties but also differing in significant respects arising from their geographical location, relation to surrounding ecosystems, etc. Let us suppose that modelling of individual instances proceeds in the manner discussed in the previous section. However a second level of generality is added to integrate and generalize the individual projects. Central to such organization is a system entity structure with specialization hierarchy capability (Section 11.2). Such an overall structure may be designed so that individual entity structures may be extracted from it by pruning. The bases for models and experimental frames that are organized by this overall structure, should now store generic models and frames capable of being specialized to the concrete form employed in individual projects. The parameter base associated with the overall structure should contain only parameters that are considered, or found, to be fundamental ones, i.e. invariant in value across all individual instances of a type of systems.

Engineering synthesis of a family of systems

Systems are often designed as a family of possible configurations of generic components. For example, a computer system may consist of several versions of a central processing unit with several choices of memory sizes and peripherals. To select a configuration to meet a client's specification is a much smaller problem then the general systems design problem (Chapter 12). However, it is still not a trivial matter if many possibilities exist, it is costly to mispredict the performance of a configuration, and it is difficult to make a sufficiently accurate prediction.

The system entity structure with specialization hierarchy capability is the basic means of organizing the set of all possible configurations of the system to be synthesized. The designer of the family of systems is responsible for setting up this entity structure. The model base contains models for the atomic system components that are coupled together. The user, e.g. salesman or other system configurator, interacts with the entity structure pruner, to select a desired configuration. After configuring a system, the user should be able to specify a model for it using a model specification tool and then be able to experiment with it using a simulation tool. There are several possible approaches to the design of such a system and the time frame for implementation may depend greatly on the approach. One would have to decide whether to build a simulator from scratch in a simulation language or provide an interface to an existing simulation test bed.

Experimental frames would include standard templates for evaluation of price, space requirements, power consumption, reliability, input/output performance indexes and utilization of resources indexes (Chapter 12). The user can request evaluation of system configurations in these basic frames. Advantage can be taken of the reduced model complexity made possible in this way to obtain rapid program generation and simulation experimentation. Once system configurations have been found that satisfy specifications relating to individual frame objectives, they may be evaluated using more complex models in frames combining the basic frames for trade-off considerations.

The system may be extended to provide help in searching the configuration space. To the extent possible, Artificial intelligence (AI) methods may be employed to incorporate accepted design heuristics. System configurations not so eliminated must be sampled experimentally using the simulation component but computer assistance in the integration of this information can be provided. Graphical displays of relationships would be an important interface to the user.

Management of ill-understood non-recurrent systems

In contrast to the design context, management usually concerns the formulation of policies for intervening in existing systems rather than the building of new ones (Chapter 1). Moreover the existing system may not be well understood nor sufficiently stable to permit the long term construction of a family of validated models. Elzas [2] distinguishes between *repeatable*, *recurrent* and *unique* sets of phenomena. Repeatability of experiments is the essence of a classical science such as physics and fosters long term knowledge acquisition since it is both possible to set up the circumstances necessary to test models and also to apply them. A system characterized by recurrent phenomena may not lend itself to being manipulated into a desired experimental condition, but if phenomena recur sufficiently regularly, opportunities to test and take advantage of models still arise. Unique phenomena are those that do not naturally recur and cannot be caused to repeat at will. The ever changing environment of modern business, characterized by highly competitive drives to introduce new technologies, is an example of where unique phenomena may be the rule.

In such unstable circumstances, the emphasis will be on the "what if" mode of model application, i.e. policies are tested on models and scenarios that are largely hypothetical, representing guesses as to what the future system and the uncontrollable environment will be like. In this case, tools should be provided so that a decision maker may rapidly construct models and experimental frames that embody "what if" hypotheses. Intended for short term forecasting, such models and frames are likely to be relatively simple structures incorporating relatively simple hypothetical relationships. The model and frame bases in this application should constitute a repository of structures that can be readily disassembled into components that can be coupled together and integrated into a new structure. Due to the uniqueness of the phenomena, validation of particular models and accumulation of parameter information will be meaningless. However, one structure may be more useful than another in the sense that it may be employed more often and/or when it is employed, it participates in more successful forecasts. Such a usefulness rating should take the place of validation confidence in feeding back to guide the selection of components for model and frame synthesis.

Training and education

A well established application of simulation is in training operators to control equipment in a safe but realistic setting. Examples include the well known flight simulators and more recently, simulators of nuclear power plant control ensembles for training operators how to react to crises that may arise in plant operation. Emphasis in this context is placed on reproducing the real system in sufficient detail so that the reactions learned in the simulated environment carry over to the real one. Thus the multifacetted approach here does not lead to construction of a multiplicity of lumped models but to a comprehensive base model. However, besides realism, an important property of a simulator is flexibility to be reconfigured to represent a new version of the simulated equipment. To design for flexibility, the emphasis should be on modular and hierarchical construction of the base model and the use of formalisms that support this kind of model specification (Chapter 8).

A promising form of model-based education is found in its use for concept, rather than skill, development. In such a context, a student learns the structure and complexity of a class of real systems by interacting with a computer in a tutorial dialog. The system entity structure is the central medium to teach the various decompositions, components, specialization and generalization classifications found to be useful in conceptualizing the class of real system. The user interface should be designed so that the entities and aspects are explained in terms of their relation to the real system.

Several levels of capability are conceivable. The first is aimed only at conveying a familiarity with the dynamic interactions embodied in models rather than at an understanding of the models themselves. At this level, the "model base" should contain didactic descriptions of dynamics. The second level may aim at providing a more in-depth understanding of the models in a model base. Accordingly, tutorials should be associated with the models and experimental frames, documenting their structure, and explaining how simulation experiments can be performed. Third and fourth levels are conceivable in which a modelling methodology for the class of real systems, and utilization of modelling in design, management or control of such systems are taught respectively. Instruction of medical practitioners is an example in which such computer-aided instruction systems may be designed at one or more of such levels.

3. Software/Hardware Steps Toward Multifacetted System Architecture

We have seen an architectural scheme in which computer support of multifacetted modelling methodology may be contemplated and problem contexts which emphasize different features of such support. We gain a second perspective on the possibilities by examining designs of software and hardware systems that can be viewed as steps toward more comprehensive computer support of multifacetted modelling. No attempt is made in the following to enumerate all state-of-the-art software/hardware of relevance. Rather, examples are given of such designs that are prototypic of components in ultimate multifacetted modelling support systems.

Object-oriented high level languages

As previously described (Section 18.1) software systems for support of modelling can be viewed as providing tools for the definition and manipulation of classes of objects. A number of high level programming languages have been developed from this perspective, originating with the class concept of SIMULA, perhaps the one most widely known being SMALLTALK [3]. We shall briefly describe such an object-oriented language that is close in conception to our view of object classes as specified by formalisms (Section 3.6). GARF (Generalized Arithmetic Recursive Functions) combines the SIMULA class concept for data structuring with functional procedure definition characterized by LISP. Its conceptual innovation however is that the object classes can be parameterized so as to achieve structures of great generality as we shall explain.

In GARF, object classes are defined by induction using some basic forms of specification. A class definition sets up associated predicates for testing for membership of an object in the class, and for testing for equality of members within a class. A class definition may contain in addition to the structural definition of its objects definitions for predicates than can compute the truth of properties of objects, operators that change the state of objects, and mappings that transform objects into other objects.

The structural segment of a class definition consists of three statements: signature, object and argument. The *signature* statement declares the symbols to be used for the functions that will be composed to form the objects. The *object* statement provides the inductive definition for such composition processes together with the *argument* statement that gives the arguments to be employed in the compositions. For example, the following defines natural numbers as objects in the class NAT:

    class NAT
    signatures: O, suc
    objects: O, $suc(n)$
    arguments: (NAT|n)

Here there are two signature functions: O is a constant function (has no arguments) and suc is a unary function. The object statement declares that O is an object of class NAT (the basis step in the inductive definition). It also says that one can construct a new object by applying suc to an argument n; the arguments statement specifies that such an n must itself be from the class NAT, clearly an inductive definition of the class NAT. One may infer that NAT = $\{O, suc(O), suc(suc(O)), ...\}$. This is the classical definition of natural numbers using Peano's axioms.

The segment of a class definition that defines functions employs the *case schema*: (if P_1 then A_1, if P_2 then A_2,..., else A_n) where the P_i are predicates and the A_i are actions to be taken in case the corresponding P_i is the first predicate in the list to be true. Definition of the class of boolean objects illustrates the use of the case schema:

    class BOOL
    signatures: F,T
    objects: F,T
    arguments: (BOOL|x,y)
    defined functions:
       BOOL|or (x,y) = (if EQ(x,T) then T,
                         if EQ(y,T) then T,
                         else F)
       BOOL|and(x,y) = ...
       BOOL|not(x) = ...

Note that in this class there are only two objects T (true) and F (false). This, and any other class of finite cardinality, can be defined directly in terms of constant functions as illustrated here. The basic logic operator "or" is defined using the case statement. Others such as "and", "not", etc. can be defined in like manner. Note that the classes of arguments of these functions are specified in the arguments statement, and the class of their result is declared as BOOL. Thus such functions are indeed logical operators mapping from boolean objects to boolean objects. Such specification of the domain and range types of a function is called strong typing in modern programming languages and arises naturally from the world view of GARF as a set theory based language. In the definition, we employ the equality predicate EQ that the language provides automatically with each class.

The following definition for the class of strings illustrate class

parameterization:
   class STRING(X)
   signatures: null, ir (empty string, include on right)
   objects: null, $ir(v,x)$
   arguments: $(X \| X)$, $(X | x)$, $(STRING(X)(v,w))$
   basis functions:
      (BOOL|empty(null) = T; empty($ir(v,x)$) = F)
      (X|outer($ir(v,x)$) = x)
      (STRING(X)|tail($ir(v,x)$) = v)
   defined functions:
   NAT|length(v) = (if empty(v) then O,
                         else suc(length(tail(v)))))
   BOOL|is__in(x,v) = ...(true just in case x is in the string v)
   STRING(X)|delete(x,v) = ...(deletes the first occurrence of
                                   x from the string v)
   STRING(X)|concat(v,w) = ...(concatenates strings v and w
                                   to form new string).

The parameter of the definition is X, which is specified to be any class in by the phrase $X \| X$ in the arguments statement. Thus the definition is actually a template for defining strings over any class of objects. Thus STRING(X) may be thought of as an abstract class. To give it a concrete realization, a class name is substituted for X. Thus STRING(NAT) is that class of strings of natural numbers, while STRING(STRING(NAT)) is the class of strings whose elements are themselves strings of natural numbers.

Also illustrated in the definition is the use of *basis* functions to determine the nature of objects in the class and decompose them. Each signature gives rise to a corresponding basis function that tests whether an object begins with that signature symbol. For example, "empty" tests for the null signature which stands for the empty string. Each argument gives rise to a basis function that can extract it from its composition. For example, "outer" extracts the element x from the string $ir(v,x)$ while tail extracts the components string v. Mappings into other classes, predicates, and operators on strings are illustrated in the defined functions.

It may now be clear that one may build up complex abstract classes that can be concretized in an unlimited number of ways giving rise to a variety of structures. A compiler for such a language maps concrete class definitions into procedural form. Signature functions are mapped into procedures that operate appropriately on data structures. For example, O maps into a procedure that defines a new structure characteristic of O; "*suc*" operates on a structure representing a natural number n to extend it to represent "$suc(n)$". Abstract class definitions are mapped into templates for such procedures that can be instantiated when concrete classes for the parameters are specified.

The GARF language can be employed to provide high level definitions for system specification formalisms. For example, one can define the class of autonomous state machines with state set Q as the free parameter. Defined functions for such a class include STRAJ that maps a machine M, a state q, and natural number n, into an object in STRING(Q) that represents the state trajectory of M starting from state q of length n. Such a class can be specialized to the finite autonomous state machines by replacing the parameter Q with classes of finite cardinality. It can also be specialized to other classes, for example in which Q is STRING(N), i.e. machines whose states are strings of natural numbers. In each case however, the concept of state trajectory remains invariant and is directly implementable by applying the operator STRAJ defined at the abstract level.

Thus, GARF illustrates how the formal framework for multifacetted modelling developed earlier can be implemented through the media provided by appropriate high level object-oriented languages. To proceed along these lines however, one would want to design a language that is oriented directly to the kinds of formalisms, and the mappings between formalisms, that were characterized in Chapter 2 and which were later developed in detail in subsequent chapters. GARF is not optimal here since the desired concepts do not have direct correspondences to its features. In such a language, an operationalized version of the most general systems formalisms at each level of the hierarchy of systems specification would be defined. When a more specialized formalism is introduced, one would provide not only its class definition, but the procedural realization of its specification mapping, i.e. the translation by which its objects are mapped into systems in the general formalism. Such approach would provide both generality (due to the existence of the most general systems formalism) and flexibility (new formalisms can be readily introduced and made operational by defining the required specification mapping). It might however, suffer from inefficiency in behavior generation, i.e. simulation, and symbolic manipulation, unless provision were made to employ concepts at the least general class consistent with the desired processing.

The following descriptions of state-of-the-art software systems reveal approaches that are much more particularistic in their conception of language features for supporting multifacetted methodologies. The reader is challenged to study the issue of level of implementation language raised by these contrasting approaches.

The entity structuring program (ESP)

As discussed previously (Chapter 11), the entity structure concept is eventually intended to be employed by a modelling executive to organize and manage models and experimental frames in software support systems for multifacetted systems modelling [4]. One of the tools that already exists for interacting with a user to construct entity structures. Called the *entity structuring program* (ESP) [5], it can stand alone as a kind of conceptual scratch pad and communication tool. With it, the modeller or team of modellers, can conceptualize and record the decompositions underlying a model, or family of models, as the model development proceeds. Another current use of ESP is to teach students of simulation the ability to structure systems and models at a level more abstract that the acutal simulation code. Especially, they should acquire a flexibility of approach fostered by the ability of the entity structure to accommodate alternative decompositions for the same system. ESP is available in Pascal from the author.

In the following we shall discuss the basic features of ESP in the form of a tutorial offered to the first time user.

The following is an example of an entity structure (ES) created with ESP:
```
*1: ENTITY UNIVERSE
   2: ASPECT PHYSICAL__DECOMPOSITION
     3: ENTITY MOLECULES
       4: ASPECT MOLECULES__DECOMPOSITION
         5: ENTITY MOLECULE
```
The asterisk (*) indicates the current item occurrence (ITOC) which is the focus of the program's attention. Many of the operations refer implicitly to the current ITOC. The FIND command may be used to change the location of the "*". For example,
FIND MOLECULES
LIST STRUCTURE
will cause MOLECULES to be the current ITOC. The LIST STRUCTURE command displays a representation of the ES so that the following would result:
```
 1: ENTITY UNIVERSE
   2: ASPECT PHYSICAL__DECOMPOSITION
    *3: ENTITY MOLECULES
       4: ASPECT MOLECULES__DECOMPOSITION
         5: ENTITY MOLECULE
```
The situation upon logging in to ESP, call it a "tabula rasa", is that of an ES consisting of a single fictitious item of mode aspect called ROOT__ITEM. To construct a desired ES one must begin by attaching an entity to this aspect. This can be done via the ADD ENTITY command. For example:
ADD ENT UNIVERSE
LIST STRUCTURE
results in an ES displayed as
   *1: UNIVERSE

Once such an elementary ES has been established one can proceed to extend and modify it as will now be indicated.

The ADD commands are the basic ones available to build up an ES. With their help one can add items, item occurrences, variables, and synonyms to the existing structure.

1. ADD ENT/ASP "item_name"

creates an item called "item_name" if such an item does not already exist and attaches it to the ES in the vicinity of the current ITOC. Just where attachment takes place depends on the mode of the current ITOC: if this mode is aspect then an entity will be attached as a son and an aspect will be attached as a brother. The converse applies if the ITOC is an entity. The attached item becomes the current ITOC. For example the following sequence will result in the ES discussed above:

ADD ASP PHYSICAL_DECOMPOSITION;
ADD ENT MOLECULES;
ADD ASP MOLECULES_DECOMPOSITION;
ADD ENT MOLECULES;

We see that this sequence is entirely natural provided that we remember that an aspect must always intervene between an entity and its subordinate.

To add another aspect to UNIVERSE, either of the following sequences would suffice:

FIND UNIVERSE
ADD ASP BIOLOGICAL_ASPECT

or

FIND PHYSICAL_DECOMPOSITION
ADD ASP BIOLOGICAL_ASPECT

In the first case, the new aspect is attached as a son and in the second, as a brother. Issuing the LIST STRUCTURE command (or just STRC for short) displays:

   1: ENTITY UNIVERSE
     2: ASPECT PHYSICAL_DECOMPOSITION
       3: ENTITY MOLECULES
         4: ASPECT MOLECULES_DECOMPOSITION
           5: ENTITY MOLECULE
    *2: ASPECT BIOLOGICAL_ASPECT

2. ADD VAR "variable_name"

causes the attachment of a variable name "variable_name" to the current ITOC and by the uniformity axiom to all occurrences of this item. ESP responds at the terminal with a sequence of prompts for information on the specification of the variable. For example, to add a variable WEIGHT to MOLECULE the following dialog might occur:

FIND MOLECULE

ADD VAR WEIGHT
RANGE? (ESP response)
non negative reals
MEANING?
measures the mass of the molecule
PHYS.DIMENSION?
mass units
UNITS?
milligrams
   3. ADD SYN "given__name" "new__name"
causes the new__name to be synonymous (interchangeable in all ESP commands) with the given__name which must either be an existing synonym of the current item or of one of its variables. For example to call the UNIVERSE also the WORLD or to abbreviate WEIGHT to WT:
ADD SYN UNIVERSE WORLD
ADD SYN WEIGHT WT
This feature permits the use both of long descriptive names and short comfortable names. Moreover it accommodates the use of different names for the same entity or variable in different model, program and data files.

Modification, either to correct errors or to suit a new purpose, is accomplished in ESP by deleting undesired structure and adding desired structure. Deletion is effectuated by the DELETE commands which are roughly the inverse of the ADD commands. For example,
   DEL ENT/ASP "item__name"
causes the current ITOC to be detached from the ES leaving its father as the new current ITOC. This detachment does not destroy the item or any of its substructure — attaching this item to the ES will bring with it all this structure.

The DELETE command contrasts strongly with the PURGE command which destroys the current item (hence all its occurrences) without any trace. However, the substructure of the item is not destroyed but is detached from the ES as in the DELETE case.

While the uniformity axiom enables one to duplicate the structure of an item in any of its occurrences it does not permit any of these occurrences to be modified independently. This independent variation can only be achieved if a new structure is created in the image of the existing one. The command
   ADD VAR DUPL "source__item__name"
attaches a duplicate of each of the variables of the source item to the current item. Since the duplicates belong to a current item they may be altered independently of the originals which belong to the source item. For example, the sequence:
   ADD ENT ION

ADD VAR DUPL MOLECULE
ADD SYN WT MASS

will create an entity ION which has all the variables of MOLECULE including WEIGHT whose description is then modified as desired. Note that the DUPL subcommand achieves an approximation to the concept of variable type in which a variable_name automatically carries with it a fixed structure no matter to which item it is attached (Section 11.1).

In ESP, upwards extension of the ES to represent expanded system boundaries (Chapter 14), is accomplished with a sequence of add and delete commands. The entity desired to be the new root is added first as a brother to the old one. An aspect is then added to it and finally the old root entity is deleted and attached to this new aspect. The old root entity thus becomes a subentity of the new root entity.

There are a variety of commands to display various features of the ES. LIST STRUCTURE, used above, is an example. Among other such LIST commands are LIST SUBSTRUCTURE and LIST PATH to display the substructure below the current ITOC and the path to it from the root entity, respectively.

The SAVE command writes a complete representation of the ES on a file which is human readable form and serves also to direct a desired resurrection of the ES (see below). In addition the user can request that the LIST commands write their displays on a documentation file in addition to appearing at the terminal.

As indicated above, the representation of the ES produced by SAVE can be employed by ESP to reconstruct the ES at a later session. To do this the RESTORE command is issued at the beginning of the session and the file name containing the representation of the desired ES is given in response to ESP's request.

Steps toward model based decision support

The architectures proposed for decision support systems by Sprague and Carlson [6] contain a model base as one of the three defining components (the others being a data base and dialog user interface, Section 1.1). Although these authors define some of the functions that would be performed by a model base management system (analogous to a data base management system) they provide no in-depth theory for the design of such a system.

Spiguel [7] recently presented a conceptual framework for a model base system in the business support system that takes into account the cognitive processes underlying the activities of a decision maker/model builder. The business models dealt with in this framework are discrete time arithmetic structures typically formulated in simulation languages such as IFPS[38] for

---

[38] Execucom Systems Corp. (1980). *IFPS (The Interactive Financial Planning System)* User's Manual.

forecasting business performance indicators over limited time horizons.

Spiguel's model specification, called a schema, consists of two sub-schemata: a *skeleton subschema* and a *function subschema*. The skeleton subschema specifies a set of descriptive variables of the model (with associated range and units) and a set of *interactions*. An interaction is specified by its sets of input and output variables and may be *structural* (static) or *temporal* (dynamic) the difference being that the input and output variables of a structural interaction belong to the given set of descriptive variables while at least one input variable of a temporal interaction has a non-zero integer i coupled to it. The integer, called a *lead* if positive, and a *lag* if negative, signifies that the value of the associated descriptive variable i units in the (future/past) is to be employed rather than its current value in computing the current values of the output variables. The function subschema supplies, for every interaction, the function that maps value assignments to the input variables into value assignments for the output variables.

The separation of skeleton and function schemata leads to a model base in which model skeletons may be stored independently and in which a model provides a reference to both a model skeleton and an associated function subschema. The functions employed may be synthesized within a given language or may be stored in a function base.

In the methodology underlying Spiguel's framework, model construction is based on the skeleton abstraction. In a process called *model fabrication*, parts of skeletons existing in the model base may be coupled together with new interactions to synthesize a desired model skeleton. Such fabrication is supported by a formal deductive system for manipulating interactions analogous to those developed for manipulation of relational data dependencies [8]. The deductive system provides rules for inferring the existence of interactions implied by a given set of interactions. The following are basic questions that arise in fabrication and that can be answered with the deductive system:

1) Given sets of input and output variables, and a model skeleton in the model base, is there a submodel of this skeleton that determines the output variables, given the input variables?
2) Given a set of variables desired as output and a model skeleton, what are the input variables whose values are needed to in order to compute the values of the output variables?
3) Given a set of variables and a model skeleton, what variables are determined by this set of variables in this structure?

These questions arise as the modeller probes the model base for existing skeletons that might be exploited to build the current model. Following the objectives driven methodology of Chapter 12, the modeller starts with an experimental frame that specifies available input variables and desired output variables. A search is performed on the model base for a skeleton that meets this requirement: this is an iteration on Question 1. If no such skeleton is found, the modeller may decide to work backward from the output variables to find input variables needed to determine them, and/or forward from the input variables to find output variables they determine. These are searches through the model skeletons iterating in Questions 2 and 3, respectively. Having a new set of inputs and output variables, the search loops back to an iteration on Question 1 for a skeleton that meets the input/output requirement. The search process continues until a set of subskeletons is extracted whose union generates the original output variables from the original input variables, or the modeller decides that additional interactions must be constructed to meet the requirements.

Of course, once such skeletons are discovered they must be modified and coupled together, not a trivial matter, since input variables may include lead/lag terms and since skeletons may overlap in the variables they employ leading to potential inconsistencies. This problem awaits further work. However, taking another approach, that of modular model construction it is readily solved as illustrated in the language GEST81 to be described below.

GEST-language for modular and hierarchical model construction

GEST (General Systems Theory Implementor) as originally conceived [9] departed radically from extant simulation languages. It has since undergone a number of revisions as its definition becomes more aligned toward service as the central model and experimental frame specifying language within a comprehensive modelling and simulation system.

A GEST81 program is highly descriptive and acts as documentation, as well as specification, of a simulation study. Such a program is intended to be developed in conversational form with the computer providing assistance and prompts to satisfy the syntactical and semantical requirements of the language. (Syntax editors are already in existence for some common programming languages.) Thus help is provided to minimize the effort required to supply the information required by GEST program (which exceeds that required by a conventional simulation language).

A GEST81 program consists of *model, experimentation* and *output module* segments. We shall briefly describe each of these segments. More details are available in [10].

Model specification segment. A model specification consists of a *parametric model specification* and any number of *labelled parameter assignment sets*. The idea is that to specify a particular model from the parametric class only the label of a parameter assignment set need be supplied.

A model may be atomic or the resultant of a coupling of components, each of which are themselves models. This gives rise to modular and hierarchical model construction.

An atomic model may be continuous (differential equation), discrete time, or memoryless. Thus models containing components specified by different formalisms are possible. Extension to discrete event and other formalisms would be desirable.

An *atomic model specification* consists of static and dynamic structure segments. The *static structure* declares descriptive variables in the following categories: input, state, output, auxiliary variable, constant, parameter, auxiliary parameter. The conventional type (integer, real, array, etc.) of each variable must be declared and a subrange specification may be given. Auxiliary variables are those variables employed in the dynamic structure section that are not state or output variables. Similarly auxiliary parameters are variables that mediate the computation of parameter set assignments but are not themselves parameters of the model.

The *dynamic structure* segment consists of transition and output function segments. The *output function segment* relates the output variables to input, state and auxiliary variables. The *transition function segment* specifies equations for the derivatives of the state variables (continuous case) or for the next values of the state variables (discrete time case). A memoryless model contains only an output segment and lacks state variables and a transition segment.

A *coupled model* consists of a global interface segment, an externals segment, a list of component models, and a coupling specification. The *global interface* declares the input and output variables of the coupled model while the *externals segment* identifies these input and output variables with input and output variables of the component models.

An example is shown in Figure 18.1a, where model Z is the resultant of coupling models M and N. The externally declared input variable and output variable of Z are A and H, respectively. Thus $Z.A$ and $Z.H$ are the names by which these variables will be known to all nested couplings in which Z is a component. However, the externals declaration identifies $Z.A$ with $M.B$, an input variable of M and $Z.H$ with $N.H$ an output variable of N.

The *coupling scheme* specifies for each input variable (except those identified with external inputs), an output variable that is to be connected to it. This implements the Wymore coupling scheme of Appendix 1, Chapter 15. The conversational system generates the list of input variable names of the components and waits for an output variable name for each (other than the external ones). Documentation of the coupling for each component is produced as illustrated in Figure 18.1b.

A *nested coupling* (or hierarchical construction) is one in which at least one component model is a coupled model (non atomic). Since the input/output

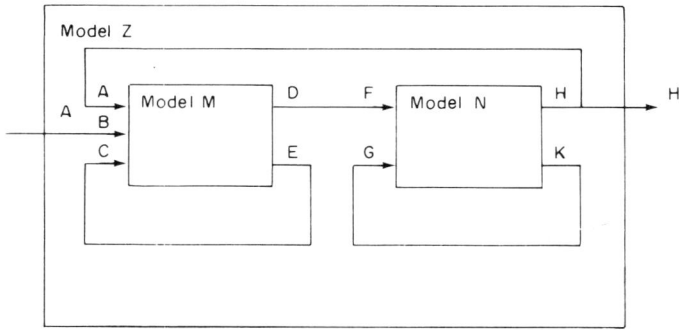

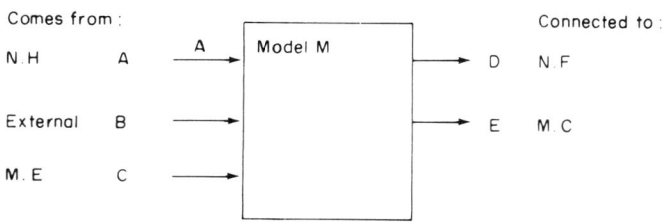

Fig. 18.1. Documentation of the input/output relationships of one of the component models of a coupled model.

interface of the latter model is declared it can be employed in a coupling to other components even when its internal structure has been completely specified. Such would be the case in top down design (Section 3.5).

GEST81 incorporates a *class generation* mechanism. Replicas of any parametric model M may be created and given distinct names. Once created the components may be assigned different parameter sets and subjected to different specific experimental frames (see below).

Experimentation specification segment. Experimentation is specified in experimental frame, model/frame pair and post study segments, the latter being optional. The *experimental frame segment* consists of two parts: a global frame that applies to a coupled model and specific frames that apply to each of its components. A global frame specifies experimentation intended for the resultant model of a coupling. Specific frames apply to atomic models in such a coupling. To start with the latter, they include initialization of state variables, data collecting requirements, etc. We shall see that the GEST concept of experimental frame assumes applicability to a particular parametric model so that initialization is specified in terms of its state variables. Thus it does not provide the more abstract run control variable concept that is required if a frame is formulated to apply to more than one model (c.f. Chapter 15). A coupled model is initialized by providing specific frames for each of its atomic components. A global frame for a coupled model specifies such experimentation conditions as time unit, input segments, and output sampling.

A *model/frame pair* specifies the combination of parametric model, parameter set, and experimental frame to be used in a simulation run. Note again, in contrast to Section 12.4.2, GEST81 does not separate experimental frame and execution control functions, both being combined within the experimental frame concept. Thus it is sufficient, in GEST81, to specify a model/frame pair and a model parameter set in order to completely specify a simulation run. Such a run may be specified by the following kind of statement:
  RUN 1 to OBSERVE MODEL population__growth
   WITH PARAMETER SET high__birth__rate
    IN FRAME low__employment.

Output display and analysis for a single run can be specified by appending a statement indicating the output module and device (printer, crt, etc.) that are to be used for this purpose. Such *output modules* are defined in the final segment of the GEST81 program. They specify the report form to be used, variables to be plotted, etc.

A *post study segment* specifies any processing that is to be performed on the trajectories produced by a set of simulation runs.

Although GEST81 does not implement the full model/experimental frame/execution control modularization advocated in Section 16.1, it is the simulation language realization closest to this conceptual view. As usual, there is a trade-off to be considered. In the GEST81 approach, it is easier to specify experimentation directly in terms of a particular model, to check for consistency of model/frame pairs, and to convert such a specification into code. However, the ability to apply the same frame to many models, necessary for test of model simplification (Section 13.4), frame based model retrieval (Chapter 15), and the other "modelling in the large" activities that constitute the multifacetted modelling methodology are not supported.

Internal language for discrete event model representation
While GEST aims at modular and hierarchical model specification, it accepts the user's model description and makes no attempt to transform it into some other form. Overstreet [11] has developed a representation language for discrete event models that is amenable to such manipulation. The language is neutral with respect to the three world views (event, activity and process) and models specified within it can be transformed into each one of these formalisms. In contrast to the user-oriented languages given in Chapters 7 and 8, Overstreet's language is intended to be less abstract and

governed by a stricter syntax so that symbolic analysis is facilitated. Examples of such analysis problems are:
1) Given a model description check it for completeness, consistency, and absence of ambiguity.
2) Given a model description reduce its complexity as far as possible (with respect to given complexity measures).
3) Given a model description decompose it into component models such that a measure of complexity of the interaction between components is minimized.
4) Given a model description, a particular world view, transform it into an equivalent model within this world view having minimum complexity (in some measure).

While as Overstreet shows, ambitious forms of these analyses are recursively unsolvable for discrete event models (see for example Problem 1 of Chapter 6), or are combinatorially intractable, there are less ambitious forms for which feasible software tools can be provided.

We present a brief discussion of Overstreet's language for discrete event model specification (for more detail, see [11]). Overstreet generally conforms to the systems theoretic view of models so that input, output and state are distinguished and the principle of model/experimental frame separation is largely adhered to. The state transition structure of a model is given in the form of a finite set of *condition-action pairs*. Such a pair has a *condition* component that is a boolean expression and an *action* component that is intended to be executed while the condition is true. The action type of particular interest is the one in which the values of model variables are changed. Each such so-called *value change description* has the following structural components: *input variables, output variables,* and an *evaluation procedure* which defines new values for the output variables determined from the input variable values (input and output variables in this sense are the equivalents of the influencer and influencee concepts respectively of Chapter 7).

To introduce time sequencing, a variable type called a *time-based signal* is introduced. The statement: SET_ALARM (<alarm name>, <alarm time>), employed in a value change description, sets <alarm name>, the name of a time based signal, to "go off" at the time given by <alarm time>. To make use of such alarms, the statement: WHEN (<alarm name>) can appear in the condition part of a condition-action pair. For example, consider condition action pairs:

(WHEN(MACHINE_FAILURE),
SET ALARM(MACHINE_REPAIR), T1)) (1)
(WHEN(MACHINE_REPAIR),
SET ALARM(MACHINE_FAILURE, T2)) (2)

The action part of Pair (1) will cause the alarm called MACHINE__REPAIR to be set for time T1. This variable will become true at time instant T1 and is false otherwise. Pair (2) will have its condition part become true when MACHINE__REPAIR becomes true and will the schedule the alarm called MACHINE__FAILURE to be set for T2 at which time the condition of Pair (1) will be true instantaneously, and so on.

More powerful primitives can be defined in terms of the WHEN construct. For example,

(AFTER(<alarm name>) & <boolean expression>, action)

causes the action to be carried out as soon as the <boolean expression> becomes true after the <alarm name> has gone off. The effect of the AFTER can be achieved with the WHEN primitive with the following sequence of condition-action pairs:

(WHEN(<alarm name>), flag := true)
(flag=true & <boolean expression>, (flag := false,action)).

CANCEL(<alarm name>), which nullifies a SET__ALARM, is a second operation that is convenient but can be synthesized from the basic primitives.

Overstreet does not include two features that are apparent in the formalism developed in Chapters 4, 7 and 8: the SELECT function for tie breaking of simultaneous events, and mechanisms for querying the state of an alarm clock. For example,

ELAPSED__TIME(<alarm>)

would be a V-function (Section 4.4) associated with the O-function SET__ALARM (<alarm>) that would return the time elapsed since the alarm has been set. Similarly, TIME__LEFT (<alarm>) would return the difference between the scheduled time and the elapsed time. The importance of these functions in expressing the state of the model justifies their inclusion in the set of primitives although their effect can also be simulated with the available primitives.

While such a language could be implemented to constitute yet another simulation language, Overstreet argues that complexity, both computational and psychological, mitigates against this approach. The computational efficiency of simulation would be low since much condition testing has to be done that is not necessary if the model is reformulated in one of the traditional world views. The understandability would be low since the description suffers from lack of "locality" (having the complete description for an object or all related actions in one location). Indeed, Overstreet found that he tended to make many errors working within the language, that he attributed to the locality problem (personal communication).

Overstreet discusses procedures for converting model specifications in the base language into equivalent ones in each of the world view formalisms. The *action cluster digraph* serves as the central abstraction underlying such transformations. An *action cluster* is defined for each condition and consists of the set of condition-action pairs whose condition is the cluster condition. Thus an action cluster represents all the actions in the model that are enabled by the truth of a particular condition. An *action cluster digraph* has as its nodes the set of action clusters in the model and has two types of directed edges (so it is actually a multigraph). As illustrated in Figure 18.2, the first type of edge, depicted as a solid line from cluster A to B, say, represents the ability of A to cause the immediate activation of B; the second type of edge (depicted as a dotted line) represents the ability of A to cause the *delayed* activation of B.

We interpret the phrase "A can cause the activation of B" as "there exists a transition of the model from some state s to some state s', such that in s, the conditions for A and B are true and false respectively, while in s' the condition for B is true." Overstreet shows that under this interpretation, the digraph is not effectively constructable. However, a super digraph based on *rebuttable successors* can easily be constructed. Call an action cluster *determined* if its condition contains a WHEN statement (or equivalently a time-based variable); let it be *contingent*, otherwise. Then there is a dotted line from action cluster A to action cluster B if B is a determined action cluster and A SETS its time-based alarm; there is a solid line from A to B if B is contingent and an output variable of A also appears in the condition part of B. The term "rebuttable successors" refers to the fact that edges in this last digraph may not be actual activation successors, that is to say, might not appear in the digraph representing causal activation as just defined. While elimination of non-actual rebuttable successors is not decidable, algorithms that detect common special cases could be constructed. Such simplification is important since it is made possible by the grouping of action clusters that occurs in world view transformation.

Such transformations are based on the following digraph operations:

next event: identify an Event with each determined action cluster. The subgraph of all contingent action clusters accessible via solid edges from a determined action is expressed in the case for the Routine to be associated with its Event. In other words, determined clusters are scheduled, contingent clusters are tested and performed within the code of scheduled events.

activity scan: identify an Activity with each contingent action cluster. The subgraph of all determined action clusters accessible from a contingent cluster is expressed in the code for the Routine to be associated with its Activity. Thus, the conditions of contingent clusters are continually scanned for and their associated activity codes express the determined action clusters.

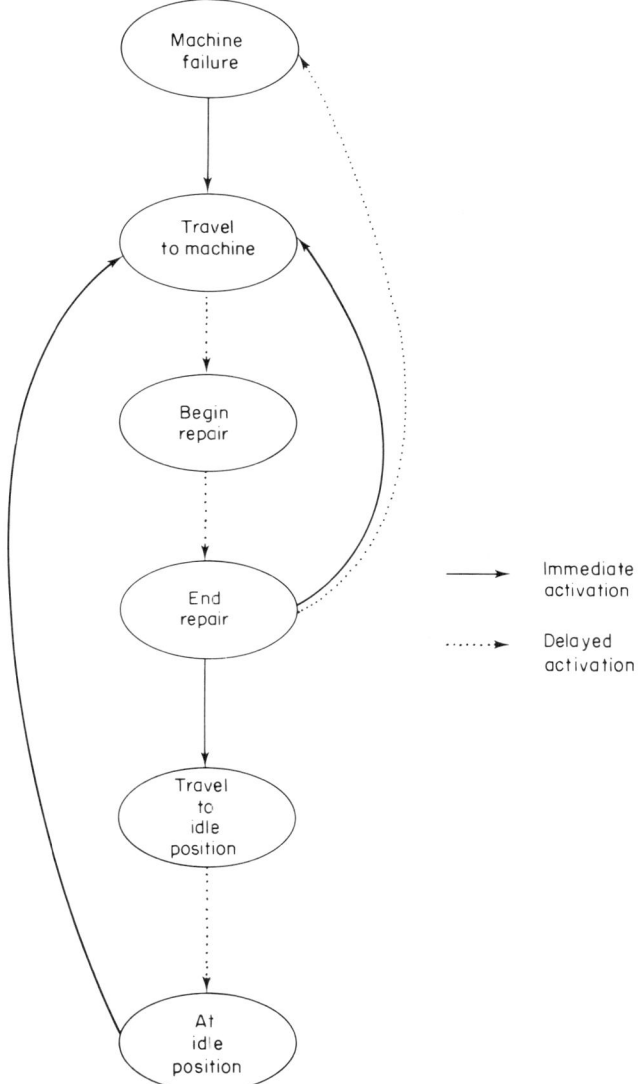

Fig. 18.2. Action cluster diagraph for travelling repairman model (from Overstreet [11]).

process interaction: identity a Process with each component in the model. The subgraph of all action clusters containing variables attached to this component is expressed in the code for its Process.

Simplification is induced by such coverings due to the fact that the status of a condition may be entirely predictable once a prior context is established. For example, the action of a cluster A may always be to make the condition of cluster B true. If B follows A in one of the above subgraphs, then the subgraph code need express only the action part of B. On the other hand, suppose the action of another cluster A' is always to falsify the condition of B. If B follows A' in a subgraph, then its action may be entirely eliminated from the code of the subgraph.

It should be commented that the coverings for the cluster digraph chosen by Overstreet represent particular choices in a class of possibilities that could be explored.

Support of general systems methodology

A conceptual framework for systems problem solving called *General Systems Problem Solver* (GSPS) has been evolved by Klir and his students [12]. It sets out taxonomies of *system types* and *problem types* with the goal of eventually supplying a set of tools for each of the recognized types and an interface to assist the user in identifying and formulating his problem within the framework. A working system has been implemented for a subclass of problem types supporting the process of *inductive modelling*. We shall briefly describe the general framework and then its current operationalized version.

The taxonomy of system types in GSPS is very similar in conception to the hierarchy of system specifications described in Chapter 3. It is called a *hierarchy of epistemological levels*: the term "epistemology" referring to "knowledge", highlights the point of view that the modeller is attempting to acquire knowledge about a particular real system, and increasing levels represent increasing knowledge about the internal structure of the system. *Inductive modelling* is the approach to knowledge acquisition in which one starts with an observation frame, collects data within it, and attempts to climb up the hierarchy by inferring successively higher level structures from the data alone. This contrasts with the deductive or *postulational* approach which characterizes simulation modelling [13].

Some key differences in the hierarchy of espistemological levels as defined by Klir and that of the hierarchy of systems specifications are:

(a) The hierarchy of systems specifications is based on the orientation of a system (distinction between input and output variables) and on the concept of internal state. Klir's hierarchy allows both oriented and non-oriented systems. It restricts the use the state concept to essentially the finite memory case (Section 4.5.1): only variables that can be observed at level 0 and lagged versions can be employed at higher levels. This restriction is appropriate in the case of strict inductive modelling where only the data acquired in the observation frame can be employed. It is not appropriate however in the case of postulational modelling where appropriate state variables might be suggested from centuries of previous experience in modelling systems of the same kind (e.g. using Newton's laws for mechanical systems) or where models may be synthesized from components extracted from a model base.

(b) The hierarchy of systems specifications distinguishes time as a fundamental variable (the systems are dynamic, i.e. change over time), and systems with internal structure (level 2 and above) are formulated as deterministic (stochastic systems however can be accommodated, Section 16.2.2). Klir's objects are essentially probabilistic in form, and the probability measure may be replaced by various fuzzy set measures. Time does not occupy a privileged support status, other variables can be chosen, e.g. space or individuals of the same kind, ordered in some fashion.

(c) Klir's hierarchy includes the meta-system level which is the one above the structured system level. The metal-system level specifies changes in structure at the next lower level as a function of the support variable. Moreover, a meta-meta-system may specify changes in the meta-system, and so on upward an indefinite number of levels. The hierarchy of system specifications does not have a concept for change in structure, although such a concept could be added to it. Such structure modification capability has been advocated by Oren [14] and Dekker et al. [15] in the simulation context.

Problem types in the GSPS framework are conceived as triples of the form: (initial system type, terminal system type, requirement type). Problems may take two forms:
(a) *transformation* problems in which the initial system and the requirements are given, the problem being to transform the initial system into a terminal system (of the specified type) that meets the requirements.
(b) *comparison* problems in which the initial system, terminal system, and requirements in the form of questions are given, the problem being to answer the questions concerning relationships between the two systems.

Inductive modelling is characterized by a class of problems of the transformation type. These problems take the form: given a system, say S_i, at level i, and set of systems, S_{i+1} at level $i+1$, called admissible systems, determine a subset of S_{i+1} that satisfies given requirements, R.

The software system, SAPS (System Approach Problem Solver) [16] provides tools for solving such problems. The user specifies a list of variable names and formats (in Fortran notation), called the *source* or level 0 system. Data gathered for such a system is in the form of a *list of cases*, each case being an assignment of values to the list of variables. This list constitutes the data *or* level 1 system, *called the* data array. Actually, the range sets of the variables at the data level must be subsets of integers. A crucial decision required of the user is the how to code ranges at the source level to those at the data level (tools are provided to execute the coding once defined).

The level 2 structure of GSPS is called the *generative* system. It consists essentially of a support-invariant (e.g. time-invariant, space-invariant, etc.) relationship, called a *mask*. The mask is translated along the data array so that the frequency distribution of the mask states is computed, called its *basic behavior*. From the basic behavior, the conditional frequency distribution of the predicted variables conditioned by the generating variables, called the *generative behavior*, is computed.

The *generative uncertainty* of a mask is the Shannon entropy applied to the generative behavior. The ability of a mask to regenerate the original data array is inversely related to this uncertainty (for an uncertainty of zero, each successive shift of predicted variables can be predicted exactly from the values of the generating variables). The *complexity* of a mask is taken as its size, i.e. essentially the number of generating variables that are used to do the prediction. SAPS provides a tool that given a maximum size of mask, iterates through submasks in order of increasing size, computing for each the basic behavior, generative behavior and generative uncertainty. The user may select a mask on some basis such as one that has smallest generative uncertainty, or the best uncertainty reduction to complexity ratio, etc. The selected mask constitutes a solution to the inductive inference problem for the transition from level 1 to level 2.

The level 3 concept, called a *structured system*, consists of a collection of generative systems coupled together by sharing of variables. From the basic behaviors of the components and the constraints imposed by the coupling, a basic behavior for the global system (consisting of the union of the variables of the components) is computed using a probabilistic version of the relational join operation. Klir [17] provides an interesting argument that the basic behavior so computed is optimal in the sense that it takes account of all, and only, the information provided (this is formalized by the principle of maximum Shannon entropy).

Given a generative system, S, Klir defines a family of admissible structured systems for it. Each such system corresponds essentially to a maximal cover of the set of sampling variables of S. SAPS provides a tool that will iterate through the admissible structured systems, computing the global behavior of each, and comparing the latter with the basic behavior of S. A natural metric to employ for this comparison is the difference between the amount of information in S and the amount preserved by the candidate structured system. The user may select one, or more such candidates, for which the difference falls below a desired threshold, as solutions to the inductive inference problem for the transition form level 2 to level 3.

Finally, SAPS provides a tool for inferring metasystems from data systems. Here a metasystem refers to a finite set of generative systems, each with a domain of the support variable in which it is to apply (these domains are to partition the range of the support variable). Given certain parameters, the program seeks points in the given data array, at which generative uncertainty computed for the current mask degenerates rapidly, signalling the likelihood that this mask should be replace by another for continuing through the array. The set of masks associated with these crossover points constitute a metasystem and hence a solution to the inference problem for the transition to level 4.

Inductive modelling, as described above, optimally employs all, and only, the data acquired from the particular source system under observation. Its results are system representations of this data that are unbiased by any preconceptions or prior knowledge about the system. Were it possible to construct models in this way in all cases, this would no doubt be preferred methodology. Unfortunately, the computational complexity of the search methods increases dramatically with increasing number of variables, limiting the pure form of induction to small numbers of variables in practice (typical applications have used five or six). For such small cases, SAPS provides an attractive alternative to, and can be used in combination with, conventional statistical inference packages.

However, for large systems, the pure methodology is impractical and one must resort to postulational, simulation supported, methodology. Still, hybrid combinations of the two methodologies may be feasible. Inductive modelling might be employed initially when approaching an entirely novel real system and the postulational methodology, might be phased in as experience progresses. The multifacetted approach, and its model base concept, would be crucial in such an approach. Inductive approaches may also be used to provide compact summaries of complex system behavior (real system or model) for convenient comparison in validation and verification activities.

Distributed simulation on networks of microcomputers

Microcomputers of ever increasing computing power and decreasing cost are being made available on the commercial market. Simulation software is being developed for such computers [18]. Means to interconnect such computers to form local networks are also available and there are even examples of such networks that can be purchased as a unit. Thus the time has arrived for considering networks of microcomputers as a most attractive medium for distributed simulation offered by state-of-the-art technology.

Such a network of computers can be viewed as the latest stage in the evolution of digital design departing from the original von Neumann SISD (single instruction stream-single data stream) architecture in which first memory and processing capability were distributed in arrays (SIMD, single instruction stream-multiple data stream) and later control was decentralized as well (MIMD, multiple instruction stream-multiple data stream) [19].

Simulation oriented problems have constituted a major driving force in the SIMD and MIMD hardware developments. Array-of-processor machines (e.g. ILLIAC IV), the pipelined supercomputers (e.g. STAR-100), and the array processors controlled by serial hosts have been applied to exploiting the parallelism inherent in simulation computations, for example in the solution of partial differential equations, and in the manipulation of large matrices [20].

Networks of microprocessors have more recently been designed, in which the instruction repertoire of the LSI chip components is specialized to the arithmetic and logical operations required for differential equation solution [21]. Given the relatively small computational power of the processor elements, and the application to partial differential equation simulation, such networks are characterized by large numbers of elements distributed in spatial arrays of up to three dimensions. Weisberger [22] reviews the various problems that must be dealt with in the design and implementation of such networks. Progress in the solution of these problems however has been slow.

In contrast to the limited power of the microprocessor element, a microcomputer is a complete and powerful sequential processor. A network of such devices offers new possibilities and prospects for distributed simulation. While traditionally, the emphasis has been placed in the attaining increased speed of computation by exploiting parallelism in the model, the power of the microcomputer network makes it possible to seek other objectives as well as, or in lieu of, speed.

A large model may be a coupling of components, each of which is itself large, and which may be implemented on a microcomputer by a simulation program written in a high level language. In this way, the flexibility advantages of software *vis à vis* hardware in construction and modification may be exploitable. Moreover, the moderate (over 256K) core memory owned by each microcomputer is immediately accessible to its central processor. Compare this to implementing the model conventionally on a single sequential processor, where the sum of all component memory requirements may exceed main memory, necessitating swapping in and out from auxillary stores. Similarly, the input/output capabilities of each microcomputer may be employed for interactive experimentation with its component model and display of its simulated behavior. Finally, and perhaps most significantly for the future, a network of microcomputers may come closer to permitting a "one-to-one analogy in space and time" between the model and its simulator. As interpreted by Dekker *et al.* [15], this means that the modular and hierarchical structure of a model is mapped into corresponding processors of the simulator, rather than abstractly represented in the code of a single conventional program; also, a change in this structure brought about by the model itself, is represented by a change in the corresponding processors. This transparancy contributes to improved understanding and interactivity with the model.

An experimental parallel processor was built at the Delft University of Technology to investigate the feasibility of the "one-to-one space and time analogy". It consists of eight data processing elements (DPE) controlled by a minicomputer, LSI-11. Each DPE is a microprogrammable autonomous processor, the program being translatable from a specially developed assembly language. The DPEs are interconnected, and coupled to the minicomputer, via a data transfer system. Program preparation, loading of the DPE definitions, implementation of hierarchical model structure, and state initialization is done on the minicomputer. During simulation, the latter also implements the changes to be brought about in the hierarchical model structure.

The Delft experimental processor provides one prototype for distributed simulation on networks of microcomputers where the latter substitute for the specifically designed DPEs. However, a major departure that is possible with a network of microcomputers is that of truly distributed processing as opposed to the centrally clocked timing that characterizes the Delft Processor as well as all other architectures described above. In such distributed processing, each computer is able to compute at its own pace (asynchronous processing) except for obeying certain synchronization signals that co-ordinate its processing with that of other computers. In the simulation context, the main problem is the co-ordination of computers so that states they compute are correctly related to each other and to the global clock of the model despite the unpredictable rates of local computing. Some approaches to this synchronization problem were discussed in Chapter 17.

4. Epilogue

This concludes our presentation of the concepts of multifacetted modelling methodology and the computerized systems that could make working with such concepts a reality. Throughout we have stressed the benefits of the multifacetted approach and have concentrated on the what has to be done to achieve these benefits. But it must be recognized that doing what has to be done to achieve long term benefits is an overhead cost when viewed from the short term. Concern with this overhead should sensitize us to the tradeoffs that must be examined in attempting specific implementations of multifacetted methodology. But it would do no good to become paralysed in contemplation of the magnitude of the cost. Progress is made when there is a goal out there worthy of achievement *and* where specific, concrete steps are taken towards it. The satisfaction, and benefits, lie partly in reaching modest stepping stones on the way.

References

[1] Oren, T. I. (1982). "Computer Aided Modelling Systems". In *Progress in Modelling and Simulation*, (ed, F. E. Cellier), Academic Press, London.
[2] Elzas, M. S. (1982). "The Use of Structured Design Methodology to Improve Realism in National Economic Planning". In (ed, H. Wedde) *Model Adequacy* (Proceedings of the Intl. Conf. on Model Realism, Bonn), Pergamon Press, London.
[3] Ingals, D. H. H. (1978). "The Smalltalk-76 Programming System: Design and Implementation", *Proc. 5th An. ACM Sym. on Principles of Programming Languages*, Tuscon, AR, pp. 9–16.
[4] Belogus, D. (1983), *Multifacetted Modelling and Simulation Methodology: A Software Engineering Implementor*, Doctoral Dissertation, Weizmann Institute of Science, Rehovot, Israel.
[5] Zeigler, B. P., D. Belogus and A. Bolshoi (1980). "ESP: An Interactive Tool for System Structuring". In, *Proc. Eur. Meet. Cyb. and Sys. Res., Vienna*, New York: Hemisphere Press.
[6] Sprague, R. H. and E. D. Carlson (1982). *Building Effective Decision Support Systems*, Prentice Hall, N.J.
[7] Spiguel, C. P. (1982). *Computer-Aided Modelling: An Application to Decision Support in Business Environments*, Doctoral Dissertation, University of Michigan, Ann Arbor.
[8] Ullman, J. D. (1980). *Principles of Data Base Systems*, Computer Science Press, Potomac, MD.
[9] Oren, T. I. (1971). "A Combined Digital Simulation Language for Large Scale Systems". *Proceedings of the Tokyo 1971 AICA Symposium on Simulation of Complex Systems*, Tokyo, Japan, pp. 8–11/4.
[10] Oren, T. I. (1983). "GEST — A Modelling and Simulation Language Based on System Theoretic Concepts". In *Simulation and Model-Based Methodology: An Integrative View*, (eds, T. I. Oren et al.), Springer-Verlag, NY.
[11] Overstreet, C. M. (1982). *Model Specification and Analysis for Discrete Event Simulation*, Doctoral Dissertation, VPISU, Blacksburg, VA.
[12] Cavallo, R. E. and G. J. Klir (1979). "A Conceptual Foundation for Systems Problem Solving". *Int. J. Systems Science* 9, 219–226.
[13] Zeigler, B. P. (1974). "A Conceptual Basis for Modelling and Simulation". *Int. J. Gen. Sys.* 1, 213–228.
[14] Oren, T. I. (1975). "Simulation of Time-varying Systems". *Proc. Int. Cong. of Cybernetics and Systems*, Oxford, 1972, Gordon and Breach, England, pp. 1229–1238.
[15] Dekker, L. E. J. H., G. C. Kerckhoffs, J. C. Vansteenkiste and J. C. Zuidervastat

(1981). "Outline of a Future Parallel Simulator". In *Simulation of Systems '79*, North Holland, Amsterdam.
[16] Uyttenhove, H. J. J. (1983). "SAPS – A Software System for Inductive Modelling". In *Simulation and Model-Based Methodology: An Integrative View* (eds, T. I. Oren *et al.*), Springer-Verlag, NY.
[17] Klir, G. J. (1983). "Reconstructability Analysis: An Overview". In *Simulation and Model-Based Methodology: An Integrative View* (eds, T. I. Oren *et al.*), Springer-Verlag, NY.
[18] Ellison, D., I. Herschdorfer and J. T. Wilson (1982), "Interactive Simulation on a Microcomputer". *Simulation Journal* **38**, No. 5.
[19] Enslow, P. H. (1977). "Multiprocessor Organizations — A Survey". *ACM Computer Surveys* **9**, 103–129.
[20] Thurber, T. J. (1979). "Parallel Processor Architectures, part 1: General Purpose Systems". *Computer Design* **18**, 89–97.
[21] Ortega, J. M. and R. G. Voight (1977). *Solution of Partial Differential Equations on Vector Computers*, Report 77–7, ICASE.
[22] Weissberger, A. J. (1977). "Analysis of Multiple-Microprocessor System Architectures". *Computer Design* **16**, 151–163.

SUBJECT INDEX

A

Abstract simulator of a DEVS, 318, 321
Abstraction, sets and set operations, 21
Acceptance, experimentation by, 218
Acceptor, 72, 224
Accommodates (see also applicability), 247
Action cluster, 348
Activatable component, 104
Activate, 65, 124
Active component, 65
Activity scanning, 24, 118, 356
Admissible segment, 211
Aggregate, 160–4
Aggregation, 14, 160–4, 283, 293
Aggregation, uniformity of influence, 162–3
Alternating mode axiom, 196
Amalgamation, 264, 294
Applicability, experimental frames to models, 83, 237, 247
Aspect, 15, 196, 273, 346
Association, in hierarchy, 28, 42–6
Asynchronous distributed simulation, 327, 361
Atomic entity, 263, 273
Atomic model specification, 351
Attached variables, 196
Attribute (SIMSCRIPT), 191
Axioms, entity structure, 196

B

Base model, 81, 163, 248
Basic behavior, 360
Behavior, 36, 38, 211, 360
Behavior data base, 316
Behavior processing, 334

Behavioral equivalence of systems, 48
Belongs relation, 198
Binary relation, 22
Block of components, 49, 163
Bottom-up construction, 284

C

Calibration, 369
Cellular automaton formalism, 92
Change detectable variable, 307
Change detector, 307
Class generation, 185, 341, 352
Climbing up the specification hierarchy, 53
Closure of formalisms under coupling, 60, 326
Co-ordinate, 41
Combined continuous/discrete models, 287, 351
Compatible specifications, 47
Competitive models, 13
Complementary models, 13
Complexity measure, 24, 31, 354
Component, 15, 42, 281
Component names, 42, 123, 134
Component nodes in decomposition tree, 179
Component type, 179
Composite model, 26, 43
Composition property, 40
Composition tree, 27, 281
Computable model, 108
Computational complexity, 32
Computer aided modelling and simulation, 4, 294, 334
Computer networks, 327, 361
Concerns relation, 262
Concretization, 22

Condition-action pairs, 354
Condition, initialization and termination, 220–1
Congruence, 160
Consistency of coupling scheme, 283
Contexts for multifacetted methodology, 336
Continuation conditions, 222
Continuous time, 37
Control function of experimental frame, 248
Control objectives in modelling, 41, 17, 257
Control segment, 221
Controllable input variable, 17, 83, 218
Controllable part of real system, 4
Correctly simulates relation, 82, 321
Correspondences, parameter, 170
Correspondences, state, 48
Correspondences, variable, 48
Cotyledons and experimental frames, 226–32
Coupled system models, 25, 41, 134, 283, 351
Coupling constraint, 280
Coupling of system models, *see* Coupled system models
Coupling recipe, 282
Coupling scheme, 41, 278, 282, 351
Crossproduct, 21

D

Data, real system, 80
Decision maker, 3, 339
Decision support system, 5, 348
Decomposition tree, 15, 175
Depth of formalism, 28
Derivability of experimental frames, 237, 243–55
Design objectives in modelling, 4, 226
DEVS formalism, *see also* DEVS models, 62
DEVS interpretation, 65
DEVS models, 66, 97–107, 118–27, 142–45
DEVS segment, 63
Differential equation system specification, 56
Directed graph (digraph), 23

Discrete event system specification (see DEVS), 4
Discrete time model, 59, 93
Distributed simulation, 327, 361
Distribution (generic variable) 207, 253
Diversity of formalism, 28
Documentation of models, 11, 345, 350
Dynamic structure, 37, 174, 350

E

Elaboration, 11, 276
Elapsed Time, 62, 319, 325
Entity structure, 15, 194–8, 261, 272, 344
Entity structure, amalgamation, 261
Entity structuring program, 344
Equivalence of entity structures, 266
Equivalence of models, 46, 251
Event times, 56, 217
Event-like input segments, 167
Events, 46, 356
Execution control program segment, 88, 309–15
Experiment, 218
Experimental frame program segment, 88, 301–3, 350
Experimental frame, definition, 82, 210–1
Experimental frame, examples, 16, 206, 276
Experimental frame, realization of, 224, 301
Experimental frames and validity, 84, 250
Experimental frames for isolation of components, 248
Experimental frames for model simplification, 251
Experimentation specification, 352
Expressive power of model formalism, 108
External event, 63, 167
Extraction from entity structure, 263

F

Filling out procedure, 179
Final value transducer, 73, 74
Finite memory DEVS, 74

Finite memory machine, 30
Finite state machine, 23
Formalism, 21
Formalism for discrete event cell space models, 97–107
Formalism for discrete event systems, see DEVS formalism
Formalism for hierarchical structuring, 25, 127, 149–51
Formalism, adequacy of, 107, 151
Formalization of modelling assumptions, 248
Frame-to-frame processing, 247
Function, definition of, 22

G

General Systems Problem Solver, 358
Generative behavior, 359
Generator of segments, 71, 218, 224
GEST modelling and simulation language, 350
Global transition function, 93
Greatest lower bound, 291

H

Handling of collisions, 113
Hierarchical complexity measure, 24
Hierarchical DEVS model, 127, 149–51, 297, 356
Hierarchical DEVS simulators, 326
Hierarchy of epistemological levels, 358
Hierarchy of system specification morphisms, 46–51
Hierarchy of system specifications, 35–42
Homomorphism, 48

I

I/O data space, 211
I/O function, 39
I/O relation, 38
I/O system, 39
Identifiable system, 54
Ignored input event, 66
Implementation, definition of, 29
Indecomposable system, 9
Indistinguishable systems, 48

Induced complexity measures, 31
Induced segment-to-segment mapping, 247
Inductive modelling, 358–9
Influencees, 134, 301
Influencers, 25, 42, 208
Informal model description, 10
Information neighbors, 93
Initial state, 39
Initialization, 220, 312
Initialization conditions, 220
Input segment, 38, 64
Input segment generation and acceptance, 216
Input value set, 37, 216
Input variables, 17, 41, 212, 280, 351
Input/output function observation (IOFO), see IOFO function
Input/output segment pair, 38, 80, 211
Interactive complexity measure, 31
Interactive parameter exploration, 313
Interest variable, 208
Interface map, 42, 134, 278
Interface, man-machine, 336
Interior nodes of tree, 24
Internal role designation, 280
Internal structure, 39
Interrupt external event, 66
Intervariable relations, 312
Inverse neighborhood, 96
Isomorphism, 46
Item, 98, 265, 346
Item occurrence, 98
Iterative methodology for model construction, 288

J

Job description structure, 66
Justifying conditions, 53

K

Knowledge acquisition, 358
Knowledge of internal structure, 81

L

Lag, 358
Language for DEVS model specification, 123–4, 137–9

Lead, 348
Leaf of tree, 23
Least upper bound, 291
Legitimacy of DEVS, 64
Level of detail, 14
Levels of system specification, 37–42
Linkage types, 279
Links in coupling recipe, 279
Local map, 49
Local transition function, 93, 97
Lower bound, 15, 291
Lumped model, 163, 250, 283

M

Management objectives in modelling, 5, 258, 339
Mapping, definition of, 22
Mask, 360
Maximal element, 291
Meaning of variable, 175, 203–4
Mediating variables, 208
Methodology for modelling "in the small", 10
Methodology for modelling "in the large", 11
Methodology, iterative model construction, 288
Methodology, objectives driven, 274
Minimal state variable set, 311
Mixed formalism models, *see* Combined continuous/discrete models
Mode of item, 98
Model base, 12, 262, 290, 348
Model class, 54, 281
Model construction, *see* Methodology for iterative model construction
Model fabrication, 348
Model generation/referencing, *see also* Model-base, 334
Model input variables, *see* Input variables
Model output variables, *see* Output variables
Model quality assurance, 334
Model specification program segment, 88, 350
Model transformation, *see* Translation of formalisms
Model/frame pair, 88

Modelling and simulation as a process, 12
Modelling formalism, *see* Formalism
Modular realization of model/frame pair, 10, 132–3, 300
Modularity, 146
Modularization, 132–147
Module, 132
Monitored variable, 302
Morphism, observation frame, 47
Morphism, structured system, 49
Morphism, system coupling, 49
Morphism, I/O function, 48
Morphism, I/O Relation, 47
Motion, representation of, 113
Moving down the morphism hierarchy, 51
Multicomponent models, spatially distributed, 92, 107
Multicomponent DEVS in hierarchical form, 127–9, 149–51
Multicomponent DEVS in modular form, 133–37
Multicomponent DEVS in structured system form, 117
Multifacetted model building methodology, 272–95
Multifacetted modelling approach, 8–14, 272–95
Multifacetted systems and decomposability, 9
Multiple decomposition, 183–5
Multi-objectives approach, 5
Multiple-step transition function, 59

N

Name of variable, 181
Nested coupling, 351
Network of computers, *see* Computer network
Next event cell space models, 97–8
Next event formalism, 118
Non-atomic entity, 263
Non-event closure, 66, 216
Non-event symbol, 66, 216

O

O-function, 66
Object, 22, 339, 341

370   Index

Objectives, 5
Objectives and expanding system boundaries, 257–61
Objectives-driven methodology, 207–15
Objectives-guided pruning, 275
Observation frame, 37
Off-line realization of experimental frame, 316
On-line realization of experimental frame, 301
Operations on variables, 188–90
Organization role of entity structure, 261
Orientation of model, 280
Output function, 40
Output function program segment, 40
Output module, 353
Output trajectory, 46
Output value set, 38, 216
Output variables, 17, 41, 212, 280, 351
Overdetermination of state, 312

P

Parameter base, 255, 267, 337
Parameter correspondence, 171, 268
Parameter identification, 269
Parameter structure, 268
Parameterized experimental frame, 218, 309
Parameter, definition, 22, 171
Parameters in execution control, 309
Parametric model specification, 253, 268, 278
Partial decomposability, 9
Passivate, 65, 124
passive component, 65
Patches in ecosystem, 284
Performance index, 208
Pertaining to relation, 198
Piecewise continuous segments, 58
Pointwise segment-to-segment mapping, 240
Population distribution objective, 258
Port, 295
Ported input/output specifications, 295
Post study program segment, 353
Postulational approach, 358
Predictively valid model, 85

Prestructure, 179
Principle of derivability, 240
Process formalism, *see also* DEVS models, 124
Process interaction, 118, 124
Propagation of parameter information, 267
Proper morphism, 283
Pruned entity structure, 272
Qualified variable, 198
Queue, formalization of generalized, 66
Quiescent state, 98

R

Random phase-random space approximation, 164–7
Range set, 175
Rate of change, 58, 351
Re-assigning variables, 266
Read access, 1407
Real system, definition, 80
Real system experimentation, 334
Realization, definition, 29
Realization of experimental frames, 224, 301
Realization of model/experimental frame modules, 301
Realization of the DEVS formalism, 120, 301, 325
Rebuttable successors, 350
Reconstitution of system from components, 36, 44
Recurrent environment, 339
Reduction, 54, 160
Relationally equivalent systems, 47
Relative labelling, 189
Relevant model, 291
Removing redundant aspects, 266
Repeatable environment, 339
Replicatively valid model, 85
Representing states, 85
Representing systems in DEVS formalism, 167–70
Resource utilization measure, 208
Resultant of coupling, 26, 43
Role designation of variables, 280
Run control segments, 221
Run control variables, 17, 220

S

Sample of activity, 164
Scenario, 17
Schema, 349
Scope experimental frame, 246, 254
Select function in DEVS, 97, 117, 134
Semantic relation, 202
Semantic structure on variables, 202–4, 338
Separation of model and experimental frame, 215
Sequential states, 62, 117
Set of state variables, 310
Set theory, 21
Signature function 341
Simple morphism, 52
Simplification, 14, 86, 159–72, 281, 293
Simulation complexity, 86, 354
Simulation program, 88, 302, 319, 322, 350
Simulation relation, 79, 320
Simulator, 79, 318, 321
Skeleton subschema, 349
Source system, 359
Specialization hierarchies, 185, 199
Specification, *see also* System specification, 27, 30
Specification using prestructure, 179
State space reduction, 160
State trajectory, 46
State transition function, 35, 175, 344, 349, 351
State variables, 278, 280, 310
State variables in execution control, 310
Static structure, 35, 175, 344, 349, 351
Strict hierarchy, 180, 196
Structural complexity, 24
Structural inference, 85, 358
Structural levels and volatility, 314
Structural validity, 85
Structure, definition, 22
Structured function, 41
Structured set 41
Structured system, 41, 360
Structured system morphism, 49

Sub-coupling scheme, 49
Sub-frame, 219
Subclasses of systems, 54
Submits relation, 36
Subordinate relation, 198
Substructure, 198
Successor relation, 23
Summary mappings, 211, 225
Synonyms, 263
System boundaries, 263
System entity structure, *see* Entity structure
System input variables, *see* Input variables
System output variables, *see* Output variables
System problem types, 358
System specification, 35–42, 179
System theory, 35, 358
System types, 358
Systems and models, 80

T

Temporal relation, 349
Termination conditions, 229
Theory construction, 22
Tie-breaking rules, *see* Select function in DEVS
Time advance function, 97, 114, 134
Time base, 37
Time invariance, 57, 326
Time left, 97, 319, 325, 355
Time step, choice of, 113
Time-based signal, 354
Top down modelling, 10, 50, 195, 284, 352
Total state, 63, 326
Trade-off experimental frame, 231
Transducer, 73, 225
Transformation, *see also* Translation of formalisms, 73, 225
Transition function, *see* State transition function
Transitory state, 65
Translation of formalisms, 30, 104, 139, 167
Tree formalism, 23, 281
Tricotyledon theory of system design, 22, 281

U

Uncontrollable component, 231
Underdetermination of state, 312
Uniform influence principle, 161, 164
Uniformity axiom, 181, 196
Unique phenomenon, 334
Unique solutions, 44
Units of measurement, 175
Unqualified variable, 198
Useful abstractions, 33
User oriented DEVS language, 138, 151

V

V-function, 66, 335
Valid brothers, 194
Valid coupling scheme, 283
Valid model, 82–5
Valid simplification, *see* Simplification
Validation, 13, 82
Value change description, 354
Variable, 17, 41, 175, 179, 198, 208
Variable type, 179, 198
Variables: rights of access, 198
Verification, 13, 82
Volatility, 314

W

Wait until statement, 119, 121, 124
Well-described model, 311
Workload, 233
World view of formalisms, 107
Write access, 140

OHIO UNIVERSITY LIBRARY

this   as soon
         rder